CAMBRIDGE LIBRARY COLLECTION

Books of enduring scholarly value

Life Sciences

Until the nineteenth century, the various subjects now known as the life sciences were regarded either as arcane studies which had little impact on ordinary daily life, or as a genteel hobby for the leisured classes. The increasing academic rigour and systematisation brought to the study of botany, zoology and other disciplines, and their adoption in university curricula, are reflected in the books reissued in this series.

Types of British Vegetation

Distinguished plant ecologist A.G. Tansley (1871–1955) is widely considered to be the father of British ecology. During his career he edited two important journals on the subject: *The New Phytologist* and the *Journal of Ecology*, and he was one of the founding members of the British Ecological Society. He was also part of a committee formed in 1904 to survey systematically the vegetation of the British Isles. This book, edited by Tansley and first published in 1911, is the result of that survey. It contains contributions by leading botanists of the early twentieth century, and detailed maps, photographs and figures. The physical characteristics and climate of Britain are outlined early in the book and later the plant communities of particular areas such as moors, fens and the coast are discussed. This is a significant work that will appeal to both plant ecologists and natural historians.

Cambridge University Press has long been a pioneer in the reissuing of out-of-print titles from its own backlist, producing digital reprints of books that are still sought after by scholars and students but could not be reprinted economically using traditional technology. The Cambridge Library Collection extends this activity to a wider range of books which are still of importance to researchers and professionals, either for the source material they contain, or as landmarks in the history of their academic discipline.

Drawing from the world-renowned collections in the Cambridge University Library and other partner libraries, and guided by the advice of experts in each subject area, Cambridge University Press is using state-of-the-art scanning machines in its own Printing House to capture the content of each book selected for inclusion. The files are processed to give a consistently clear, crisp image, and the books finished to the high quality standard for which the Press is recognised around the world. The latest print-on-demand technology ensures that the books will remain available indefinitely, and that orders for single or multiple copies can quickly be supplied.

The Cambridge Library Collection brings back to life books of enduring scholarly value (including out-of-copyright works originally issued by other publishers) across a wide range of disciplines in the humanities and social sciences and in science and technology.

Types of British Vegetation

By Members of the Central Committee for the Survey and Study of British Vegetation

EDITED BY A.G. TANSLEY

CAMBRIDGE
UNIVERSITY PRESS

CAMBRIDGE UNIVERSITY PRESS

Cambridge, New York, Melbourne, Madrid, Cape Town,
Singapore, São Paolo, Delhi, Mexico City

Published in the United States of America by Cambridge University Press, New York

www.cambridge.org
Information on this title: www.cambridge.org/9781108045063

This edition first published 1911
This digitally printed version 2012

ISBN 978-1-108-04506-3 Paperback

Tansley's Vegetation

ERRATA

Plate IX. *Omit* Heath Moor of Pennine (August)

„ XXXI, *for* Phot. F. F. Blackman
read Phot. T. G. Hill

p. 51 footnote, *for* tracks *read* tracts

TYPES OF BRITISH VEGETATION

CAMBRIDGE UNIVERSITY PRESS
London: FETTER LANE, E.C.
C. F. CLAY, Manager

Edinburgh: 100, PRINCES STREET
Berlin: A. ASHER AND CO.
Leipzig: F. A. BROCKHAUS
New York: G. P. PUTNAM'S SONS
Bombay and Calcutta: MACMILLAN AND CO., Ltd.

TYPES OF BRITISH VEGETATION

BY MEMBERS OF THE CENTRAL COMMITTEE FOR THE SURVEY AND STUDY OF BRITISH VEGETATION

EDITED BY

A. G. TANSLEY, M.A., F.L.S.

UNIVERSITY LECTURER ON BOTANY IN THE UNIVERSITY OF CAMBRIDGE

With 36 Plates
and
21 Figures in the Text

Cambridge:
at the University Press
1911

Cambridge:
PRINTED BY JOHN CLAY, M.A.
AT THE UNIVERSITY PRESS

TO

PROFESSOR EUGENIUS WARMING

THE FATHER OF MODERN PLANT ECOLOGY

AND TO

PROFESSOR CHARLES FLAHAULT

WHO THROUGH HIS PUPIL

ROBERT SMITH

INSPIRED THE BOTANICAL SURVEY OF THIS COUNTRY

THIS FIRST ATTEMPT

AT A SCIENTIFIC

DESCRIPTION OF BRITISH VEGETATION

IS DEDICATED

IN ALL GRATITUDE AND ADMIRATION

BY THE AUTHORS

LIST OF CONTRIBUTORS

1. *Members of the Committee*

    F. J. Lewis
    C. E. Moss
    F. W. Oliver
    Marietta Pallis
    W. Munn Rankin
    W. G. Smith
    A. G. Tansley

2. *Other contributors*

    Professor Grenville Cole
    Dr R. W. Scully
    Professor G. S. West

PREFACE

THIS book offers an account of British vegetation from a standpoint which has not hitherto been adopted in any general treatment of the plant-life of this country. An endeavour is made to recognise and describe the different types of plant-community existing in the natural vegetation of these islands, and to trace their relations, so far as these have been elucidated, to climate and soil, and to one another—in other words to present a scientific classification and a balanced picture of British vegetation as it exists to-day.

The work of systematically surveying vegetation and recording the results on vegetation maps was begun in Scotland by the late Robert Smith in the closing years of last century, and continued by his brother, W. G. Smith, and various other workers. In 1904 these workers formed a committee, with the somewhat ponderous title of "The Central Committee for the Survey and Study of British Vegetation," to organise and facilitate work on these lines.

The memoirs and vegetation maps published by these workers, of which the titles will be found in the Bibliography on p. 367, have formed the nucleus of the material dealt with in the present book, which is thus a direct outcome of the work of the Central Committee. Its composition would have been impossible but for the close co-operation secured by the existence of such an organisation. Since it was clearly out of the question for a committee as a whole actually to write a book, it was necessary for someone to undertake the task of correlating the contributions of various authors and securing uniformity of treatment. The editor undertook this task on the understanding that he should be allowed a free hand in editing the work of the authors, and this was in nearly every case most willingly accorded. The result of this loyal co-operation has been, it is hoped, to secure a consistent treatment of the whole subject. It is not to be expected that all the members of a large committee should agree in every detail, and each author is of course responsible only for his own contributions. The parts of the book to which authors' names are not attached have been written by the editor.

To the following gentlemen, not members of the committee, who have most kindly contributed sections or helped in other ways, the editor expresses his cordial thanks: to Professor Grenville Cole, F.R.S., of the Royal College of Science, Dublin, who contributed the section on the Soils of Ireland; to Professor G. S. West, of the University of Birmingham, who contributed the section on the British Freshwater Phyto-plankton; to Dr Horne, F.R.S., and Mr Crampton (the latter a member of the committee), of the Scottish Geological Survey, who revised the section on the Soils of Scotland; to Dr J. E. Marr, F.R.S., of the University of Cambridge, who read the proofs of the section on the Soils of England and Wales and contributed a paragraph on the East Anglian Heaths as representing a survival of steppe-conditions; to Dr Reginald Scully, of Dublin, who contributed the information on the vegetation of the Killarney woods; and to Mr J. A. Wheldon, of Liverpool, who kindly contributed information on the bryophytic flora of the Lancashire sand dunes.

The following gentlemen, other than members of the committee, have very kindly contributed photographs: Messrs W. Ball, F. F. Blackman, C. J. P. Cave, R. H. Compton, F. H. Graveley, F. F. Laidlaw, S. Mangham, R. Welch, and A. Wilson.

The information on the distribution of crops is mainly taken from the official "Returns of the Board of Agriculture and Fisheries" and from the Agricultural Statistics of the Irish Department of Agriculture and Technical Education; to the officials of the Board the editor is indebted for most courteously given supplementary information, involving considerable trouble, with regard to changes in the area of wheat land.

The editor also desires to express his special indebtedness to his colleague, Dr C. E. Moss, who, besides contributing important sections of the book, has given constant help and advice throughout[1]; and to Dr W. G. Smith, the Secretary of the Committee, who has read a considerable part of the proofs and contributed many valuable suggestions. The editor is indebted to his wife for undertaking the laborious task of constructing the general index.

The treatment adopted in Part II is first to give a general account of the various formations and associations recognised, with lists of species occurring in them. The plant-communities are then, in some cases, illustrated by descriptions of special instances.

[1] It is right also to mention that the concept of the plant-formation and of its relation to the association, which forms the logical basis on which the scheme of classification embodied in the book is constructed, owes its present form to Dr Moss. Cf. "The Fundamental Units of Vegetation," *New Phytologist*, Vol. IX. p. 18, 1910.

The imperfections of the book will be sufficiently obvious to the reader. Some are at present unavoidable owing to the unequal development of our knowledge of the vegetation of different parts of the British Isles. While some regions have been carefully explored and their vegetation analysed, others have been only "reconnoitred," and of others again, such as Wales and considerable parts of Ireland, the plant-communities, as distinct from the flora, are almost unknown.

Most of the continental work on plant-communities similar to or identical with British ones has been deliberately neglected. A serious attempt, which will eventually have to be made, to correlate British with continental plant-communities would have involved a task extending far beyond the immediate object in view.

The Bryophyta, and the lower plants generally, are, for the most part, mainly through want of knowledge, very inadequately treated, and in the case of many plant communities, ignored altogether. This is greatly to be regretted, since these plants are frequently of the first importance in differentiating plant-communities. The algal associations of the freshwater aquatic formation are scarcely touched upon.

In a book of this nature, involving the correlation of a great mass of unequally developed material from many sources, it is inevitable that there will be numerous errors and omissions. Great improvements will be possible in a second edition, especially having regard to the fact that the subject is advancing very rapidly; and the editor will be grateful for all criticisms and suggestions.

The treatment of the ecological effect of the different soils almost wholly lacks the basis which can only be afforded by careful quantitative analysis and comparison. Data of this kind are now being accumulated, but are not, as yet, sufficiently extensive to be of use. It is believed, nevertheless, that the statements made about the relation of soils to the different types of vegetation, which are partly based on analysis and partly inferred, are essentially accurate, and afford a sound basis of classification. It is believed, in fact, that the classification of vegetation and the treatment of the genetic relations of plant-communities are in advance of anything hitherto attempted, though modifications are of course to be expected as knowledge advances.

The field of analytical and experimental ecology lies widely open to workers able and willing to devote themselves to the laborious tasks involved in the attack on the various problems underlying the phenomena of vegetation.

The chief obstacle to the rapid development of ecology on fundamental lines is the laborious and time-consuming nature of the work and the chemical and physical training required for its prosecution. It is hoped that this book will serve to call attention to many of the more immediate problems.

Nomenclature of specific names. The nomenclature of species and the order in which they are placed in the general lists for the most part follow the 10th edition of the London Catalogue (1908). To continental readers the adherence to the system of Bentham and Hooker will seem an anachronism, but the convenience of following a recognised standard list of British plants dictated this course.

Symbols denoting frequency. The symbols denoting the frequency of species in the general lists of the composition of associations are as follows: *d* = dominant, *ld* = locally dominant, *lsd* = locally subdominant, *a* = abundant, *la* = locally abundant, *f* = frequent, *o* = occasional, *r* = rare, *vr* = very rare. It is to be understood of course that these symbols refer to frequency within each association as a whole, and have no reference to frequency in the country at large.

A. G. T.

GRANTCHESTER, CAMBRIDGE,
July, 1911.

CONTENTS

Introduction.

Part I. The Conditions of Vegetation in the British Isles.

Part II. The Existing Vegetation of the British Isles.

LIST OF PLATES

Folding diagrams

INTRODUCTION

THE UNITS OF VEGETATION—THEIR RELATIONSHIPS AND CLASSIFICATION

IT has been well said by the distinguished plant-geographer Professor Drude of Dresden that plants may be studied in pure science from three points of view, and from three only:—the physiological, the phylogenetic and the geographical. Geographical Botany or Plant-geography may be defined as the study of the facts and causes of the distribution of plant-life over the surface of the earth.

When we consider the distribution of the plant-life of any given region or country, whether large or small, we find that it may be regarded from two distinct points of view. In the first place we may study the distribution of the species, and, in the case of the larger areas of the earth, the distribution of the genera and families of plants. The list of species, arranged taxonomically, is called the *flora* of the region or country, and the study of their distribution is *floristic* plant-geography.

But there is another way of regarding the distribution of plant-life. If we consider the general plant-covering or vegetation of any geographical region we find that it is naturally divided into units, each of which has a characteristic appearance or physiognomy and consists of characteristic species. These *vegetation-units* have been recognised from the earliest times, since not only do they form outstanding features of all landscapes, but man absolutely

Vegetation-units or plant-communities.

depends upon them in his primitive relations to nature. In their more obvious forms, therefore, the units of vegetation have common names in all languages, and these are everywhere preserved in the names of places. *Wood, moor, heath, marsh,* are some of the commonest and most obvious examples. Now it is to be observed that most of these names refer to something more than the vegetation alone. A moor is an area supporting certain kinds of plants, but it also implies the presence of a peat-soil on which those plants flourish. Heathland nearly always involves a relatively poor and dry soil. A marsh is an area where the soil is always wet. In other words a vegetation-unit is always developed in a *habitat* of definite characteristics.

To look at the matter from the opposite standpoint, synthetically instead of analytically, certain kinds of plants are always found associated together under definite conditions of life, and such groupings may be called *plant-communities.* A plant-community is simply a vegetation-unit regarded as an aggregation of species and individuals instead of as a division of the whole vegetation of the region. Whichever standpoint we adopt, the connexion with habitat is a fundamental part of the conception. This way of considering the distribution of plant-life is called *ecological plant-geography,* from the Greek οἶκος, a house (habitat), in contradistinction to the floristic plant-geography, referred to above, which is concerned primarily with the distribution of species.

Ecology includes more than the study of vegetation-units or plant-communities; it deals with the whole of the relations of individual plants to their habitats. This latter branch evidently cannot be sharply separated from physiology; and may in fact be justly considered as a part of that subject. It has been aptly called by Professor Schröter of Zurich *autecology,* to distinguish it from *synecology* or the study of plant-communities.

The present book attempts to describe the principal kinds of plant-communities or units of vegetation met with in the British Isles. During the last twelve years a great deal of work has been done in observing and describing these vegetation-units, and their relationships. The work is far from complete even in the matter of mere description, and this book necessarily carries many evidences of the incompleteness. Yet enough is known to justify the attempt at a preliminary sketch of the subject, which it is believed will interest botanists and lovers of nature, as well as students of scientific geography.

It may be said that we ought not to occupy ourselves with synecology till we have a complete or an approximately complete knowledge of autecology, but this is a mistaken notion. It might as reasonably be contended that we ought not to study the phenomena presented by the nations and races of men before we know all about the physiology and psychology of the individual man. As a matter of fact the study of synecology is considerably in advance of autecology (which is indeed still in a very backward state of development), and the progress made has amply justified the attention devoted to the wider though less fundamental branch of the subject. The plant-community, in fact, offers a convenient mode of approach to the study of plant-life in relation to habitat. The systematic description and classification of vegetation affords a natural framework in which autecological studies will find their proper place.

Nomenclature and classification of vegetation-units.

There has recently been a great deal of discussion as to the nomenclature of ecological phytogeography or synecology, and nothing like general agreement yet exists as to the proper naming and classification of the different categories of vegetation-units or plant-communities. One of the main subjects of contention is the use of the term *plant-formation,* which we originally owe to Grisebach[1]. It is not intended to enter here into a discussion of this question, for which the literature must be consulted[2]. The uses of the terms *plant-formation*, *plant-association* and *plant-society* in this book are those adopted by the Central Committee for the Survey and Study of British Vegetation. Whatever may be the ultimate fate of these terms it will be generally conceded that their consistent use in the description of the types of vegetation of a given region will afford a useful test of their appropriateness.

The plant-formation, the plant-association, and the plant-society.

The Plant-formation

A plant-formation is the natural vegetation occupying a habitat with constant general characters which determine the communities of plants occurring in that habitat[3]. The characters, or as they are often called the *ecological factors*, of the habitat, which influence vegetation, are often classed as *climatic* and *edaphic*.

Ecological factors.

Climatic factors are those primarily dependent on the climate of the region in which the plant-formation is developed, such as temperature, precipitation (i.e. rainfall, snowfall, dew, etc.), humidity of the air and wind.

[1] Grisebach, H. R. A., "Ueber den Einfluss des Climas auf die Begränzung der natürlichen Floren." *Linnæa*, XII. 1838.

[2] See especially Schimper, A. F. W., *Pflanzengeographie auf physiologischer Grundlage*, Berlin, 1898 (English translation, *Plant Geography upon a Physiological Basis*, Oxford, 1903–4); Clements, F. E., *Research Methods in Ecology*, Lincoln, Neb., U.S.A., 1905; Warming, E., *The Œcology of Plants*, Oxford, 1909; Moss, C. E., "The Fundamental Units of Vegetation," *New Phytologist*, IX. 1910 (issued separately by the *New Phytologist*, Botany School, Cambridge); Ch. Flahault and C. Schröber, *Phytogeographical Nomenclature*, Zurich, 1910.

[3] This of course is not a logical definition but only a preliminary description. The conception of a plant-formation, as understood in this book, can only be made fully clear by reference to the examples dealt with in Part II.

Edaphic factors, on the other hand, depend upon the features of the soil in which the plants grow, such as water-content, food-content, aeration, the presence and amount of humus and of certain minerals, acidity, and so on[1].

Not all the features of a habitat can be reckoned as effective ecological factors. Thus two soils, alike in many respects, may differ in others, such as the presence or amount of certain salts. But unless the differences have an actual effect on the vegetation they cannot be considered as ecological factors. Similarly one locality may have twice the rainfall of another, and yet the same type of soil may bear the same plant-formation in both.

The effective ecological factors which separate one formation from another we shall call the determining, differentiating or *master-factors*.

In the British Islands the most striking effect of climate is seen on the higher mountains. The various climatic factors involved in differences of altitude determine for instance a limit above which woodland does not extend; and the rock-vegetation near mountain summits is another effect, though partly an indirect effect, of increased altitude (see Chapter XIV). There is also a considerable difference in the climate of the eastern and western portions of Great Britain, and this is probably related to the prevalence of moors in the west as opposed to heaths in the east.

Characteristic but subordinate features of other plant-formations also depend on climate, for instance the abundance of ferns in the woods of the west and north as contrasted with their paucity in the east and south-east. Certain species of plants are confined to parts of the country characterised by a definite climate, and their absence elsewhere may often be due to climatic factors,

[1] It must be understood that nothing approaching a detailed treatment of ecological factors, and of their relations to one another, is possible in the present work.

though a strict determination of the various cases is lacking. But these phenomena, interesting as they are, do not count in the determination of the fundamental units of vegetation, the plant-formations, and we may fairly say that, on the whole, the plant-formations of the British Isles are *mainly* determined by edaphic factors, *i.e.* by soil.

Determination of plant-formations.

Many plant-formations are easy to determine, because they are limited by quite obvious habitats to which very definite and characteristic plant-communities correspond. Such are for instance the two chief "maritime" formations, those of the sand dunes and of the salt marshes.

The moor-formations are also well characterised, if we take the word "moor" in its current German sense to apply to types of vegetation developed on deep peat[1]. Two principal formations with essentially different plant-communities may be distinguished, that on peat relatively rich in mineral salts and neutral or alkaline in reaction—"fen" as it is called in this country (*Niedermoor* of Weber, *Flachmoor* of various authors)—contrasting strongly with the formation on peat poor in mineral salts and acid in reaction—the moor proper (*Hochmoor* of German writers). Whether we should distinguish a third formation—the "transition-moor" (*Uebergangsmoor*)—is open to question.

The vegetation of chalk and limestone, of the non-calcareous clays and loams, of the sands relatively rich in plant-foods, and of the sands and gravels relatively poor in plant-foods and developing an acid humus, fall into different plant-formations. Each of these types of soil bears vegetation with characteristic features and characteristic species of plants, though many of the species are common to several. There are however certain

[1] This use is not, however, in accord with the current use of the common English word "moor." For a discussion of the subject see Chapter IX.

superficial difficulties in determining the nature and limits of these formations, difficulties mainly due to human interference with natural vegetation.

In the days before the face of the country was changed by the activity of man woodland prevailed over practically the whole area of the types of soil mentioned, with the exception, probably, of parts of the first and last, *i.e.* of the chalk and limestone and of the poorer sands. At the present day, cultivated crops, "permanent pasture" which has been "laid down" and plantations, occupy, of course, the greater portion of the country, but apart altogether from these the natural vegetation of each type of soil is actually varied. Besides woodland each type has a characteristic "scrub" or bushland, and a corresponding grassland or, in the case of many sands, a heathland.

Retrogressive changes in formations.

These different types of plant-community on the same soil have no doubt originated mainly from the clearing of the original woodland and the pasturing of sheep and cattle. This prevents the regeneration of the woodland, and of most of the shrubs also, if the pasturing is sufficiently heavy and continuous, while it encourages the growth of grasses. Thus the plant-formation determined by the particular soil, and once represented by woodland, shows a series of phases of degeneration or retrogression from the original woodland, brought about by the activity of man. The intimate relationship of the various phases are clearly seen in the associated plants. The woodland proper has of course a ground vegetation consisting of characteristic shade plants, but the open places, and the "drives" or "rides" of the wood, are occupied by many of the species found among the scrub and in the grassland, while those true woodland plants which can endure exposure to bright light and the drier air outside the shelter of the trees often persist among the grasses of the open. In some cases where grassland is not pastured the shrubs and trees of the formation recolonise the open land, and woodland is regenerated[1].

Besides these degenerative processes, due to human interference, there are others due to "natural" causes, which are for the most part little understood.

Primary development or primary succession in a formation.

The natural process of the development of formations on new soils has also to be taken into consideration. The formation which is to occupy a new soil rarely springs into being fully constituted. It normally passes through a series of phases of development (primary succession), each phase exhibiting a definite *plant-association* (see below). Clements has given a very interesting account of this primary succession of associations (called by him "formations") in the case of several American formations; while the primary successions met with on sand-dunes, on salt-marshes, and on the borders of inland lakes where peat accumulates, have long been familiar to European plant-geographers. Clements[2] has shown that primary succession follows definite laws, and has pointed out that the pioneer species generally form an association in which but few species occur and which does not completely cover the soil (open association), while the succeeding associations are mixed in character and the final association is closed and often dominated by a single species, or by a few similar species.

Primary succession must take place universally where new land habitable by plants is formed, but in a country like our own, where geological changes are slow and insignificant, there is little opportunity for observing it

[1] This very brief summary of the relation of woodland, scrub and grassland or heathland omits all reference to complicating factors. Some of these are discussed under the different formations in Part II.

[2] *The Development and Structure of Vegetation*, Lincoln, Neb., U.S.A., 1904; *Research Methods in Ecology*, Lincoln, Neb., 1905. See also Moss, *op. cit.*

except on the sea-coast where new land is constantly being formed in certain places, on the edges of certain lakes where the land is gaining on the water, and to some extent on the talus of cliffs, and on the detritus of mountains.

The colonisation of waste ground, such as derelict building sites, of ballast heaps, etc., furnish, it is true, very interesting and instructive studies in the succession of plant-communities; but owing to the peculiar nature of the substratum in many of these cases, and also owing to the supply of seed available, these successions do not often lead to the establishment of communities belonging to the normal plant-formations of the country, and such land forms a battle-ground for aliens and casuals.

In the normal primary development of a formation, as in the retrogressive processes before mentioned, the associations involved show intimate relations and transitions one to another, and the whole set of associations has a definite flora dependent on the type of soil. It is for these reasons that we consider the entire set of plant-communities on a given type of soil, in the same geographical region, and under given climatic conditions, as belonging to one formation, in spite of the diversity of the dominant plant-forms in the different associations.

The plant-formation thus appears as the whole of the natural and semi-natural[1] plant-covering occupying a certain type of soil, characterised by definite plant-communities and a definite flora. The "wilder" formations, those less modified by man, for instance sand-dunes, salt-marshes, heaths and moors, are quite easily determined, but where human agency has been extensively at work, a careful study of the flora and vegetation and of their relations to the soil is needed before the formation can be accurately determined. In some cases we cannot as yet

[1] *i.e.* owing its present form to human activity though not planted by man.

decide with certainty, in others we may never be able to decide, where to draw the limits of the original formations. For instance, much of our heathland is undoubtedly degenerate forest, but in other cases it is probably primitive, existing on soil which does not naturally bear woodland at all. Let us suppose that a tract of primitive heathland (which must be counted as a separate formation) adjoined a tract of woodland on a light sandy soil. If the woodland is extensively cleared the heath community will undoubtedly invade its area, and the original limits of the two formations will be obliterated. The same is true of chalk pasture. There is evidence that some of our chalk downs are primitive grassland, that is, were never covered by forest; but there are many tracts of chalk pasture which certainly occupy the sites of old beech and ash woods that have disappeared under the axe or from other causes. It is difficult or impossible in the present state of our knowledge to draw the limit between the two cases, which may be represented by identical associations.

When habitats and their corresponding plant-communities are separated by characters less important than the master-factors determining formations, the term *sub-formation* is employed to designate a division of a formation, based on these less important factors. A sub-formation however exhibits the same features (of succession etc.) as a formation and is to be distinguished carefully from an association.

The Plant-association

The plant-association is the vegetation-unit next below the plant-formation. Plant-associations are in general the most obvious plant-communities that we recognise in the field. Thus each of the types of vegetation, woodland, scrub and grassland, *within a given formation*, is a plant-association, and so is each definite phase in the primary development of a formation.

Chief and subordinate associations.

The highest type of association within a formation (often woodland), to which development tends, is called the *chief association* of the formation. In the absence of disturbing factors, such as the interference of man, landslips, and so on, the chief association will ultimately occupy the natural formation-area to the exclusion of the other associations, which may be collectively designated *subordinate associations*[1]. But other associations may attain to a considerable degree of stability as the result of an interference with the normal course of development, even apart from continuous human interference.

Sub-associations and facies.

In some cases smaller differences of habitat, of scarcely sufficient importance to warrant the recognition of distinct associations, may nevertheless bring about certain changes in the vegetation. Communities thus differentiated may be called *sub-associations*. In other cases where there is no definite segregation of distinct communities, but the vegetation varies in character, we may speak of different *facies* of the association.

A plant-association, especially a chief or other stable association, is generally dominated by a single social species (*e.g.* in many woodland associations), or by several social species of the same or similar plant form, as in grassland. These *dominant species* determine the physiognomy of the association and largely also (by their habit and mode of growth, by the protection they afford, the shade they cast, and so on) the *associated species* found among or beneath them[2].

[1] Moss, "The Fundamental Units of Vegetation," *New Phyt.* Vol. IX. p. 38.

[2] The technical name of an association is formed by the addition of the suffix *-etum* to the stem of the generic name of the dominant species, followed by the specific name in the genitive. Thus a wood dominated by the pedunculate oak is a *Quercetum Roburis*, a name which is really an

The association has a very definite and usually a much more restricted flora than the formation, largely owing in the case of closed associations to the special conditions determined by the dominants. Transitions between different associations of the same formation are constantly met with, and in these a mixture of both vegetation and flora occurs, a fact which serves to knit the various associations together and to emphasise the unity of the formation to which they belong.

THE PLANT-SOCIETY

Locally within an association there occur more or less definite aggregations of characteristic species or of small groups of species, and these, which appear as features within the association, may be recognised as smaller vegetation-units or *plant-societies*. Sometimes their occurrence may be due to local variations of the habitat, at other times to accident and the gregarious habit originating from a general scattering of seed in one place or from the social growth of a rhizomic plant[1]. While a plant-formation is always made up of associations, an association is not necessarily or even generally made up of societies, which are essentially local discontinuous phenomena.

Thus of the three orders of vegetation-unit or plant-community which we recognise, the largest, the formation, is absolutely determined by habitat. Transitions between

abbreviation of *Quercetum Querci Roburis*. A qualifying adjective is sometimes tentatively added.

[1] It is a question whether it would not be better to separate these two causes of the formation of societies within an association, and to restrict the term society to the aggregations due to the latter alone. In this way we should obtain a more logically coherent conception. But the more detailed analysis of vegetation has scarcely progressed far enough at present to justify a finer classification of plant-communities.

formations are found only where intermediate habitats occur, and are, on the whole, rare and restricted in area, because the great habitats depending on soil are essentially distinct units, except in those cases in which a fundamental and progressive change in the habitat is going on, *e.g.* in the gradual conversion of an aquatic into a terrestrial habitat. In such cases we have to recognise the gradual supplanting of one habitat by another. Man has, in general, no power to change the formation, except by changing the whole character of the habitat, *e.g.* by draining marsh land, or by substituting planted crops for the natural vegetation.

The association, on the other hand, is determined by minor differences of habitat or by the course of succession, and is often completely changed when human interference takes place. Transitions between associations are very frequent, owing to the gradual nature of progressive and retrogressive changes, and to the variation from place to place of the minor features of the habitat.

Finally, plant-societies are minor features of vegetation, and their presence in certain spots is generally determined by some biological peculiarity of a species, not by the habitat as such.

The classification of the units of vegetation here adopted cuts across the superficial classification based on plant-form, physiognomy, or mere topography. Thus we do not divide vegetation into groups of tree-formations, shrub-formations and herb-formations, because species belonging to all these plant-forms, and associations dominated by these species, actually occur in one and the same general type of habitat.

Habitat must be the basis of any natural classification of vegetation because it is on similarities and differences of habitat that vegetation ultimately depends. It is true that plant-forms have an important relation to habitat, a relation which depends on the phenomenon of adaptation, but it is not true that plant-form runs everywhere *pari passu* with habitat. The net of adaptation is not so close-meshed, nor is it drawn so closely round the manifold of organic form, as to bring about a rigidly exact correspondence of this kind. In broad outline the correspondence exists, and we may classify the sum-total of the plant-formations of the world by the plant-form of the dominant species of their chief associations. Such a classification is the expression of the fundamental climatic zonation of the world and of the great mountain complexes.

But a classification of this kind lies outside the scope of a book devoted to the vegetation of a comparatively small country. Our primary task in the present work is to determine the formations themselves, and if we used plant-form as the basis we should associate units of vegetation which are naturally distinct, and divorce others which are naturally associated. Thus we should bring together all deciduous tree-associations in this country and separate them from the naturally related grass- or heath-associations. Such a classification would be quite barren, leading nowhere, because it would ignore habitat.

PART I

THE CONDITIONS OF VEGETATION IN THE BRITISH ISLES

CHAPTER I

THE PHYSICAL CHARACTERS AND CLIMATE OF THE BRITISH ISLANDS

THE British Islands stand on the "continental shelf" of north-western Europe. The depth of the seas (English Channel and North Sea) separating them from the neighbouring countries of the Continent nowhere reaches 200 metres and very rarely exceeds 100 metres. On the west of Scotland and Ireland the floor of the Atlantic Ocean slopes rapidly down to a depth of 2000 metres and in the course of a few hundred miles depths of over 4000 metres are reached.

Similarity of flora and vegetation to that of the Continent.

Thus the British Isles are physiographically simply a slightly detached part of north-western Europe, and their separation took place in very recent geological times. This fact is reflected in the flora, which is for the most part identical with that of the adjacent parts of the continent from which it is derived.

Floristically the British Isles are comparatively poor, for many continental species fail to reach them, and they have few, if any, endemic species[1]. Their isolation has not been sufficiently prolonged nor are their climatic conditions sufficiently distinct for the evolution of a distinctive flora. The most interesting floristic groups are certain species confined to the east and south-east, which are mid-European forms not found in northern Europe; the "arctic-alpine" species of the Scottish mountains, of which a few occur also on the mountains of northern England, North Wales and Ireland; the species confined to Ireland and south-west England, some of which are found in western France, while others (notably *Arbutus Unedo*) are Iberian or Mediterranean plants; and finally a small group of American species found in Ireland. It will be seen that all these groups are simply extensions (occasionally discontinuous extensions) of continental floras to the nearest parts of the British Isles. Into the question of their mode of migration we cannot enter here.

With regard to vegetation too, it is doubtful if we have any plant-formations unrepresented on the Continent, though it is impossible to be certain till a more systematic comparison has been made.

Geology and orography.

In spite of their comparatively small size the British Isles show a considerable variety of physiographical features corresponding with the very extensive series of geological formations represented.

Considering the island of Great Britain (Scotland, England and Wales) alone it may be said that the older and harder (Palæozoic and metamorphic) rocks, forming

[1] *i.e.* species in the aggregate sense; in the last (10th) edition (1908) of *The London Catalogue of British Plants* there are however about 30 species and 100 varieties of flowering-plants and ferns, exclusive of the genera *Rubus* and *Hieracium*, described by modern British systematists and not recorded from other countries. Besides these there are about 30 "endemic" *Rubi* and 70 "endemic" (mainly arctic-alpine) *Hieracia*.

by far the greatest bulk of the high land and all the hills over 1500 feet (*c.* 500 metres) in height, occupy the northern and western part of the country, while the midland, southern and eastern portions are composed of the newer and softer (Mesozoic or Secondary, Tertiary, post-Tertiary and Recent) deposits, which form flat or undulating plains broken only by comparatively low hills (Figs. 1 and 2).

If we draw a straight line from Hartlepool on the Durham coast of the North Sea southwards to Leicester, and another from Leicester south-westwards to the Dorsetshire coast of the English Channel we find that the country lying to the west of these lines is very largely composed of Palæozoic rocks, that about three-quarters of it has an elevation of over 500 feet (*c.* 150 metres), while practically the whole of the Scottish "Highlands," forming more than one-sixth of the total area, considerable regions of the "Southern Uplands" of Scotland, of northern England, of North Wales, and small areas in south-western England reach a height of over 1500 feet (*c.* 460 metres).

Elevations of more than 2500 feet (*c.* 760 metres) are rare except in the Scottish Highlands, where there is one considerable *massif* (the Cairngorm group in Perthshire) which has an extensive area exceeding that altitude, while there are several summits reaching above the same level in the Southern Uplands, in northern England and in North Wales.

The highest altitudes attained are Ben Nevis (4406 ft. = 1343 m.) in the Western Highlands, Ben Macdhui (4296 ft. = 1309 m.) and Braeriach (4248 ft. = 1294 m.) in the Cairngorm group, Ben Lawers (3984 ft. = 1214 m.) in the Central Highlands, and many other Highland peaks over 3000 feet; Snowdon (3560 ft. = 1084 m.) and Scafell Pike (3210 ft. = 978 m.) in North Wales and the Lake District respectively (both districts possessing several

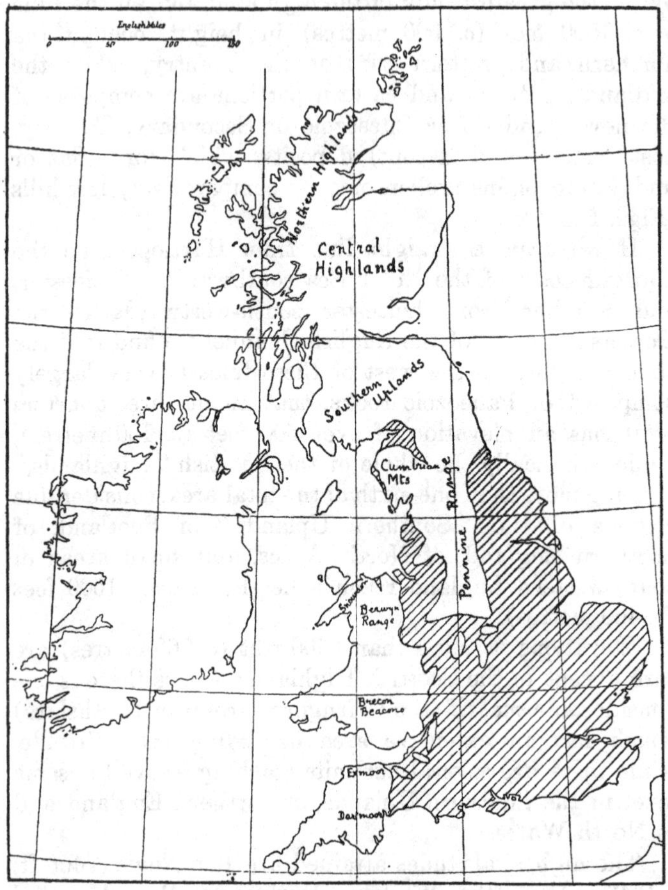

Fig. 1. The British Isles: showing the distribution of Archæan and Palæozoic (harder) rocks (unshaded), and Secondary, Tertiary and post-Tertiary (softer) rocks (shaded). Glacial deposits are ignored.

Comparison with Fig. 2 shows that the great mass of the ground over 500 feet and all of that over 1500 feet is formed of the older rocks.

Fig. 2. **The British Isles: showing the distribution of high ground. Below 500 feet (unshaded); between 500 and 1500 feet (diagonally shaded); between 1500 and 2500 feet (cross-hatched); above 2500 feet (black). The extent of high ground has been somewhat exaggerated owing to smallness of scale; not all the isolated summits above 2500 feet are indicated.**

other peaks above 2000 ft.); several hills in the Pennine range over 2000 feet; and High Wilhays on Dartmoor in Devon (2039 ft. = 621 m.).

To the east of the imaginary line mentioned the contrast is extremely striking. Palæozoic rocks are not represented at the surface, the gently undulating plains are broken by comparatively few ranges of hills exceeding 500 feet (*c.* 150 m.) in height; only the Cleveland *massif* in north Yorkshire exceeds 1000 feet (*c.* 305 m.), and there are no elevations reaching 1500 feet (*c.* 460 m.).

Ireland, which is mainly composed of Palæozoic rocks, has its hilly regions in the north, west and south, enclosing an extensive central plain of an elevation less than 500 feet. The highest altitudes are found in Macgillicuddy's Reeks (3414 ft. = 1040 m.) in the south-west, but there are several other mountains exceeding 2000 feet (610 m.).

Causes of the British climate.

The mild "oceanic" climate of the whole of the British Islands is determined in the first place by the great length of coast-line compared with the total area, and the consequent general dampness of the air, which tends to lower the temperature in summer and to raise it in winter. This effect is of course a direct function of proximity to the sea, as can be seen very strikingly by an examination of maps recording the mean maximum temperatures for January and the mean minimum temperatures for July. The lines indicating equal values are in both cases concentric about the land masses[1].

The second great factor is the direction and character of the prevailing winds. Owing to the nearly constant position throughout the year of the north-Atlantic area of low barometric pressure, in the neighbourhood of Iceland, the prevalent winds during the whole year are from the south-west, *i.e.* they are warm moisture-laden winds coming from the warm regions of the mid-Atlantic and

[1] Bartholomew's *Physical Atlas*, Part III. *Meteorology*; 1899, p. 11.

traversing a great extent of ocean. These winds, impinging on the colder elevated land of the west coasts, are diverted upwards into colder strata of air, forming dense mists and clouds and leading to considerable rainfall on the western slopes and summits of the hills.

Spatial distribution of rainfall.

Thus much of the west coast of Great Britain, most of the Scottish Highlands and Southern Uplands, and most of western Ireland receive between 40 and 60 inches (*c.* 1000 to 1500 mm.) of rain *per annum*, while there are large areas in the western Highlands, in Wales, and in south-west Ireland, and smaller ones in the Southern Uplands and the Cumbrian mountains, which have a rainfall of over 60 inches (*c.* 1500 mm.). Considerable areas in the western Highlands, and smaller patches in most of the other regions mentioned, receive more than 80 inches (*c.* 2000 mm.), while the annual rainfall in particular stations reaches very high figures. Thus Seathwaite in Cumberland has 177 inches (*c.* 4500 mm.), the highest record in Europe, while Ben Nevis in the western Highlands has 151 inches (*c.* 3835 mm.).

On the eastern side of the mountains the air descends again to lower and warmer levels, and the annual rainfall rapidly falls off. The central plain of England, much of the south-east, a narrow strip of land running along the north-east coast as far as the Firth of Tay, and other strips bordering the Moray and Pentland Firths, have a rainfall between 25 and 30 inches (*c.* 625 to *c.* 750 mm.). Thus Brighton has 28·33 inches, Birmingham 26·54, Newcastle 27·81 inches, St Andrews 28·45 inches. Finally the Fenland extending inland from the estuary of the Wash, as well as adjoining country to the north and south, a strip of east Essex and south-east Suffolk north of the Thames estuary, and various small strips further north along the eastern coast of Great Britain, have an annual rainfall below 25 inches. The lowest recorded

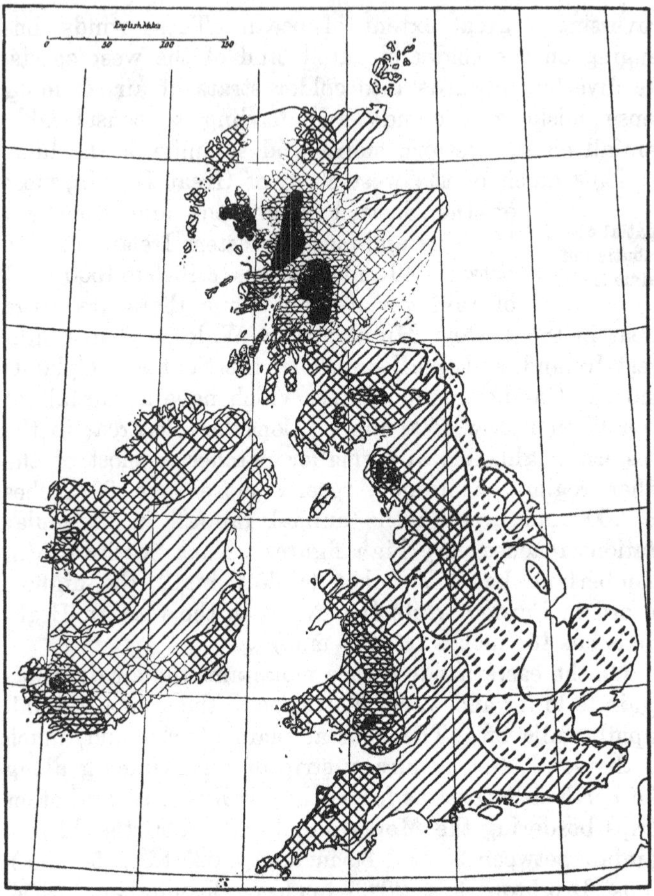

FIG. 3. Distribution of mean annual rainfall in the British Isles.

Below 25 inches (625 mm.)

25—30 inches (625—750 mm.)

30—40 inches (750—1000 mm.)

40—60 inches (1000—1500 mm.)

60—80 inches (1500—2000 mm.)

Over 80 inches (2000 mm.)

means are 19·1 inches (485 mm.) at Spurn Head on the Yorkshire coast, 20·6 inches (523 mm.) at Shoeburyness on the Essex coast, 22·4 inches (569 mm.) at Ely, etc. These areas of lowest rainfall all lie on flat eastern coastlands and the immediately adjoining flat country (Fig. 3).

Seasonal distribution of rainfall.

The seasonal distribution of rainfall in the British Isles shows no very marked features, the rain in all stations being fairly well distributed between the different seasons. The autumn months (September, October and November) are the rainiest months over most of the country. In the mountainous regions of the west however the heaviest rainfall occurs in the winter months (December, January and February). The driest months are February and March in many parts of the east, April and May over most of the country and June in the east of Ireland, south-west England and some parts of eastern Scotland.

Throughout the year the climate of the country is characterised by the very frequent passage of a series of barometrical depressions of mild cyclonic type, mostly coming from the west, and the anticyclonic periods of dry weather are comparatively few and quite uncertain in occurrence.

Sunshine, cloud and mist.

Cloud and mist are very prevalent. An examination of a sunshine or cloudiness map of Europe shows at once the deficiency of mean annual sunshine in the British Isles as compared with western Europe generally. Only the south, and parts of the east and south-west, coasts receive more than 1500 hours of bright sunshine per annum, while nearly the whole of France and much of central Europe receives more than 1750 hours per annum. Nearly the whole of Scotland and parts of England and Ireland get less than 1250 hours in the year, an amount exceeded everywhere else in Europe except in Scandinavia. In May nearly the whole of the British Isles receives less than 200 hours in the month, while western Europe generally, except Denmark and Scandinavia, receives more than 200. A comparison of these figures with those of rainfall shows that the British Isles as a whole are considerably cloudier in proportion to their rainfall than is the case in most of western Europe.

Distribution of temperature. Winter isotherms.

The effect of the warm south-west winds on the islands as a whole, and especially on the west coasts, is also well seen in the distribution of the isotherms (lines of equal temperature), especially during the winter months (Fig. 4). The course of the January isotherm of 5° C. (41° F.) is particularly instructive. This isotherm, which has a mainly east and west direction through southern Europe, turns northwards in south-western France, crosses the English Channel, bending north-westwards to the Dorsetshire coast, continues into South Wales, pursues a northerly course to a point between the Isle of Man and the Scottish coast, makes a great bend to the south-westwards including the whole of central and north-eastern Ireland, and then returns to the islands off the western Scottish coast, through which it runs due north into the northern seas.

Thus the whole of south-western England, the western coasts of Wales, the whole of southern and western Ireland and the western islands of Scotland have a mean January temperature as high as that of western and parts of southern France, while Cornwall and the extreme south-west of Ireland has a mean January temperature as high as that of the Mediterranean coast of Provence (Riviera). It is this fact of course that accounts for the successful cultivation in the open air of many Mediterranean and subtropical plants, particularly of many evergreens, in these parts of the country, and probably also for the occurrence of *Arbutus Unedo* in the Killarney woods. Many native herbaceous plants too,

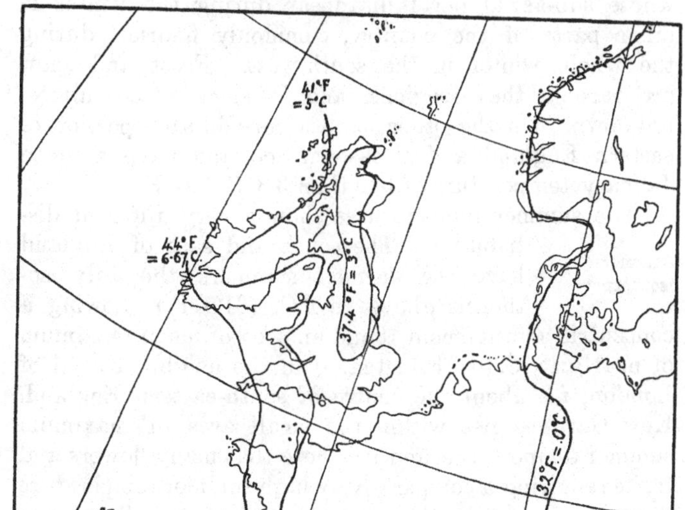

FIG. 4. Course of January isotherms in western Europe, showing the influence of the warm air current from the Atlantic on the western coasts. Data taken from Bartholomew's *Physical Atlas*, Part III. *Meteorology*, 1899.

whose sub-aerial parts disappear during the winter in other parts of the country, commonly flourish during the whole winter in the south-west. Frost and snow are rare in these regions, and in some places almost unknown. On the other hand a considerable portion of eastern England and of eastern Scotland have a mean January temperature of less than 3·3° C. (36° F.).

Summer isotherms.

The summer temperatures show a very different distribution. The south and east of England have the warmest summers, the July isotherms above 16·5° C. (61·5° F.) showing a concentric arrangement round an area of mean maximum of nearly 18° C. (64° F.) situated in the neighbourhood of London, *i.e.* about the centre of south-eastern England. Kew Gardens lies within the small area of maximum summer temperature, and it is here that many flowers and fruits requiring a comparatively high summer temperature can be brought to the greatest perfection. The lower July isotherms—east and west in general direction—bend very sharply northwards over the land both in England and Ireland, and correspondingly southwards as they approach the North Sea, the Irish Sea and the Atlantic Ocean, showing precisely the opposite behaviour in this respect to that exhibited by the January isotherms. The temperature falls off steadily northwards, the northern Hebrides showing a mean temperature of 12·78° C. (55° F.), which gives a mean range between January and July of only 7·8° C. (14° F.) as opposed to a range of 14·5° C. (26° F.) in London.

Summary of climate.

The general features of the climate of the British Isles may be summed up by saying that while the general type of climate is "insular," *i.e.* showing in a marked degree the moderating effect both on summer heat and on winter cold of the close proximity of the sea, this effect is far more marked on the western coasts, which are also very much wetter than the eastern and central portions of the country. The east central district of England (say from London northwards to the Wash) shows the nearest approach to a continental climate: the lowest rainfall, the hottest summers and the coldest winters.

Effects of climate: on distribution of plant-formations.

Some of the general effects of the geographical variation of the British climate on vegetation in relation to temperature have already been alluded to. With regard to rainfall one general effect on the distribution of plant-formations is the prevalence of moors dominated by various peat-forming species of Cyperaceæ, such as cotton grass (*Eriophorum*), and *Scirpus cæspitosus* (as well as the "grass moors" dominated by *Molinia cærulea*, *Juncus squarrosus* and *Nardus stricta*), on the western side of the country, and their absence in the regions of low rainfall on the eastern side, where deep peat is only formed under special edaphic conditions. On the other hand heaths on shallow, dry, peaty soils, dominated by the common ling (*Calluna vulgaris*), are commonest in the regions of lower and medium rainfall in the centre, south and east.

Grassland.

The agricultural effects are equally marked. The mild climate of the British Isles with its comparatively cool summers and well distributed rainfall is particularly well suited to grass, and indeed the country is deservedly famous for the richness and beauty of its pastures. A direct result of this is the flourishing condition of the pastoral industries, and the fine quality of the British beef and mutton. The damper climate and milder winters of western England and Ireland are the best suited to grass growing, and it is here as well as in the western midlands and parts of the north that the bulk of the pasture is found, and the cattle-raising industry particularly flourishes. In several of the western and midland counties there are more than twenty head of cattle and in Ireland twenty-three to the hundred acres of land of all kinds.

The great sheep-regions are the hillside pastures, such as the Southern Uplands of Scotland, the hillsides of Wales and the Chalk Downs of southern England. In these areas there are from 50 to 100 sheep to every hundred acres of land.

In some of the midland and western midland counties, such as Leicestershire, Warwickshire, Shropshire, Worcestershire and Gloucestershire, the area of permanent pasture exceeds half the total area of the county, and is more than double the total area of arable land. In Somerset the permanent pasture occupies about two-thirds of the area of the county and is more than four times the area of arable land.

Arable land.

These figures contrast very strikingly with those for the eastern counties, *e.g.* Cambridgeshire, where the proportions are reversed, the arable land exceeding two-thirds of the whole area and being more than three times that of the permanent pasture[1]. In Norfolk, Suffolk, Essex, Huntingdonshire, and Lincolnshire the area of arable land exceeds, and sometimes considerably exceeds, half of the total area of the county, while it approaches or exceeds twice the area of permanent grass.

Wheatland.

This eastern district with the large proportion of arable land is also the region where most wheat is grown, the area under wheat in any given year varying from one-sixth to one-quarter of the total area under arable crops. Eight of these eastern counties including Lincolnshire provide about 42 per cent. of the total wheat area of Great Britain, while the total of

[1] If the small patches of permanent pasture round the villages, which it is necessary to maintain for the grazing of farm animals and of cows for the local supply of milk, etc., were subtracted, the total would be very greatly reduced.

their land areas is only 11 or 12 per cent. of the whole island.

Considerably more than half the rest of the wheat area, *i.e.* about 33 per cent. of the whole, is found in the eastern midland, the western midland and the south-eastern groups of counties, whose total land areas make up between 19 and 20 per cent. of the land area of the island. Here again the average area under wheat in a given year is about one-sixth or one-fifth of the total area under arable crops, but, as already pointed out, the total arable area is very much smaller proportionally than in the eastern counties. A great deal more wheat used to be grown in the midland and south-eastern counties, but with the great decline in prices which occurred between 1875 and 1895 this region suffered a more severe reduction in its wheat area, as in its total arable area, than the eastern counties[1]. Though not perhaps quite so favourable for wheat, the climate of much of this region does not differ very markedly from that of the eastern counties. The midlands, however, contain a large proportion of heavy clay land, and this "strong soil" though quite good for wheat, and at one time considered as typical wheatland, requires a great deal of labour and does not pay to plough when the price of wheat is low. Consequently much of this land has been gradually laid down to permanent pasture.

Of the 25 per cent. of wheatland remaining to be accounted for the bulk occurs in the north and south-west of England, while the west and north of Scotland show areas which are practically negligible, the Orkneys, and Shetlands, Caithness and Sutherland (extreme north),

[1] The decrease in the area under wheat between 1875 and 1895, which was 42·75 per cent. in the eastern and 52·56 per cent. in the north-eastern counties, was nearly 60 per cent. in the south-east and over 60 per cent. in the midlands, though the wheat area was much smaller in the latter regions than in the former at the beginning of this period.

Argyll and Bute (extreme west), as well as some of the counties situated in the Southern Uplands, growing no wheat at all. According to W. G. Smith[1] "wheat ceases to be a regular crop of the farm at an altitude where the mean July temperature is below 56° F. [13·3° C.] and the rainfall exceeds 32 to 34 inches (about 800 to 850 mm.) per annum."

The total amount of wheat grown in Ireland is insignificant, but the province of Leinster, containing the driest part of the country and having the warmest summers, possesses the largest acreage.

It is thus seen very clearly that most of the British wheat is grown in the regions of low rainfall and comparatively warm summers, while the regions of high rainfall and the coolest summers grow no wheat at all.

Other cereals.

Barley, of which the total British acreage is now rather below that of wheat, is not so dependent on climatic conditions, but a large proportion is grown in the wheat-producing counties of England, though Wales and the east of Scotland have a good deal more barley than wheat, and no Scottish county shows absolutely none.

Oats, the third great cereal crop, mature well in far damper and cooler summers than either wheat or barley. In some recent years the acreage in Great Britain has been nearly double that of wheat, but in 1909, owing to a considerable increase in the wheat area and a slight decrease in the area under oats, the proportion was not quite so high. The distribution of the crop is very different. While the eastern counties grow considerable quantities of oats, the acreage is well below that of either barley or wheat. In the south-west, on the other hand, the acreage under oats approximately equals that under wheat and barley together, while in the north-west it is

[1] W. G. Smith, 1904, p. 627.

nearly three times as great. In Wales and south-eastern Scotland it is nearly twice as great, in eastern and central Scotland it is more than three times as great, while in northern and western Scotland the oats crop, though absolutely smaller than in the east and centre, is the only important cereal crop. Its limits are only fixed at the higher altitudes in the north by the early autumn frosts, which kill the plant before the grain is mature.

In Ireland the crop of oats enormously exceeds all the other cereal crops put together. The Irish acreage under potatoes is also very large, greatly exceeding the acreage of all the cereals together, except oats.

The rye crop is quite insignificant compared with the other cereals, though a little is grown in nearly every county except those of the extreme north of Scotland.

CHAPTER II

THE SOILS OF THE BRITISH ISLES

THE geological structure of the British Isles and particularly of England, is extremely varied, practically all the great geological formations being represented. The soils derived from these rocks are of course likewise very various, but for practical purposes, so far as they determine vegetation, they may be divided into comparatively few categories, according to their outstanding physical and chemical characteristics. Thus geological formations of widely different age, and often differing also in lithological constitution, weather to form soils of the same essential physical and chemical type, bearing the same kind of vegetation.

Types of soil. The main types of soil and their general characters are well known, but a short summary may be given here.

1. *Sand.* The typical sand is composed of relatively coarse particles of silica more or less mixed with other constituents. It is derived either from a recently formed loose sandy deposit that has never been consolidated, of which an extreme case is the blown sand of the sea-coast and of some inland regions, or from a coarse-grained sandstone, or from a crystalline rock, such as granite, containing much quartz. Sand forms an open light soil, little retentive of water. It is typically poor in lime,

though calcareous sands occur. Where surface-moisture accumulates, *e.g.* owing to the occupation of certain kinds of vegetation, acid peaty humus is formed, apparently largely owing to the deficiency of lime. Very fine-grained sandstones weather to a soil indistinguishable physically from clay, and bearing the same kind of vegetation.

2. *Loam.* This is formed of a mixture of sand and clay and results from the weathering of impure sandstones and similar rocks and from mixed alluvial deposits. It is more retentive of water than the purer sands, and where a fair proportion of lime is present forms the best agricultural soils.

3. *Clay.* The purer clays are mainly formed of hydrated aluminium silicate, are typically fine-grained, and very retentive of moisture. They crack when dry and form close heavy soils. Clay deposits, bearing typical clay soils, form the "rock" of much of the midlands of England. When clay is very deficient in lime it may, like sand, if kept wet, accumulate an acid peaty humus.

4. *Siliceous soils of the older rocks.* The older, non-calcareous rocks, which have become indurated and compacted by pressure, though composed of the same lithological constituents as the more recent sediments, often yield soils which have somewhat different characters from those of the newer sediments. Though technically they may be classified into sands, loams and clays, according as the particles are mainly coarse, a mixture of coarse and fine, or mainly fine, yet they often differ in respect of the vegetation they bear from the soils produced by more recent deposits. This is true of the majority of the Palæozoic and also of the metamorphic and igneous rocks. Thus the older Palæozoic shales and slates, which correspond lithologically with the more modern clays, do not bear the same vegetation as the clays, and the same may be said of the quartzites and many of the older sandstones and grits, which correspond

with the newer sands and friable sandstones. One feature in which these old rocks generally differ from the newer is that the soils they form are on the whole shallower, owing to the greater resistance to disintegration of the harder rocks, and this certainly affects the vegetation in an important manner. But we are still very ignorant of the effects of these soils upon plants, and a great deal of work is required before we shall be in a position to formulate a scientific classification of soils in their relations to vegetation.

The problem is also complicated by the fact that the older rocks are exclusively developed in the north and west of the British Isles, where the climate differs decidedly, as we have seen, from that of the south and east, and we cannot as yet present a satisfactory analysis of the factors due to climate as it affects the vegetation, not only directly, but through the soils. The combination of soils very poor in lime with high atmospheric humidity undoubtedly leads to the marked acidity which is so characteristic a feature of many of the soils of the north and west and has a most important influence in determining the vegetation. The natural drainage channels become clogged with mosses, impeding free drainage and tending to keep the soil constantly damp, while aeration is hindered, and acid peaty humus accumulates.

5. *Calcareous soils*, *i.e.* soils containing a fairly large proportion of lime, are mainly derived from limestones, calcareous sandstones, and highly calcareous marls[1]. They are mostly light open soils, except in the case of the heavier marls, but become sticky when wet, somewhat like clay. The influence of lime on the soil is certainly very great, though its effect upon vegetation and its relation to other factors are still very obscure.

[1] Marl is properly a mixture of clay and lime, but the term is sometimes loosely used in geology for ferruginous clays which may contain very little lime, *e.g.* the Keuper marl and other Triassic marls.

Calcareous soils accumulate "mild" as opposed to acid humus. Calcareous loams and marls are very favourable for agriculture; the lime is said to cause the clay particles to flocculate and thus makes the soil lighter and more open.

Very shallow-soiled limestones possess a characteristic dry soil, supporting a characteristic grass-vegetation.

6. *Organic soils* derived from plant-remains are of various types and differ very much in their effect on vegetation:—

(*a*) *Mild humus* such as is normally found on the floor of woods, and in meadows, has a neutral or alkaline reaction, and is well aerated by earthworms, etc.

(*b*) *Mild peat.* When there is excess of water, as on the margins of lakes, and a deficiency of oxygen, but lime and other bases are present in quantity, the humus accumulates and forms a purely organic soil or peat, whose water gives an alkaline reaction. This peat may reach a great depth and is the characteristic "black soil" of the fens (see p. 211).

(*c*) *Acid humus* is formed on soils deficient in lime, where the surface is moderately damp and oxidation is deficient.

(*d*) *Acid peat.* Under certain conditions of moisture and deficiency of oxygen, acid humus accumulates so that a pure peat soil is produced, and if the conditions are maintained this acid peat soil may reach a great depth, giving the characteristic "moor" soil (in the narrower sense, Chapter IX) and supporting the characteristic moor plants[1].

[1] The distinction here given between "mild" and "acid" humus and peat corresponds to a difference well known to agriculturalists, and having a most important effect on vegetation. It is not by any means certain however that the acid and alkaline reactions of the soil water which is found in extreme cases is the really important differentiating factor of the two classes of habitat. The researches of Van Bemmelen and the recent

7. *Salty soils* containing a considerable percentage of sodium chloride (common salt) or certain other soluble mineral salts, are very unfavourable for the majority of plants. They are characteristically inhabited by a special ecological class of plants—the *halophytes*. In this country they occur mainly in the mud flats of tidal estuaries which are covered by the high spring tides. Sand in immediate proximity to the sea, and spray-washed cliffs, are also often impregnated with salt and bear a halophilous vegetation.

8. *Rock, i.e.* hard rock not covered by loose earth, can hardly be called a soil in the ordinary sense. Where it forms horizontal or comparatively slightly inclined surface, the products of weathering, largely through the agency of lichens, mosses and algæ, quickly cover it with soil in most cases, but vertical or very steep rock faces which cannot hold soil are inhabited only by these plants, which can adhere to the bare rock face, and by an occasional higher plant which has succeeded in colonising a small thin patch of soil formed by a lichen or alga. Rock clefts in which soil accumulates often have a special vegetation of their own. The vegetation of rocks is influenced very much by the chemical and physical constitution of the rock, and also by such factors as available moisture, exposure, etc.

The talus of cliffs and rocks and "mountain top detritus" (p. 302) form other classes of substratum allied

work of Baumann (*Mitteil. d. bayrischen Moorkulturanstalt*, 1909 and 1910) has shown that humus is to be regarded as a colloid complex, and that its properties are determined by the special laws governing the behaviour of colloids to other bodies. The old conceptions of "humin" and "humic acid" do not, apparently, correspond with real chemical bodies, and the most important character of humus in relation to other chemical substances in the soil and to plants is whether it is "absorptively saturated" (mild humus) or "unsaturated" (acid humus). On these lines we may expect a real increase in our knowledge of the all-important relations of humus to plant life, a subject which has hitherto been very obscure.

to rock but with special characters influencing the vegetation (see Chapter XIII).

A detailed account of the distribution of British soils would be out of place in this book, even if the necessary data, which would have to be derived from a systematic soil-survey, were available. But some account of the general distribution of soils in relation to the geological formations may be given.

General distribution of geological formations.

As we have already seen (p. 17) the east and south-east of England, as well as much of the Midlands, are entirely composed of Secondary, Tertiary, post-Tertiary and Recent deposits, while the north, west and south-west, Wales, and nearly the whole of Scotland and Ireland are formed of metamorphic, Palæozoic, and intrusive igneous rocks. Very considerable areas of these older rocks however, especially in Scotland and the north of England, as well as some of the newer ones in the Midlands and east, are covered by deposits of glacial drift left by the retreating ice. This drift may form a heavy "boulder clay" or it may be sandy or loamy.

The archæan and Palæozoic rocks and the igneous formations of various ages, include all the hardest rocks, and they have been far more elevated and folded by earth-movements than any of the later ones. For these reasons they form all the mountainous land in the country, while the more recent formations, including the glacial deposits, are for the most part comparatively soft and yielding, have been but little elevated or disturbed, and form flat or undulating plains broken only by ranges of low hills.

Soils derived from rocks of different ages.

Many of the hard older rocks, such as limestones, sandstones, flags, slates, etc., and also many of the crystalline, metamorphic and igneous rocks, weather with difficulty, and consequently form very shallow or comparatively shallow soils, unsuitable for agriculture. On the other hand, the soft "rocks" (many of them really clays and sands, and therefore not rocks at all in any other sense than that of technical geology) of the south and east give rise to deep soils. It is largely for this reason, apart from the climatic factors already described (pp. 20–31), that the east, the south and the Midlands of England are the great agricultural regions of the British Isles. Many softer rocks, such as shales, are however interbedded with the harder members of the Palæozoic formations. These often occupy the valleys and sometimes form quite good agricultural land.

Considered as a whole, however, the Palæozoic, metamorphic and igneous rocks, with their thin soils, are very largely under natural or semi-natural grassland used for grazing, or bear woodland of definite and characteristic types, or are strictly "waste" land, *i.e.* covered with heath or moor on peat or peaty soil (see Chapters IV, XIII). Waste-land is not nearly so common on the newer and softer rocks, being mainly confined to the poorer sands bearing heath or heathy woodland. The loamy and clay soils are used for agriculture or laid down in permanent pasture. The woods are generally patches left mainly for the local supply of "small wood" and are almost invariably used for game preserves.

THE SOILS OF SCOTLAND

By W. G. Smith

The metamorphic, Palæozoic and igneous formations, which constitute nearly all the rocks of Scotland, contain no extensive deposits of limestone covering wide areas and comparable with the Mountain Limestone, Oolites or Chalk of England. Localised bands of limestone are, it is true, fairly abundant and widely distributed, but taken as a whole both the rocks and the widespread glacial drift derived from them are non-calcareous, and carry types of vegetation characteristic of "siliceous" soils.

Glacial deposits.

The soils of the lowlands up to about 1000 feet (*c.* 300 m.) have been almost entirely deposited by moving water or ice, and the solid rock is exposed only here and there. Glacial drift also frequently occurs above this altitude up to 2000 feet (*c.* 600 m.) in depressions or on terraces among the hills, while the summits and steeper slopes are largely covered by local rock-débris. It is probably safe to assume that about three-quarters of the surface-soil below 2000 feet is composed of this transported material, an assumption supported by observations on the vegetation, since the limits of the different vegetation-units by no means follow, at the lower elevations, the limits of the rock-formations.

Most of the glacial "boulder clay," and to some extent also the sands and gravels, show a tendency to form an acid humous soil or an acid and relatively pure peat. These humous and peat soils occur at all altitudes up to about 3300 feet (*c.* 1000 m.), and according to their different characters give rise to the different prevailing types of "moorland" vegetation so characteristic of the country. Thus *Calluna*-heath occurs on the drier humous soils, *Calluna*-moor on the drier peats, moors characterised by *Eriophorum*, *Scirpus cæspitosus* or *Sphagnum* on deep wet peat, wet grassland or grass moor characterised by *Molinia cærulea*, *Nardus stricta* and *Juncus squarrosus* on the peaty soils with more mineral content but still forming acid peaty humus, and dry grassland on the less peaty and drier soils. On the whole the wet moors and wet grasslands are typical of the wetter climate of the west, while the heaths, *Calluna* moors and dry grasslands prevail in the drier climate of central and eastern Scotland[1].

The Highlands.

Something like two-thirds of the surface of Scotland lies more than 1000 feet (*c.* 300 m.) above sea-level, and therefore above the agricultural area, which only in a few places reaches this altitude and rarely exceeds it. The great bulk of this elevated area is known as "the Highlands," the great hill complex which contains the highest summits in the British Isles (p. 17). The Highland area occupies practically the whole of the northern half of Scotland, except the broad shelf of low country fringing the North Sea (Fig. 2, p. 19).

The Highlands fall naturally into two districts, separated by a chain of lochs[2] lying in a deep trench (Glenmore—the big glen) which cuts across the country from north-east to south-west and marks the line of a great fault.

The North-western Highlands form a wild, remote, and mountainous region, though its average altitude is not so great as that of the Central and Eastern Highlands lying south and east of Glenmore. This northern region is much cut up by deep fjord-like sea-lochs, and by deep inland valleys, so that areas exceeding 2000 feet (610 m.) are small and detached, though numerous. The country is built up of schists, quartzites and granites; the high plateaux are in places covered with glacial deposits, and these with deep peat, from which the isolated ranges and summits rise abruptly. Though lime-containing soils are by no means absent, and sometimes influence the vegetation, the mantle of deep, highly acid, peat passes

[1] Cf. the vegetation map accompanying M. Hardy's "Esquisse de la Géographie et de la Végétation des Highlands d'Écosse," *Scottish Geographical Magazine*, 1906.

[2] These lochs are connected by the Caledonian Canal, which joins the North Sea and the Atlantic.

uniformly over everything alike and is uninfluenced by changes in the rock types. In the extreme north and along the east coast the rainfall is comparatively low (30 to 40 inches = 76 to 101 cm.), and here as in the Central and Eastern Highlands *Calluna*-moor (the driest type) and heath are more prevalent. In the west, which has a very high rainfall, mostly over 60 inches (152 cm.) and with large areas over 80 inches (203 cm.), the moors are largely dominated by *Scirpus cæspitosus.*

The whole area is naturally very thinly populated, and is difficult of access, except along the east coast where there is a fringe of cultivation. From what is known of its botany it appears to be poor floristically.

The Central and Eastern Highlands, lying south and east of the valley of Glenmore, likewise consist mainly of schistose metamorphic rocks, more variable in kind than those of the North-west Highlands. Granites occupy large areas. The district is characterised by possessing extensive continuous areas exceeding 2000 feet (*c.* 600 m.) in altitude, while considerable tracts above 3300 feet (*c.* 1000 m.) rise ultimately to several summits of 4000 feet (*c.* 1200 m.) or more. In this mountain mass glacial deposits are limited to the valleys, so that the influence of the lithology on the vegetation is much more evident than in the north-west. The western part of the area has a similar climate to that of the North-west Highlands, and bears similar vegetation, but the central and eastern portions are somewhat drier, and *Calluna*-moor and heath and the drier grasslands are prevalent. *Calluna* is commonly dominant below 2000 feet (*c.* 600 m.) and *Vaccinium Myrtillus* at higher levels.

Locally the quartzose schists are replaced by limestones, and by chloritic and mica-schists, etc., which introduce edaphic and topographical features favourable to the presence of "arctic-alpine" species at the higher altitudes. The Central Highlands include the richest localities in Britain for these species; among them may be mentioned the Cairngorm mountains on the north, and the Lochnagar group on the south of the Dee valley, forming the eastern region of the Central Highlands. To the south-west of these regions Ben Lawers and the neighbouring hills in Breadalbane have attracted many botanists by the variety of their flora. Robert Smith pointed out that these floristically rich areas are accompanied by considerable tracts of mountain grassland of a comparatively dry type, which occurs on soils formed from rocks richer in minerals than the quartzitic schists which form the basis of the ericaceous "moors[1]" (heaths).

Central Lowland Plain.

The Central Lowland Plain of Scotland is an extensive tract of comparatively low country, only occasionally exceeding 650 feet (*c.* 200 m.) in elevation, and stretches from the Firth of Clyde, on the west coast, eastwards over a low watershed to the eastern coastlands of the Firths of Forth and Tay, with a considerable extension north-eastwards along Strathmore (the great valley). This great transverse depression, separating the Highlands on the north from the Southern Uplands in the south, includes the greater part of the agricultural and industrial areas of Scotland, and contains the largest cities, Glasgow and Edinburgh, as well as many other important centres such as Perth and Stirling.

The soils are almost all of glacial or alluvial origin, and although very fertile near the coasts, there are extensive tracts of glacial "till" or boulder-clay which are generally unproductive as farmland and are now left uncultivated as wet grassland or peat-bog. The best areas of farmland are largely on the lower ground near the coasts; in the valleys among the lowland hills the farmland is restricted to the bottoms and lower slopes of the valleys, though sheep-grazing extends all over the

[1] Robert Smith, 1900.

grassland of the lower hills. The most fertile lands, where some of the best farming in Great Britain is carried on, are largely on loamy glacial drifts derived partly from the Old Red Sandstone, as in the Lothians round Edinburgh, especially at Dunbar, where the narrow strip of coastland is probably the most expensive and most highly farmed land in the world, in the Fife peninsula between the Firths of Forth and Tay, and in Strathmore. These allow of the inclusion of wheat, the finer strains of barley, potatoes, oats and turnips in the regular rotation of crops. This land is practically restricted to the relatively dry east coast, and forms a band of varying width, extending northwards along the eastern edge of the Central Highlands as far as the shores of the Moray Firth, at the north-eastern extremity of the Caledonian Canal.

On the west coast the Ayrshire or southern side of the Firth of Clyde is almost the only tract of high fertility, owing to the favourable topography and the mild spring; early potatoes are here an important crop. Outside these fertile areas a zone is reached in which wheat is grown to a small extent only, but barley of the hardier varieties and potatoes are still included. On the less productive soils and amongst the hill-valleys, these quickly disappear and only the hardier oats and turnips, and sometimes rye, are left; the cultivation of these last, combined with cattle-raising, marks the limit of the arable land. On the fertile coastlands grass and forage crops are as a rule laid down for short periods only, but as the soils become less productive the permanent grassland increases, till at the higher altitudes beyond the arable limits the enclosed farmlands consist of grazing grounds alone.

Above the enclosed grassland comes unenclosed natural grassland, or moorland. Even the latter is sometimes grazed, but it is mainly used as shooting moors for grouse (*Lagopus scoticus*) and other hill game, while in the Highlands the extensive areas of the higher mountain moors are entirely reserved as "deer-forests" or ranges for the red deer (*Cervus elaphus*).

Southern Uplands. The Southern Uplands extend from the Central Plain southwards to the English border. They consist of Ordovician and Silurian siliceous rocks with local intruded igneous masses. The hills have steep slopes and broad rounded summits, rarely exceeding 2000 feet (*c.* 600 m.) and generally covered with grassy or heathy moorland which provides grazing for large numbers of sheep. This region and the somewhat similar Welsh hills are the two most important sheep-grazing areas in Great Britain, sustaining in both cases a proportion approaching or exceeding one sheep to every acre of land of all kinds. The Southern Uplands are girt by broad lowland coastal belts on both east and west. In the valleys of the Tweed, Nith and Clyde and their tributaries agriculture and manufactures connected with wool are largely carried on.

THE SOILS OF IRELAND

By Grenville A. J. Cole

Geological structure. The general geological structure of Ireland is easily comprehended. A great low-lying plain, based upon Carboniferous Limestone, forms the entire centre of the country. This plain reaches the Irish Sea on the east, at Dublin, and touches the Atlantic here and there on the west; but much of the west coast, as well as the north and south of the country, is mountainous. In the extreme west and north-west this mountainous country is formed of very old metamorphic rocks pierced here and there by granite. In the south-west Millstone Grit forms some of the hilly country, but most of the southern mountains are formed of Old Red Sandstone, which occupies a large area, and also rises from under the limestone of the southern portion of the central plain to form isolated mountain ranges. In the north-east and south-east there are considerable areas of Ordovician and Silurian rocks, in the centres of which rise lofty mountains of granite. Finally, the extreme north-east is a plateau country of basalt overlying chalk and Trias; these are almost the only rocks of later date than the Palæozoic existing in the country, with the exception of glacial deposits which overlie the older rocks in many places.

Carboniferous Limestone Plain.

In sharp contrast to Scotland, with its almost uniformly siliceous soils, by far the greater part of the rock on which the soils of Ireland are based is the Carboniferous Limestone, which forms the great central plain, extending from Dublin on the east coast right across the country to Galway on the Atlantic, and from Fermanagh on the north to the borders of Co. Cork in the extreme south. Most of the glacial drift with which this plain is largely covered contains much lime, and it is to this calcareous soil, together with the equable moist climate, that the excellent Irish grasslands, justly famed for cattle-raising, are due[1]. Conspicuous among the glacial accumulations are the dry grass-covered gravel ridges known as "eskers." In Co. Cavan and other places "drumlins" of boulder-clay with steep rounded flanks furnish loamy and often stony soils. In spite of their form, the drumlins are often selected

[1] In 1910 there were 23 head of cattle to every hundred acres of land of all kinds in the whole country, while in some counties this number was considerably exceeded; thus Co. Meath in Leinster had 38 head to the hundred acres. These figures compare with (in 1909) less than 16 for England and only 6 for Scotland. Some of the English counties however, in the Midlands and west, show high figures, *e.g.* Cheshire 28, Leicester 27, Cornwall 25; Anglesey in Wales had nearly 32.

for arable land, rising as they do above the moister levels of the plain.

The direct decay of the limestone often produces a stiff clay soil on which water lies. Where the limestone is very pure, it is liable to yield only a sparse mud or sand, which is carried away as soon as it appears on the surface, leaving barren slabs and terraces of limestone. These rise into very striking regularly terraced limestone hills in Clare and Galway. On the wetter lowlands, where water does not run off, broad peat bogs resting on old lake-marl or water-logged limestone have accumulated.

Old Red Sandstone.

Ridges of Old Red Sandstone, such as the Slieve Bloom and Galty ranges (summit of Galtymore, 3015 feet = *c.* 920 m.), frequently rise through the limestone plain. These usually form barren heaths, but support good coniferous plantations. In the cores of the Old Red Sandstone hills Silurian shales often appear, weathering out into hollows occupied by farm lands and surrounded by the Old Red Sandstone scarps. Throughout the southern counties, from the west of Kerry to the centre of Waterford, long east-and-west ridges of Old Red Sandstone and Carboniferous slate prevail. Fertile clay-covered belts of Carboniferous Limestone occupy the long depressions of the northern part of this area, and these support finely developed woods by the sides of the rivers. The sandstone and slate uplands, in contrast, commonly form moorland, and culminate in the rugged scenery of the Reeks of Kerry, of which the highest summit reaches 3414 feet (*c.* 1040 m.).

Granites and Ordovicians.

The granite of the Leinster chain along the south-eastern coast of Ireland forms thin soils bearing moorland[1], in marked contrast with the clays worn from the Ordovician foothills

[1] In the north of this chain the moorland is dominated by *Calluna* on the slopes and by *Scirpus cæspitosus* on the summits. See Pethybridge and Praeger, 1905.

on either side, which support good arable land. Along both flanks of this chain the soils are much modified by the presence of glacial boulder-clay, which has imported limestone blocks into regions otherwise poor in lime.

In the north-east a broad hummocky Ordovician and Silurian area stretches from the centre of Co. Longford through Co. Down, and farmers have taken full advantage of its loamy soils.

Old metamorphic rocks.

On the Atlantic coast, in the extreme west of Co. Galway and Co. Mayo, and through a large part of Co. Donegal in the north-west, the old quartzitic metamorphic rocks, pierced here and there by granite, provide very little soil, and acid peat is a common feature of the impermeable surfaces. Where mica-schist prevails, however, as in eastern Donegal and Co. Londonderry, rich orange loams are formed, which are much more favourable for agriculture.

Basalt and Chalk.

Finally the red-brown soil of Co. Antrim in the north-east is largely derived from basalt, and is locally modified by glacial drift. The quarries in the Chalk underlying this basalt easily supply lime where it is required on the arable land.

THE SOILS OF ENGLAND AND WALES

The range and variety in the geological structure, and therefore in the soils, of England and Wales, which form the southern portion of the island of Great Britain, are much greater than in those of Scotland and Ireland.

The Palæozoic rocks of England and Wales.

"Broadly speaking the northern and western parts of England and the greater part of Wales are formed of the older rocks known as primary or Palæozoic. These were considerably folded and disturbed before the newer rocks were laid down. Resting on their upturned edges, or abutting against them, lie the Secondary strata[1]."

The geological structure of the north of England is mainly dominated by the great Pennine anticline, forming a broad ridge of high land running north and south, and sharply cut off on the south by a great east and west fault. The centre of this anticline is formed by Carboniferous or Mountain Limestone, on whose flanks lie the Millstone Grit, a coarse sandstone, and then the Coal Measures, which largely determine the position of the great industrial districts of the north. To the north-west of the Pennine axis lies the mass of still older rocks forming the Cumbrian (or Lake District) group of mountains.

The west of Wales consists mainly of Cambrian and Ordovician rocks, flanked on the east by Silurians, while in the south is a great basin of Carboniferous rocks, edged by a rim of Mountain Limestone and containing the important South Welsh coalfield. The south-western peninsula of England (Devon and Cornwall) consists of rocks of Devonian and Carboniferous age with great masses of intrusive igneous rock (largely granite).

Isolated patches of ancient rocks occur here and there in the midlands, *e.g.* the midland coalfields and the rocks of Charnwood Forest. Here and there rocks of pre-Cambrian (Archæan) age appear at the surface, but their total extent in England and Wales is comparatively insignificant.

The general effect on the physiognomy and soil-characters of the country produced by the older rocks of the north and west has already been described, and it would not repay us to consider the different formations in detail.

Limestones.

The limestones form an outstanding series with very marked soil-characters and vegetation. By far the thickest and most extensive is the

[1] H. B. Woodward, *The Geology of England and Wales*, Second edition, p. 26, 1887.

Carboniferous or Mountain Limestone, developed in the north of England, in North and South Wales, and in the Mendip Hills of Somerset; the vegetation of the English areas of this has been carefully studied[1].

The Cambrian, Silurian and Devonian series all have numerous limestone bands occurring at intervals between other types of rocks; the vegetation of these however has been scarcely investigated at all. The Magnesian Limestone of Permian age occurs in Nottinghamshire, Yorkshire, and other northern counties.

The limestones generally form elevated land masses or ridges, the valleys and slopes being usually occupied by adjacent shales. On very steep slopes and crags, where the products of chemical erosion by water containing CO_2 are at once carried away, the rock is left bare with characteristic plants occupying any spot where a little soil can collect. On less steep slopes it is covered by a very thin soil occupied by a special type of natural grassland ("calcareous pasture"). Where the insoluble products of chemical erosion are able to accumulate, as on flat summits, on gentle slopes and in valleys, a deeper soil results. On flat ground the lime is often almost completely washed out of this surface soil, with the result that a heath vegetation occupies the area. Sometimes "limestone pavement" (p. 160) is formed on flat ground. On slopes and in valleys scrub and woodland of characteristic type occur.

"Siliceous" rocks.

The rocks other than limestones (sandstones, grits, mudstones, slates, shales, schists and various crystalline rocks) are of extremely various physical character and lithological composition, and the vegetation they bear is modified not only by these factors, but also by the exposure and slope of the ground, water supply, drainage and other factors. In general, as already mentioned, they are characterised

[1] Smith and Rankin, 1903; Lewis, 1904; Moss, 1906.

by comparatively shallow or very thin soils, and these are frequently poor in mineral plant-food, but numerous exceptions occur, *e.g.* in soils produced by the disintegration of rocks rich in certain minerals. At high altitudes, apart from crag and cliff, the soil is covered by various types of "siliceous pasture" or by moorland; at lower altitudes, where trees and shrubs can successfully establish themselves, by woods of the oak and birch series, or their corresponding scrub.

Generally speaking the Palæozoic, metamorphic and igneous rocks are not fertile (apart from climatic conditions), but various important exceptions occur. Thus the Old Red Sandstone of Herefordshire yields a strong loamy soil which is fertile, and largely under good permanent pasture, with many orchards and hopyards. The limestone bands (called "cornstones") which occur in it form however the richest land. The diabases, tuffs (volcanic ash) and slaty volcanic rocks, forming red soils, of the Ashprington series of the Middle Devonian near Totnes in south Devon, support some of the best arable and grassland in the district. Again the Magnesian Limestone of Permian age forms a good light dry arable soil[1]. This comparatively soft limestone is however sometimes included in the secondary (Mesozoic) series.

Secondary rocks.

The secondary rocks which rest on or abut against the primary groups enumerated, form a much more regular series. Speaking generally they dip regularly to the south-east, one under the other, so that their outcrops are arranged in a succession of bands crossing England from north-east to south-west, and occupying more than half the area of the country. The north-eastern ends of these bands bend northwards, and occupy the eastern part of northern England to the east of the Pennine ridge, running out successively to the north-eastern coasts of Yorkshire and Durham.

[1] H. B. Woodward, *op. cit.*

The oldest member of the series is the Trias. This forms a broad area, broken by islands of older rocks, in the very centre of the country. Besides its long northward outcrop on the eastern flank of the Pennines, extending to the Durham coast, and its southward outcrop—narrow and somewhat interrupted—to the coast of Devon, it has an important northwestward extension, separating the Palæozoic rocks of the Pennines from those of North Wales, and reaching the Irish sea on the coasts of Cheshire and Lancashire[1]. Speaking broadly the Trias forms low ground: to the north the plain of York, to the north-west the plain of Cheshire, and in the centre the Central Plain of England. It has two main divisions in England, the Bunter, consisting of sandstones, and the Keuper, mainly of marls, the "Muschelkalk" of the Continent being absent from this country. The Bunter sandstones form a poor soil, woodlands of the oak-birch series and heathland, as in Sherwood Forest in Nottinghamshire and Delamere Forest in Cheshire, occurring on these beds. The Keuper marls form fine meadow and pasture land, for instance in Cheshire.

Trias.

The Lias comes next, extending from the coast of Dorset to the coast of Yorkshire. Like the Trias, its greatest extension is in the Midlands, where it forms part of the plain of central England. The Lias consists of clays, marls, shales, and more or less pure limestone bands, and, locally, some sand and sandstone. It forms mainly flat country, with low escarpments, representing the limestone bands, in places. The clays and marls give fertile agricultural country, with rich pasture-lands.

The vegetation of the Keuper and Liassic marls of Somerset has been studied by Moss[2]; he found that (ash-)oak-hazel wood (see p. 182) and calcareous pasture,

[1] With small isolated tracks occupying the Eden valley in Cumberland and the Vale of Clwyd in Wales.

[2] Moss, 1907, p. 50.

with ash wood on the limestone exposures, were the natural types of vegetation.

Oolites.

The Oolites, which follow the Lias, also run from the coast of Dorset to the coast of Yorkshire. Their outcrop occupies a broader band of country than the Lias, the greatest breadth being in the east midlands, where however the eastern division is overlaid by the alluvium and peat of the Fen District.

The Oolites are partly composed of comparatively hard rock forming hill-ranges, the most important of which, the Cotswold Hills in Gloucestershire, and the Cleveland Hills of north-east Yorkshire, attain a height of over 1000 feet (305 m.). The Oolite hills of the south, notably the Cotswolds, are very pure limestones, but as the outcrop is traced northwards the limestones partly give place to rocks of a different lithological character, while the north-east Yorkshire representatives of the series are mainly non-calcareous.

The vegetation of these last has been studied by W. G. Smith, whose results are not yet published, but it may be stated in a general way that the valley sides of the Cleveland district are clothed with ash-oak woods (indicating the outcrops of limestone) and with oak woods, while the plateaux of the summits bear extensive heaths. The limestones of the Cotswolds carry beechwood and calcareous pasture. The agricultural value of the Oolites varies very much according to the very variable local character of the rock. In the stratigraphically upper portion of the Oolite series are extensive belts of clay (Oxford clay and Kimeridge clay) which are very heavy soils, expensive to work, and are largely under permanent pasture.

Cretaceous series.

The Oolites are followed by the Cretaceous series, which occupies a broad belt of country extending from the Dorsetshire to the Norfolk coast, and bending north-westwards, appears again in Lincolnshire and south-east Yorkshire.

The Lower Greensand, whose outcrop is a narrow interrupted belt at the base of the series, forms in places a distinct sandstone escarpment. This rock varies much in lithological character, and its vegetation is complicated. It is followed by the Gault, a heavy clay, sometimes calcareous, which may bear the damp type of oakwood, and is now largely pasture-land.

The Chalk.

The Chalk, the most important member of the Cretaceous series, is on the whole a very pure soft limestone forming more or less elevated masses of characteristic gentle rolling contours (downs) and fairly steep escarpments. The Chalk occupies a large area in southern and eastern England. In no case does the Chalk attain a height of 1000 feet (305 m.), though it often exceeds 500 feet.

The Chalk plateaux of the south are often overlaid by large areas of a deposit known as "clay-with-flints," which is of variable composition, usually a loam, sometimes a heavy clay and sometimes sandy, and typically contains numerous flints. It is supposed to represent the insoluble residue of the dissolution of a great depth of overlying chalk, though in some cases it may be the remains of Tertiary formations which once lay on the top of the Chalk. The Clay-with-flints is generally very deficient in lime. It is largely arable land, which is much improved by liming or treatment with the artificial manure known as basic slag[1]. The natural vegetation is oakwood of various types corresponding to the various soils. Heath is sometimes developed on the lighter poorer soils.

Where the chalk itself forms the immediate subsoil, the soil itself is extremely thin, often only 2 to 5 inches (*c.* 5 to 12 cm.) in thickness, and is frequently extremely poor in lime. In the south-east the natural vegetation is beechwood on the slopes of the escarpments and valleys; towards the south-west this passes into

[1] Mainly phosphate of lime.

ash-wood (see p. 167). Pasture, supporting large flocks of sheep, covers large areas of the summits and many of the slopes of the chalk downs.

Agriculture of the Chalk.

The arable land in which wheat is regularly included in the crop-rotation forming the "wheat-country" mentioned on pp. 28–29, is largely situated on the Chalk and the overlying thin soils already described. The characteristic agricultural method is based on what is called the "Norfolk four-course system," in which wheat is followed by a "root-crop" (turnips or swedes), this by barley, and the barley by a "seed-crop," generally leguminous, such as alsike (*Trifolium hybridum*) and white (*T. repens*) or red clover (*T. pratense*), or a mixture of clover and rye-grass (*Lolium perenne*). Oats are often substituted for wheat on the lighter lands, and numerous other variations exist. This rotation essentially depends on the "folding" of sheep on the "green" crops and the manuring of the ground by their droppings for the following cereal crop. This is the characteristic feature of the agriculture of southern and eastern England. The typical pasture of the South Downs of Sussex, and of Salisbury Plain in Wiltshire, supports great numbers of sheep, and in the down farming the sheep feed sometimes on the pastures, sometimes in the folds. The Chalk of the North Downs (Surrey and Kent), and of Cambridgeshire, Norfolk, Yorkshire and Lincolnshire, has very much less (often practically no) open "sheepwalk," the Chalk country in the four last-mentioned counties being almost entirely arable.

The Wealden anticline.

The outcrop of the Cretaceous rocks is not confined to the main belt running from south-west to north-east; there is an important extension caused by a line of uplift running eastwards from Wiltshire, through the counties of Hampshire and Sussex, and across the English Channel to the neighbourhood of Boulogne in France. This line of elevation, known as the Weald anticline, determines the geological structure of the whole of south-eastern England. It brings up the chalk on its northern and southern flanks to form the North Downs running through Surrey and Kent, and the South Downs in Sussex, respectively, while exposed by the removal of the Chalk, which once connected the two ranges of downs, are the Lower Cretaceous beds of the Wealden area surrounding the axis of the uplift. These beds are largely different from and much thicker than the strata occupying the corresponding stratigraphical position along the main outcrop. The Gault is similar, but the Lower Greensand is much more developed, at least on the north side of the Weald, than it is on the main Cretaceous outcrop. At the western end it consists of coarse massive sandstones, forming hills of some height (in one place nearly reaching 300 m.), which bear woodland of the oak-birch series and extensive heaths, and are very poor agriculturally, while further east the sandstone contains more fine particles and also thin beds of limestone, and is good agriculturally, as is also the narrow belt of the same formation fringing the southern edge of the Weald.

The Wealden beds.

Below the Lower Greensand come the Wealden beds proper, developed only in that area. First a broad belt of Weald Clay forms a plain overlooked on the northern side by the bold escarpment of the Lower Greensand. The Weald Clay gives a heavy clay soil for the most part, though it contains narrow bands of limestone and some loamy beds. It bears the damp type of oakwood characteristic of clays and loams.

In the centre of the Weald come the so-called Hastings beds, divisible into the Tunbridge Wells Sand and the Ashdown Sand, separated by the Wadhurst Clay, and often themselves interbedded with narrow bands of clay. These beds form relatively high ground, particularly the

Ashdown Sand, which makes the so-called "Forest Ridge," a heathy ridge in the very centre of the Weald, still largely covered with oak and birch wood. The Tunbridge Wells Sand is partly very fine grained, like a clay, and largely bears the damp type of oakwood.

The Chalk of the main outcrop, from southern Oxfordshire north-eastward, dips below the overlying Tertiary beds and appears again at the surface in the North Downs. In the same way the Hampshire Chalk dips below the Tertiaries in the south of the county and the north of the Isle of Wight to reappear in the Central Downs of the island. These two great chalk basins, known as the London and Hampshire basins respectively, contain all the English Tertiary deposits, except those which skirt the east coast overlying the Chalk of Norfolk and Suffolk, and which really form a north-eastward extension of the London basin.

Tertiary series.

Of the Eocene beds which form the bulk of these deposits the London Clay is the most important. It yields a stiff soil with the usual clay characters. Below it are the alternating sands and clays which give variety to the fringe of country between the chalk and the London Clay, while above it, forming extensive tracts on the borders of Berkshire, Hampshire and Surrey, and isolated patches to the north of London, is the very light Bagshot Sand, bearing heath, and now very largely subspontaneous Scots Pine. The Bagshot beds of the Hampshire basin show a greater complexity of alternating sands and clays.

Eocene beds.

The Oligocene beds which come next are confined to south Hampshire and the Isle of Wight, while the Miocene are absent from this country altogether.

Oligocene and Pliocene beds.

The Pliocene strata are the Tertiary beds which fringe the chalk on the east coast, and are thus entirely East

Anglian. These deposits are nearly horizontal and occupy a considerable extent of country, though they have no great thickness. They are mainly composed of sand and gravel ("crag") with very thin bands of clay, and are largely covered with heath of various types. It is doubtful if this part of the country was ever occupied by forest.

Quaternary deposits—Glacial Drift.

"The Quaternary deposits form a distinct group scattered irregularly over the country, and resting indifferently on any of the rocks from the oldest upwards[1]." Very important among these quaternary deposits, just as is the case in Scotland and Ireland, are the extensive sheets of ice-formed drift dating from the glacial period, and of very variable composition, consisting sometimes of non-calcareous clay (as in many parts of the north of England), sometimes of highly calcareous clay (such as the "chalky boulder-clay" of eastern England), and sometimes, though less frequently, of sandy and gravelly material. These sheets of drift completely cover the older rocks over large tracts of country, particularly in the north, centre, and east, and of course entirely determine the character of the soil, and therefore the vegetation, of these regions. In places the nature of the drift, and therefore of the soil, changes within very short distances.

In the north of England the Coal Measures of Northumberland and Durham are largely covered by drift (stiff clay or sand and gravel), and the same is true of the Coal Measures and Trias of south Lancashire and Cheshire. In the vale of York there are deposits of boulder clay resting on the Trias, and also extending up the coast from Flamborough Head northwards. Throughout Wales there are deposits of glacial drift in various places, and in the south-west of England there are various deposits of local occurrence which seem to contain evidence of ice-transport.

[1] Woodward, *The Geology of England and Wales*, p. 26.

On the midland Trias there are extensive deposits of drift containing erratics, some coming from the north, and some from the Jurassic and Cretaceous rocks further east. Eastwards this drift passes into the chalky boulder clay, which covers a wide extent of country, reaching northward along the coast into Lincolnshire and Yorkshire, and occupying wide areas of the surface in south Norfolk, Suffolk, Essex, Cambridgeshire, Bedfordshire and Huntingdonshire. Nowhere however does it come so far south as the Thames valley. This heavy calcareous clay forms wide sheets on the higher ground, and sometimes extends to the valleys on a level with the alluvium, but it has been largely removed from the valley sides and escarpments. It forms a heavy soil and is largely under cultivation. The natural woodland is of the ash-oak type owing to the relatively high lime-content of the soil (see p. 181) and is treated as (ash-)oak-hazel copse.

Other Quaternary deposits.

There are numerous other Quaternary deposits of various age and the most various composition, some of glacial and many of unknown origin[1]. Perhaps the most important of these is the "clay-with-flints" already referred to (p. 53), which very frequently covers the chalk plateaux and, being generally non-calcareous, quite alters the character of the vegetation. The derived sands and gravels resting on the chalk, and covering considerable areas on the borders of Norfolk and Suffolk round Brandon, Thetford and Mildenhall, are largely occupied by heath.

River alluvium.

Recent alluvium, laid down by existing rivers, again varies very greatly in composition. In mountain country it often consists of coarse sand and gravel, while in the flood plains of the level country it consists typically of fine clayey silt. It is

[1] There is no evidence of glacial action in south-eastern England south of the Thames valley.

often difficult or impossible to obtain any direct evidence as to the natural vegetation of such soils because they are generally extremely fertile, and entirely under cultivation or in permanent pasture. In the neighbourhood of great cities, which are frequently situated on the flood plains of the lower courses of big rivers, these rich alluvial soils are extensively used for market gardening.

Maritime soils.

Travelling seawards down the course of such a river, we pass from the plain of river alluvium to the salt mud associated with the tidal estuary. The natural vegetation is halophilous (p. 332), but when the tide is kept out by sea walls, and the land is drained, it makes admirable pasture. "The Marshland," occupying the northern (seaward) portion of the old estuaries of the Ouse, Nen, Welland and Witham and lying round the Wash, their common estuary, now much reduced in size, is the most extensive area of this kind in the country.

Blown sand, forming sand-dunes, which support a very characteristic vegetation (p. 339), occurs wherever a constant supply of sand is thrown up by the sea above highwater mark, and thus comes under the action of the wind.

Shingle or pebble beaches thrown up above high-tide mark also have a vegetation of their own, allied to that of sand-dunes but with distinct characters. The best known and most extensive are the Chesil Bank in Dorsetshire, Dungeness in Kent, and Orfordness in Suffolk.

Peat soils.

Finally, among recent deposits, we have the sharply characterised organic soils of which the peat soils, formed where vegetable remains accumulate in the presence of water and with deficient oxidation (p. 35), are by far the most important. The kind of peat formed on the edges of water comparatively rich in mineral salts is often accumulated in the upper parts of tidal estuaries which are silting up, and of which the lower parts are accumulating tidal mud. The most extensive area of this kind of peat forms the region known as "the Fenland," lying to the south of "the Marshland" and corresponding with the upper portions of the old estuaries of the Ouse, the Nen, the Welland, the Witham, and their tributaries. The upper parts of the old estuaries of the Bure, Yare and Waveney in east Norfolk, form other areas of the same kind, much less extensive, but less altered from their original condition.

The present surface-peat of the upper parts of other old estuaries, *e.g.* the peat of the Somersetshire "levels" and round Morecambe Bay (north Lancashire), is of the heathy type (*i.e.* much poorer in mineral salts), though in some cases this surface-peat is built up on old peat of the former type (see p. 246).

The hill peat, so common on the badly drained slopes and plateaux of the Pennines, in Wales and in Scotland and Ireland, is of the acid type, poor in mineral salts. Much of this peat has been formed on the sites of ancient forests.

The constant variation from place to place of the English soils will be obvious from the foregoing account. It is clearly impossible briefly to summarise their distribution. Some of the most important areas of the various kinds of soil are given below.

Organic soils. Mild peat of upper parts of old estuaries fed by waters rich in bases. Acid peat of hill plateaux and slopes in the north and west, and lowland moors.

Sands and gravels. Blown sand of various places on the coast. Valley and plateau gravels overlying various secondary and tertiary rocks. Pliocene "crag," etc.—east Norfolk and Suffolk. Bagshot sand—mainly south-west of London—and other Eocene sands of the Hampshire basin. Lower Greensand—Weald and the east Midlands.

Bunter sand—Sherwood and Delamere Forests, Forest of Arden, etc.

Clays. Boulder clay (*e.g.* of south Lancashire and Cheshire). London clay (London and Hampshire basins). Gault clay—east and south Midlands and Weald. Jurassic clays (Kimeridge and Oxford clays)—east and south Midlands.

Other "siliceous" soils (see pp. 33, 122)—Archæan, Palæozoic, metamorphic and igneous rocks (other than limestones) of the north and west.

Marls. Chalky boulder clay (mainly overlying chalk)—East Anglia and east Midlands. Liassic marls—Midlands and south-west England. Keuper marl[1]—Midlands and south-west England.

Limestones. Chalk (main outcrop and Weald)—southern and eastern England, Yorkshire and Lincolnshire wolds. Oolitic limestones—Cotswold region, etc. Magnesian limestone—Nottinghamshire, Yorkshire, etc. Carboniferous (mountain) limestone—Pennines, Mendips and Wales. Devonian limestones of Devonshire. Silurian limestones—borders of South Wales.

Alluvium. Mixed soils largely of fine texture—flood plains of rivers and on sites of old estuaries.

[1] Not always calcareous.

PART II

THE EXISTING VEGETATION OF THE BRITISH ISLES

CHAPTER I

THE DISTRIBUTION OF THE CHIEF FORMS OF VEGETATION

We can gain some idea of the actual extent of natural and semi-natural vegetation in the different countries of the United Kingdom from the annual statistics of the Board of Agriculture and of the Irish Department of Agriculture. From the returns for 1909 and 1910 we may construct the following table:

Percentages of total land-areas under cultivation and under natural and semi-natural vegetation.

	England	Wales	Scotland	Ireland (1910)
Arable land (including orchards, etc.)	32·8	15·3	17·7	16·0
Permanent grassland ...	42·9	43·2	7·8	
Mountain and heath used for grazing	7·5	27·9	47·7	68·8
Waste land and land under houses, gardens, etc. ...	11·5	9·7	22·2	13·8
Woods and plantations (1905)	5·3	3·9	4·6	1·5

Of these heads we may count the two first as "cultivation," though a great deal of permanent pasture that is not regularly manured consists of more or less modified semi-natural plant-communities. It is impossible to say how much of the land under trees is to be reckoned as plantation in the narrower sense of the word, *i.e.* plantations on previously open land, but there is good reason to believe that the great bulk of British woodland is semi-natural, in the sense of representing types of wood originally native to the country[1]. Mountain and heathland used for grazing is certainly to be reckoned as bearing essentially natural vegetation. The next heading includes two very dissimilar categories, on the one hand land so barren and remote that it is removed altogether from the pastoral industry, and on the other hand the land covered by cities, towns, and villages of all sizes and by isolated houses with their parks and gardens. It is unfortunately impossible from any published or accessible data to determine the relative extents of these two categories of land, but we shall be safe in inferring that while the latter must form a relatively large proportion in England, the former must greatly preponderate in Wales, Scotland and Ireland.

Thus we arrive at the following approximate percentages:

	England	Wales	Scotland	Ireland[2]
Cultivated land (including "permanent grass")	75	59	25	? 20 to 30
Land under natural or semi-natural plant-communities ...	15 to 20	40	70 to 75	? 70 to 80

These figures are necessarily of the roughest description.

[1] See "The Woodlands of England," by C. E. Moss, W. M. Rankin and A. G. Tansley, *New Phytologist*, Vol. IX. 1910, pp. 114—118.

[2] It is unfortunately quite impossible to say what proportion of the pasture land of Ireland (amounting to more than two-thirds of the whole area) is under approximately natural vegetation, but it is probably very large.

To a large extent in England, and probably to a much greater extent in Wales, much of the permanent grassland is not regularly manured and should be transferred to the second heading as forming semi-natural plant-communities; so we may say in a very general way that somewhere about one-quarter of England, one-half of Wales, three-quarters of Scotland and perhaps as much of Ireland, bear natural plant-communities more or less modified by the activities of man.

Comparison with other countries.

Comparing these proportions with those obtaining in the neighbouring continental countries[1] we find that France probably has a distinctly larger proportion of land with a natural or semi-natural plant-covering than England, but still a fraction of the same order, that is to say something in the neighbourhood of one-quarter to one-third, while Germany must have at least one-third, Holland decidedly over one-quarter, Belgium probably the same, and Denmark distinctly less than one-quarter. Norway, on the other hand, has more than 96 per cent. of her area uncultivated, including permanent grassland.

When we turn to the way in which these fractions are made up, however, we find very striking differences. Germany has over 25 per cent. of her area covered with woods and plantations, Norway 22 per cent., France about 17 per cent., while Holland and Denmark each has about 7 or 8 per cent., England between 5 and 6 per cent., Scotland less than 5, Wales less than 4, and Ireland only 1½ per cent. Thus Germany, Norway and France have an immensely greater proportion of their uncultivated and unmanured areas under forest than have Holland and Denmark or England, while Scotland, Wales and

[1] The statistics of the different countries are, to a large extent, impossible to compare, so that the proportions given are necessarily even rougher than those for the British Isles, and are intended only to give the most general idea of the relations.

Ireland, with their very large proportional areas of uncultivated land and very small areas of forest, have a much greater proportion of unreclaimed mountain and moor than any country, save Norway, in north-western Europe.

Former extension of forest.

The uncultivated land bearing natural or semi-natural plant-communities certainly does not, except in certain cases, bear exactly the same vegetation as it did before man began, with increasing civilisation, to modify it on a large scale. There is no doubt that by far the greater part of the British Isles was originally covered with forest: in England the whole of the east, south and midlands, except perhaps some of the chalk downs and some of the poorer sands, and of the north and west probably everything but the summits of the higher hills[1];

[1] There is still considerable difference of opinion as to the normal altitude formerly reached by woodland in the British Isles. So far north as the Central Highlands of Scotland thin wood of *Pinus sylvestris* still occurs in places up to 2000 feet (*c.* 600 m.) and even higher. There is evidence from the remains preserved in the peat that woodland once existed in many places at such levels. Thus Lewis has described a bed containing *Alnus rotundifolia* and *Viburnum Opulus* at a height of over 2000 feet in the Northern Pennines (see p. 269) and a forest bed of *Pinus sylvestris* in the counties of Banff and Inverness at a similar height (*Scottish Geogr. Mag.* 1906, p. 249). He attributes the disappearance of woodland in the latter case (as in many others throughout Scotland) to a change of climate—the reappearance of glaciation after an interglacial period—but this interpretation is not universally accepted. The altitude to which woodland extends depends of course not only on the general climate, but also on the total height of the mountain mass on which it is developed. Thus we should expect a higher altitude to be reached by woodland in the Highlands of Central Scotland, a compact *massif* with several summits exceeding 4000 feet (1220 m.), than on the narrower and lower ranges of the Pennines, and this is actually the case with the woods existing at the present time. Varying conditions of exposure also play a part in determining the local height to which woodland attains. It is certain that the upper limit of existing woodland has been considerably depressed, as in many European countries, during the historical period, largely no doubt by the browsing of animals pastured in the zone lying immediately above the forest.

in Wales the alluvial plains of the larger rivers and the valley sides up to similar altitudes; in Scotland the great central lowland plain, the flat shores of the northern firths on the east coast, and the sides of the valleys running into the heart of the Highlands; in Ireland presumably the central plain and the hill country up to a considerable altitude.

We are justified in making these statements by the climatic conditions that obtain in these areas, conditions which bring them well within the natural forest region of north-west Europe, and by the widespread and abundant evidence of the remains of trees in the peat which covers, in the west and north, large tracts of the country indicated above.

Destruction of forest.

Not only therefore has practically all the arable and "permanent grassland" been taken from the original forest area, but much also of the so-called "natural pasture," heathland and moorland, has been derived from forest. A human community, directly it passes into the first stages of civilisation and increases numerically, is the natural and inevitable enemy of the tree-communities of the countries it inhabits. Man cuts down trees for building his houses and for his firing, to make room for his crops, and to make new pasture for his flocks and herds. These in their turn by browsing in the woods eat off the tree seedlings, and may thus prevent the rejuvenation of the woods and cause their eventual death. It has been pointed out that "the western sea-board of France, the British Isles, Flanders, Holland, Schleswig-Holstein, and Denmark are practically the most poorly wooded portions of Europe." This is attributed to the mild winters of the countries bordering the Atlantic and adjacent seas, which permit of grazing and browsing throughout the year[1].

[1] A. C. Forbes, *The Development of British Forestry*, London, 1910, p. 7.

But perhaps the most widespread and rapid destruction of forests in a country like our own has been incidental to military operations. An organised force trying to crush a less organised one is obliged to destroy the cover which the latter would use to hide in after a temporary defeat, and from which its members would emerge to harass their enemies as opportunity offered. The Roman generals in the first century of the Christian era were probably the first great destroyers of our forests on a large scale, and on the Pennines, where the records of these forests are preserved in the hillside peat, there is evidence of their destructive activity[1]. The various expeditions on which the English kings of the Middle Ages embarked, when they were striving with varying success to subdue the wild tribesmen of Wales and Scotland, were responsible for much of the forest destruction in these countries, while in Ireland similar causes were constantly at work.

Many of the existing larger areas of woodland in the south of England and the Midlands are old "Royal Forests" reserved during the Middle Ages for the king's hunting. Though much restricted in area and diminished in number some of these still exist as unenclosed forest land, and a few belong to the Crown, *e.g.* the New Forest and the Forest of Dean. The name "forest" is still applied to areas at one time reserved for royal hunting, even when the areas so called are largely or entirely destitute of trees, *e.g.* Dartmoor Forest.

Fate of the land denuded of forest.

The wholesale destruction of forests in countries with a cool and wet climate, such as much of the western and northern parts of the British Isles possesses, leads very rapidly, in situations where surface water can collect, to the formation of bogs and of deep peat; this preserves the remains of many of the destroyed trees, but effectually

[1] C. E. Moss, "Peat-Moors of the Pennines," *Geographical Journal*, 1904, p. 3 of separate issue.

prevents the return of the forest, and is of little use to man in any way. On steeper and drier slopes, heath develops on the poorer soils, and grass, used for hill pasturage, on the better soils, which have been denuded of forest.

In the drier south and east, the fate of the soil from which forest has been cleared is different. Such of the land as was not actually tilled would be used, so far as it became naturally colonised by grass, for the pasturage of the numerous flocks and herds of this more thickly populated part of the country, and would thus be turned into grassland, which trees can only invade sporadically. The poorer sandy soils would be colonised by the heaths and other plants which flourish on such soils, and would come to bear a vegetation akin to that of the moors, but on a drier soil and not forming deep peat. In some cases such heaths are partially or completely re-invaded by woodland of the appropriate type. This is possible because the more barren heaths afford no pasturage, and the plant-community is of more open type and therefore more easily colonised by trees of suitable kinds.

Coppice-with-standards.

With the exception of the Royal Forests, the heathy woodlands and a few others (notably the beechwoods of the chalk), the woods which remained in the south, after the earlier and extensive clearings, took the form of regularly exploited "coppice-with-standards." This semi-artificial type of wood no doubt originated by partial felling of the trees of the original forest and the cutting of the shrubs for small wood. The increased access of light would lead to increased growth of the shrub layer, and it became the custom regularly to harvest the straight coppice shoots springing from the cut "stools" of the trees and shrubs, once every 10 or 15 years, while the few standard trees were felled occasionally. The supply of coppice-wood and timber so obtained was largely employed for the local uses of the country side, the timber for building, furniture, carts, farm implements, tools, etc., the coppice-wood for firing, fencing, hurdles, stakes, hop-poles and the like. The great majority of the existing woods, and especially the oak-woods of the south and east and of the adjoining midlands, still remain in this form, and are still used for these purposes, but the primary object for which their owners now usually maintain them is the preserving of pheasants.

The woods of the west and north, with local exceptions, are mostly in the less artificial condition of "high forest," and many of these are properly forested, *i.e.* exploited for their timber. This is especially the case in parts of the north of England and of Scotland, where modern scientific forestry is to some extent pursued, and extensive planting experiments are carried on. There are also of course wide-spread plantations in the south, very largely of conifers (such as larch), that are not native to the country, also of oak, ash, beech, etc., and in some cases planted coppices; but there is good reason for believing that the great majority of the English woods are essentially semi-natural plant-communities, the remnants, modified mainly by coppicing, sometimes by partial replanting, of the original forests which once covered the country. As we shall see in the sequel the different existing types of wood are very closely connected with the soils on which they occur.

Backward condition of forestry.

In the British Isles generally, but more especially in the south of England, the practice of forestry is in an extremely backward condition. This is largely due to the fact that very few of the woodlands are owned by the State. The great majority are in private hands, and their owners nearly always prefer the preservation of game to the proper economic development of their estates. Further, the large amount of capital that has to be sunk in the development of forests for many years before a return is forthcoming, naturally deters the landowner, whose liquid capital resources are limited, from embarking on schemes of serious forest development, the monetary return from which would only benefit his grandsons and great-grandsons. The State might, no doubt, do a great deal more than it does by properly developing the few forests it already possesses, and by acquiring suitable land for afforestation. The characteristic English distrust of the scientific expert, and an unwillingness to embark on far-reaching schemes except under the pressure of absolute necessity, are largely responsible for our failure to develop the country in this direction. There are however signs of an improvement in this respect; meanwhile the majority of our woodlands are from the forester's point of view a scandal and a disgrace.

Abundance of woodland in the south-east.

The south-eastern corner of England, one of the most backward parts of the country in the matter of forestry, is, curiously enough, by far the richest in woodland. Kent, Surrey, Sussex, Hampshire and Berkshire, comprising between them 3,886,296 acres of land and water, had in 1905, the date of the last return, no less than 448,540 acres of woodland, equal to 11·5 per cent., or more than double the percentage for England as a whole, and constituting more than a quarter of the whole area of English woodland. This preponderance in woodland is due partly, though by no means wholly, to the extensive heath woods developed on the light sands of the Weald, of the Lower Greensand, and of the Eocene series.

Poverty of East Anglia and the Midlands.

Cambridgeshire and Huntingdonshire are the English counties poorest in woodland, together having only 1·4 per cent. of their areas occupied by woods and plantations. This is partly due to the fact that these two counties contain a large part of the woodless fenland. The whole of East Anglia, Lincolnshire, the East Riding of

Yorkshire, and much of the Midlands are however very poor in woods, and this is no doubt connected with the fact that these parts of the country have the greatest proportion of agricultural land (see p. 28). The apparently well-wooded character of the Midlands, as seen for instance in passing through the country by train, is due entirely to the large number of trees in the hedgerows separating the comparatively small fields into which the prevailing pasture land is divided. Seen from a distance the trees and hedges blend and give an entirely false impression of woodland.

We may now summarise the existing natural and semi-natural vegetation of the country classified according to the physiognomy of the landscape it determines, rather than according to the natural plant-formations dealt with in the sequel.

Woodland.

1. The existing woodland is very restricted in area as compared with that of most continental countries (excepting the regions bordering the Atlantic, the English Channel and the North Sea, which are also very poor in woodland); but that which still remains represents for the most part the more or less modified remnants of the primitive forests which once covered most of the country. There is very little untouched primitive forest left, but some of the remoter hill woods, *e.g.* on the Pennines and in Scotland, may be regarded as practically primitive forest. Most of the woods of the country, at any rate of England, are however to be regarded as semi-natural woods, retaining the essential characters and the flora of the primitive woods from which they are derived. The majority of the woods of southern England, treated as coppice-with-standards, a very old form of exploitation, are at the present time mostly uneconomical and are often badly neglected. The best forestry is to be seen in the north of England, in Scotland, and in the Chiltern beech forests.

Grassland. 2. Natural and semi-natural grassland covers a very large area of the country. The strictly natural grassland, in the sense that it has never been clothed with forest, below the altitudinal forest limit, is probably confined to certain areas of the Chalk, to submaritime pasture derived from salt-marsh, and perhaps to some of the grassy heaths of East Anglia. Above the forest limit and below the arctic-alpine zone it occupies the less poor of the drier soils. The semi-natural grassland is partly derived from woodland by the removal of the trees below the wood-limit on the drier hill slopes, and is continuous with and often indistinguishable from adjacent natural grassland which probably never bore trees. Semi-derelict poor pasture, and permanent pasture which has been "laid down" from arable land and is not manured, may also be considered as semi-natural grassland, though it has been derived from manured land and not straight from woodland. The pastures characteristic of many alluvial plains is also at least semi-natural: in some cases such grassland may never have borne trees and so be strictly natural. But the grasslands of the country have as yet been quite insufficiently studied from the standpoint of their vegetation and relationships.

Heathland. 3. Heathland is developed on the poorer sandy soils, especially in the south-east of England, but also on similar areas further north, and on well-drained hillsides, over sandstones, etc., poor in nutritive salts and in lime. Some of the existing heathland is quite possibly primitive, some is probably derived from woodland by natural causes (see p. 90), but a great deal has certainly been derived from cleared woodland on the poorer soils.

Moorland. 4. Moorland (on deep peat with low mineral content) is developed in regions of relatively high rainfall, such as hill slopes and plateaux in the north and west (upland moor); also on the sites of old lakes, etc., where the water-supply is poor in mineral food (lowland moor); and generally where surface-water poor in soluble mineral salts has accumulated. The hill moors are partly formed above the wood limit, and partly replace woodland, as may be proved from the abundant remains of trees buried in the peat. Moor covers wide areas of hill country in the north of England, and in Scotland and Ireland. It is related to heathland on one side and to the damper grasslands on the other. Much of what is called moorland in common parlance is really heathland, *i.e.* developed on shallow peat over a coarse porous soil.

Fenland.

5. Fenland (on deep peat with high mineral content) is developed on the edges of lakes and estuaries where considerable quantities of nutritive salts are present. It bears a very different vegetation from moorland, though some species are common to both, and transitions occur. By far the largest area of fen was formed round the upper part of the old estuary now represented by the Wash, the common estuary of several rivers. It lies in the counties of Lincoln, Norfolk, Huntingdon and Cambridge. The soil of the lower part of this old estuary (the area near the present coast) is formed of marine silt, and was no doubt at one time salt-marsh. This is sometimes distinguished as the "Marshland" as opposed to the "Fenland" or "the Fens" *par excellence.* The whole area is almost entirely drained and under arable cultivation. Considerable areas of fen occur in east Norfolk in the old estuaries of the Bure, Yare and Waveney, which have now silted up, and these fens are still largely in a natural condition. Their formation from the shallow lakes or "broads," probably the remains of the waters of the old estuaries, may be studied in detail (Chap. X). Fenland has a special characteristic type of woodland of which the alder (*Alnus rotundifolia*) is the dominant tree.

Maritime vegetation.

6. Salt-marshes are developed on the tidal mud of protected bays and estuaries, and their vegetation, like that of fen, is land-forming. If the sea is kept out and the land drained salt-marsh makes excellent pasture and often excellent arable land.

7. Sand-dunes are developed wherever on the coast there is a constant supply of blown sand derived from the sand thrown up above the average limit of high tide. The ultimate form of sand-dune vegetation in this country is generally grassland, sometimes heath. The British sand-dunes, though they often bear shrubs, are rarely colonised by trees, perhaps because of the scarcity of suitable seed-bearing trees in the neighbourhood.

8. Shingle beaches are formed from pebbles thrown up by high tides. Considerable areas exist at Dungeness in Kent, Orfordness in Suffolk, and Chesil Bank in Dorsetshire, while various smaller areas occur at numerous other places along the coast. The shingle-beach vegetation is quite characteristic. Much of it is very closely allied to sand-dune vegetation; in other cases where it lies on silt, to which the tide has access, it bears many salt-marsh plants.

CHAPTER II

THE PLANT-FORMATION OF CLAYS AND LOAMS

THESE soils may be defined as heavy or medium in texture, containing a large proportion of fine particles and a small proportion of coarse sand. Chemically they are either very poor or comparatively poor in lime. On one side they pass into the lighter sandy loams as the proportion of coarse sand increases, and on the other into the marls and calcareous clays with an increase in lime-content. In places where the ground-water is stagnant, as in hollows or on undrained flats, they become marshy and tend to give an acid reaction.

The distribution of such soils in England is very wide and extensive, particularly in the Midlands and the south-east, where the Secondary and Tertiary geological formations are mainly developed. Some of the more extensive clays are the Oxford and Kimeridge Clays of the Jurassic formation, the Weald Clay, the Gault, the London Clay, and the non-calcareous Boulder Clays of the north. The Clay-with-flints overlying the Chalk varies from a heavy clay to a light loam. The shales, which represent the clay-formations of the older rocks, bear a vegetation which, while resembling that of the clays in many respects, differs in important characters, and is considered as belonging to a different plant-formation.

The clays and loams, forming as they do many of the

plains and valleys of the country, were no doubt among the first soils to be disforested and cultivated, so that very little purely natural vegetation remains on any of them. Nowadays they are very largely under permanent pasture.

Pedunculate oakwood association (*Quercetum Roburis*). There can be little doubt that practically the whole of the clays and loams were at one time covered with oak-forest. All the woods still existing on these soils that have any claim to be considered natural or semi-natural are oakwoods, or clearly derived from oakwoods[1], except on wet ground, along streamsides, etc., where they pass into woods of the alder-willow association. Very few, if any, of these oakwoods have been left untouched by human interference, the great majority having been heavily exploited for timber in the past. During the last four centuries oak was in special demand for shipbuilding, and the extensive oak-forests of the Weald, and the south-east generally, suffered severely during this period, which witnessed the rise of the English navy and mercantile marine, because they were within easy reach of the Thames, the Medway, and the southern harbours in the neighbourhood of Portsmouth. The iron contained in the sands of the Wealden area has also been responsible, during the whole of the Middle Ages, for much forest destruction in this region. The iron was smelted and the iron articles were wrought in charcoal furnaces, and the primitive forests were destroyed wholesale for charcoal-making. Many place-names of the Weald, "Cinderhill Wood," "Hammer Wood," "Forge Coppice" and the like,

[1] Hornbeam (*Carpinus Betulus*) is dominant on clay and loamy soils in some parts of south-eastern England, and since this tree casts a deeper shade than the oak it is possible that the ground flora is affected sufficiently to admit of the separation of a hornbeam association. But no hornbeam wood has yet been studied thoroughly enough to admit of this being done at present.

still bear testimony to this extensive industry. In the absence of a proper system of forestry the result has been that the woods have mostly degenerated very seriously, and really good standard trees are now the exception.

The dominant tree is *Quercus Robur* L. (= *Q. pedunculata* Ehr.), which is the characteristic English tree of heavy and medium damp soils, though it also occurs freely on deep sands.

In the south of England and the Midlands the great majority of the oakwoods are in the form of copse (coppice-with-standards), the standards being generally oaks, and the coppice hazel, oak, ash, birch, etc. The standard oaks are generally far apart, so that their crowns do not touch and no close canopy is formed, but each tree has a free branching habit. This is supposed to be a relic of the old method of growing oak, which aimed at producing curved timber and "knee-pieces," useful in shipbuilding, as opposed to the long straight timber produced by trees grown in close canopy. In several of the various statutes designed to check the rapid exhaustion of the English oakwoods during the later Middle Ages, it is decreed that in coppicing "twelve standels" should be left to the acre, a number which would leave ample room for each tree to develop its full crown without interfering with its neighbours. The shade cast by the standard oaks in such a wood is not nearly so deep as that cast by oaks grown in close canopy, and the shrub-layer has thus plenty of light to develop fully (Plate I b).

In many of the existing copsewoods there are less than twelve oaks to the acre—some consist simply of coppice with an occasional standard, some of coppice alone. Many of these woods are very degenerate owing to their careless exploitation. When the soil is left exposed to the sun and wind, by excessive felling of standards or clearing of coppice, the humus layer is destroyed, and the soil either becomes very weedy or cakes hard and becomes almost bare of vegetation. Under such conditions the natural regeneration of the wood from self-sown seed is checked or arrested, and unless the wood is properly taken in hand it degenerates to scrub and grassland. Such woods have often been extensively replanted with oak, ash and hazel, and sometimes with conifers. As was pointed out in the last chapter these oak-hazel copsewoods are not now commercially remunerative[1], though they are useful as furnishing a source of small wood, and to some extent of timber, for the local uses of the countryside, but their main importance in the eyes of their latterday owners is found in their use as pheasant-covers; the gamekeeper, not the woodman, is the important guardian.

Shrub-layer. The most abundant shrub of the coppice is the hazel (*Corylus Avellana*), and in very many cases the coppice consists almost wholly of this plant. It is possible that the hazel was the dominant member of the shrub-layer in the primitive oakwoods from which the existing woods are derived, and it is pretty certain it was very abundant. Thus in certain fragments of uncoppiced woodland, supposed to represent remains of the old oak-forest of the Weald, the hazel is one of the commonest shrubs, though it is exceeded in numbers by sloe and hawthorn; but *pure* hazel coppice is probably an indication of planting. Hazel is a particularly good coppice shrub, making excellent and rapid growth from the stools, and the coppice shoots are used for a great variety of purposes. Among the coppice are found however various other species of shrub, such as sallow (*Salix caprea, S. cinerea*) and dogwood (*Cornus sanguinea*), etc. Oak is often an abundant constituent of the coppice, as would be expected in copse derived from primitive

[1] It is even stated to be doubtful if oak in any form can be grown remuneratively in England in the present state of the market (Forbes, *The Development of British Forestry*, 1910, p. 192).

Phot. S. Mangham

a. *Quercus Robur* (in close canopy). *Corylus Avellana*, *Cratægus monogyna*, *Rubus* spp., *Pteris aquilina*, *Digitalis purpurea*. Staffhurst Wood, Surrey, on Weald clay.

Phot. S. Mangham

b. Oak-hazel wood; oaks in open canopy. *Corylus Avellana* (second year coppice), *Cnicus palustris*, *Euphorbia amygdaloides*, *Teucrium Scorodonia*, *Pteris aquilina*. Chevening Park, Kent, on clay-with-flints.

Pedunculate oakwood association (*Quercetum Roburis*).

oak-forest. The bark of coppiced oak formerly had a high value for tanning hides in the process of making leather. Ash (*Fraxinus excelsior*) and birch (*Betula tomentosa* and *B. alba*), frequent or abundant constituents of natural oakwood, are also often found among the coppiced shrubs, and so is maple (*Acer campestre*). Beech (*Fagus sylvatica*) and Spanish chestnut (*Castanea sativa*, planted) are occasionally met with, but are far more characteristic of the dry oakwoods.

Other very abundant members of the shrub-layer of the damp oakwood are hawthorn (*Cratægus monogyna*), sloe or blackthorn (*Prunus spinosa*), briars (*Rosa canina*, *R. arvensis*) and brambles (*Rubus fruticosus* agg.). These do not of course form coppice shoots and are largely absent from well-kept coppice.

Ground vegetation.

The ground vegetation of the damp oakwood is conditioned very largely by the amount of light it obtains during the growing season. Under the shade of oak growing in close canopy, or of old (10—20 years) and well-grown hazel coppice, the vegetation is very open and consists exclusively of species which can endure deep shade (Plate I a, II a).

In about 14 square yards (about 9·3 square metres) of oakwood, with the trees about 35 years old and growing in close canopy, in which the shrubs (hazel, sloe and hawthorn) were too deeply shaded to produce abundant foliage, the following species were found forming an open association, with much bare soil between the plants, on good mild humus above a loamy soil.

Trees.

Quercus Robur (3, and 2 seedlings).

Shrubs.

Prunus spinosa (several shrubs and many seedlings).
Cratægus monogyna (several shrubs and seedlings).
Corylus Avellana (1 shrub, 1 seedling).
Rubus *sp.* (several with trailing shoots).

Climbers.

Hedera Helix (abundant, trailing).
Lonicera Periclymenum (climbing on the oaks and trailing on the ground).
Tamus communis (one plant).

Herbs (in order of abundance).

Viola Riviniana (very abundant).	Bromus sterilis.	
Pteris aquilina.	Ajuga reptans.	
Euphorbia amygdaloides.	Lamium Galeobdolon.	
Teucrium Scorodonia.	Oxalis Acetosella.	One plant of each.
Primula vulgaris.	Lathyrus montanus.	One plant of each.
Fragaria vesca.	Dactylis glomerata.	One plant of each.
Stellaria Holostea.	Potentilla sterilis.	One plant of each.

The presence of such species as *Pteris aquilina* and *Teucrium Scorodonia* shows that the soil was distinctly light in this particular wood.

In more open situations many other species are found which cannot tolerate the deepest shade, though they may fairly be considered as woodland species.

Degenerate woods are often successfully invaded by plants from neighbouring associations, frequently by weeds from arable land and roadsides, and also show all transitions to the associations of scrubs and grassland.

The ground vegetation of regularly cut coppice shows a periodicity depending on the sudden change from the deep shade of old coppice to the brightly lighted conditions obtaining when the coppice has been cut, followed by a gradual return to shade as the coppice shoots grow up. In the first season after coppicing the vegetation generally shows no conspicuous change, but in the second and third years the ground becomes covered with an active vegetation, consisting partly of strong shoots sent up by plants whose underground parts have been lying dormant or vegetating but weakly, partly of plants produced by the germination of dormant seeds, and sometimes of new invaders from without. The continuous carpets of conspicuous social early-flowering species, such as primrose (*Primula vulgaris*), bluebell (*Scilla non-scripta*), wood anemone (*Anemone nemorosa*), dog's mercury (*Mercurialis perennis*), red campion (*Lychnis dioica*), creeping bugle (*Ajuga reptans*), yellow archangel or weasel-snout (*Lamium Galeobdolon*), which arise in this way, form one of the best-known and most attractive features of our coppice-woods. Gradually, as the coppice grows up, this vegetation diminishes in vigour, till the paucity of species and individuals characteristic of old coppice is re-established.

The following is a fairly full, though it does not pretend to be an exhaustive, list, of the *Quercetum Roburis*, or damp oakwood of clays and loams.

Tree layer.

Quercus Robur (=Q. pedunculata)	d	Fagus sylvatica (generally planted)	l
Fraxinus excelsior	f	Ulmus glabra Huds. *non* Mill.	o
Betula tomentosa	f	Taxus baccata	r to o
B. alba	f	*Smaller trees.*	
Populus tremula	o to f	Ilex Aquifolium	o to a
P. canescens	l	Pyrus Malus	f
Alnus rotundifolia	l	Acer campestre (generally coppiced)	f to a
Carpinus Betulus	l or ld		
Prunus Avium	f	Pyrus torminalis	l

Shrub layer.

Corylus Avellana	a	Juniperus communis	o
Cratægus monogyna	a	Rubus *spp.*	a
Prunus spinosa	a	Rosa canina	a
Salix caprea	f	R. arvensis	a
S. cinerea	f	Ligustrum vulgare (often planted)	o
S. aurita	l		
Cornus sanguinea	o to f	Ulex europæus (an invader)	o
Viburnum Opulus	o	Rhamnus Frangula	r

Climbers.

Lonicera Periclymenum	a	Calystegia sepium	f
Hedera Helix	a	Tamus communis	f

Ground vegetation (no single species generally dominant).

Anemone nemorosa	lsd	R. Ficaria	a
Ranunculus auricomus	lab	R. repens	f

Cardamine flexuosa	o	Veronica Chamædrys	a
C. pratensis (Kent)	la	V. montana	f
Viola Riviniana	a	V. officinalis	a
Lychnis dioica	a	Melampyrum pratense	a
Stellaria Holostea	a	Lathræa squamaria	lf
Arenaria trinervia	a	Nepeta Glechoma	a
Hypericum perforatum	f to ab	Teucrium Scorodonia	la
H. hirsutum	o	Prunella vulgaris	a
Geranium Robertianum	f	Stachys Betonica	f
Oxalis Acetosella	f	S. sylvatica	f
Vicia sepium	a	Lamium Galeobdolon	la
V. Cracca	f	Polygonum Hydropiper	la
Lathyrus montanus	a	Ajuga reptans	a
Spiræa Ulmaria	o	Rumex viridis	a
Geum urbanum	a	Euphorbia amygdaloides	l
Fragaria vesca	l	Mercurialis perennis	la
Potentilla erecta	l	Urtica dioica	la
P. sterilis	la	Daphne Laureola	l
Circæa lutetiana	a	Listera ovata	f
Sanicula europæa	f	Orchis mascula	o
Angelica sylvestris	f	Allium ursinum	l
Heracleum Sphondylium	a	Scilla non-scripta	lsd
Adoxa Moschatellina	lf	Luzula pilosa	a
Asperula odorata	o	Arum maculatum	la
Valeriana sambucifolia	o	Carex sylvatica	la
Dipsacus sylvestris	o	C. lævigata	lf
Scabiosa Succisa	a	Anthoxanthum odoratum	la
Solidago Virgaurea	o	Milium effusum	o
Arctium *spp.*	a	Deschampsia cæspitosa	a
Cnicus palustris	f	Arrhenatherum elatius	f
Lactuca muralis	o	Melica uniflora	a
Primula vulgaris	a	Dactylis glomerata	a
Lysimachia nemorum	lf	Bromus giganteus	f
L. Nummularia	o	B. ramosus	f
Vinca minor	ld	B. sterilis	f
Centaurium umbellatum (= Erythræa Centaurium)	a	Brachypodium gracile (= B. sylvaticum)	a
Myosotis arvensis (= M. intermedia)	a	Pteris aquilina (on light soils)	lsd
M. sylvatica	l	Lastrea Filix-mas	f
Scrophularia aquatica	f	L. aristata (= L. dilatata)	l
S. nodosa	o	L. spinulosa	l
		Athyrium Filix-fœmina	l

Phot. A. G. Tansley

a. Oakwood association (prevernal aspect). *Primula vulgaris.* Staffhurst Wood, Surrey, on Weald clay.

Phot. A. G. Tansley

b. Oaks and neutral pasture. *Quercus Robur* (characteristic habit grown in the open), *Taxus baccata.* Edge of wood, same locality.

Formation of Clays and Loams.

No attempt has been made in the above list to separate the vegetation of woods on heavy clays from those on light loams, although there are very considerable differences, because the data at present available are insufficient to carry out such a separation satisfactorily. The woods on light loams have something in common with the dry oakwoods (p. 92), though the fresh moist soil distinguishes them. Such plants as *Anthoxanthum odoratum*, *Pteris aquilina*, *Scilla non-scripta*, *Anemone nemorosa*, *Veronica officinalis*, *Teucrium Scorodonia* belong to these oakwoods of lighter soils.

SUBORDINATE (RETROGRESSIVE) ASSOCIATIONS (Plate II b)

On the edges of damp oakwoods which have been carelessly exploited a scattered scrub is commonly found with spaces of turf between the clumps of bushes, and an occasional isolated oak-tree. This vegetation represents degenerate oakwood from which the trees have nearly or quite disappeared, and in accordance with the principles laid down in the introduction must be reckoned as belonging to the same formation. The genetic relationship of this scrub and grassland with the oakwood association is quite obvious to any careful observer, and the floristic relationship is also clear, as will be seen immediately.

Scrub association. This may consist of any of the species of the shrub-layer of the damp oakwood, but *Cratægus monogyna*, *Prunus spinosa*, *Rubus* spp., *Rosa* spp., and *Ulex europæus* are much the most abundant. It will be noticed that all of these are spiny species, and the dominance of such species is probably due to the fact that they are protected by their spines from browsing animals; since this kind of land is very commonly used for pasturage. The bushes often form dense impenetrable clumps, which appear to possess a considerable degree of permanence, and, in the shelter of these, less protected shrubs often flourish. Shade plants of the oakwood association are also often found within the shelter of the bushes.

Neutral grassland association (*Graminetum neutrale*). This is the ultimate phase of degeneration of damp oakwood. It consists of a close turf of grasses with associated herbaceous plants. The term "neutral" is intended to express the fact that this type of grassland is free from the characteristic plants of calcareous pasture on the one hand and of acid heath pasture on the other. Neutral grassland is practically always heavily grazed, and this of course must modify its composition and character to some extent, but as yet it has scarcely been studied. Many of the "commons" and village greens of central and southern England consist of the neutral grassland (pasture) association with or without the scrub described above. The severe grazing that such land usually suffers, apart from the greater exposure to light and drought, scarcely allows of the persistence of many members of the ground vegetation of the oakwood, but in places which escape such treatment woodland plants, *e.g. Scilla non-scripta, Anemone nemorosa, Primula vulgaris* and others, may be found in such grassland.

The following is a list of species found in neutral grassland (including meadowland)[1]. As in the case of the list for damp oakwoods, it includes species found on a fairly wide range of soils in respect of texture and

[1] Manuring of various kinds has of course a very important effect on the vegetation of grassland. Pasturage itself always involves manuring by the dung and urine of the animals, while the effect of artificial manuring varies according to the kind and amount of the manure employed. Very extensive data for the study of these effects have been collected as the result of the long-continued experiments at the Rothamsted Experimental Station. In general it may be said that heavy long-continued manuring reduces the number of species occurring in the grassland, while special kinds of "one-sided" manuring sometimes lead to the introduction of fresh species. A comparative study of the British grasslands from the ecological standpoint is very much needed.

dampness. Probably the association should be divided into more than one sub-association, but the *data* to hand are insufficient to justify such a division. The words "damp" and "dry" are prefixed to the names of the species whose proclivities in regard to soil-moisture are apparently most marked.

Dominant Grasses.		
	Lolium perenne	
dry	Cynosurus cristatus	
	Anthoxanthum odoratum	
	Dactylis glomerata	
	Holcus lanatus	
damp	Alopecurus pratensis	
	Bromus mollis	
	Festuca duriuscula	
Other Grasses.		
	Agrostis tenuis (vulgaris)	f
	A. alba	f
	Poa pratensis	f
damp	P. trivialis	la
	Phleum pratense	f
	Arrhenatherum elatius	f
dry	Ranunculus bulbosus	a
	R. acris	a
	R. repens	a
damp	Cardamine pratensis	la
	Polygala vulgaris	o
	Cerastium triviale	a
	C. glomeratum	a
	Stellaria graminea	o
	Hypericum perforatum	f
	H. tetrapterum	o
	Malva moschata	o
dry	Linum catharticum	la
	Geranium pratense	o
	Genista tinctoria	l
	Ononis repens	o
	Trifolium pratense	a
	T. medium	o
	T. repens	a
dry	T. dubium	a

	Lotus corniculatus	a
	Vicia tetrasperma	o
	V. angustifolia	o
	Lathyrus Nissolia	o
	L. pratensis	a
dry	Potentilla erecta	o
dry	P. reptans	a
dry	P. mixta	l
	P. anserina	la
	Alchemilla vulgaris	r
	Agrimonia Eupatoria	f
	A. odorata	o
damp	Saxifraga granulata	l
dry	Conopodium denudatum	la
dry	Pimpinella Saxifraga	f
	Anthriscus sylvestris	l
	Silaus flavescens	o
	Heracleum Sphondylium	f
dry	Daucus Carota	o
dry	Galium verum	o
	Dipsacus sylvestris	f
	Scabiosa succisa	a
	S. arvensis	f
	Bellis perennis	a
damp	Pulicaria dysenterica	f
	Achillea Millefolium	a
damp	A. Ptarmica	l
	Chrysanthemum Leucanthemum	la
	Tussilago Farfara	la
dry	Senecio Jacobæa	a
	S. erucifolius	o
	Cnicus palustris	a
	Hypochæris radicata	a
	Crepis virens	a
	Leontodon hispidum	f

	L. autumnalis	a		Prunella vulgaris	a
dry	Hieracium Pilosella	la	damp	Ajuga reptans	a
	H. boreale	f		Plantago lanceolata	a
	Taraxacum officinale	a	dry	P. media	o
	Tragopogon pratense	la		Rumex Acetosa	a
dry	Campanula rotundifolia	l		Spiranthes spiralis (=S. autumnalis)	o
	Primula veris	la		Orchis mascula	la
	Centaurium umbellatum	la	damp	O. maculata	la
	Myosotis arvensis (=M. intermedia)	a		Narcissus Pseudo-narcissus	l
damp	Veronica serpyllifolia	a		Colchicum autumnale	l
	V. Chamædrys	a		Carex ovalis	f
	V. officinalis	a		C. flacca (=C. glauca)	a
	Euphrasia officinalis (agg.)	a	damp	C. pallescens	l
	Bartsia Odontites	a		C. panicea	f
	Rhinanthus Crista-galli	la		Ophioglossum vulgatum	o
	Melampyrum pratense	f		Equisetum arvense	o
dry	Thymus Serpyllum (agg.)	la			
	Nepeta Glechoma	f			

Juncetum communis. On low-lying clayey soils the ground-water frequently approaches the surface and where this is the case the typical pasture or meadow-land passes into an association nearly always marked by a predominance of species of Juncus, *J. conglomeratus* and *J. effusus*, frequently with *J. articulatus*, *J. bufonius*, and *J. glaucus*. Other species found in this association are:

Ranunculus flammula	a	E. palustre	f
R. repens	a	Hydrocotyle vulgaris	a
R. sceleratus	f	Galium palustre	f
R. hederaceus	f	G. uliginosum	l
Caltha palustris	f	Gnaphalium uliginosum	f
Nasturtium spp.	f	Bidens cernua	o
Cardamine pratensis	a	B. tripartita	o
Lychnis Flos-cuculi	a	Petasites officinalis	f
Stellaria uliginosa	o	Senecio aquaticus	f
Lotus uliginosus	f	Lysimachia nummularia	l
Spiræa Ulmaria	la	Scrophularia aquatica	a
Callitriche stagnalis	a	Veronica scutellata and *var.* hirsuta	l
Epilobium parviflorum	f		

Pedicularis sylvatica	l	C. panicea	f
Mentha aquatica, etc.	a	C. lævigata	o
Polygonum Hydropiper	a	C. Goodenowii	f
P. amphibium	f	C. flava (agg.)	f
P. Persicaria	a	Alopecurus fulvus	r
Orchis maculata (agg.)	f	A. geniculatus	f
Iris Pseudacorus	o	Glyceria fluitans	f
Triglochin palustre	o	Deschampsia cæspitosa	f
Eleocharis palustris	f	Poa trivialis	f
Carex vulpina	f	Equisetum maximum	o
C. remota	f	E. palustre	f
C. ovalis	f	E. limosum	o

This association forms a transition to the herbaceous marsh association (p. 207).

CHAPTER III

THE VEGETATION OF THE COARSER SANDS AND SANDSTONES[1]

Physical characters of sands and sandstones.

THE very fine-grained sands and sandstones have much in common physically with clays, since they are close-textured and retain a large proportion of water. In the case of fine-grained sands, such as alluvial sands, and soft fine-grained sandstones which weather so as to produce a considerable depth of soil, the vegetation is of the same general type as that dealt with in the last chapter. The impure sandstones, containing a large proportion of clay particles in addition to the grains of silica, give rise on weathering to sandy loams, which also bear the same general plant-covering; if, on the other hand, there is a large proportion of lime, as in the calcareous sandstones, the vegetation is altered and belongs to the type characteristic of calcareous soils. But the fairly pure, coarse-grained sands and the geologically more recent sandstones have a well-defined vegetation of their own.

Coarse sandy soil is a dry soil, because it has little water-capacity, a feeble power of holding up ground-water by capillarity and suffers from rapid evaporation.

[1] The vegetation of the harder, *i.e.* older, (Palæozoic) sandstones and grits is treated separately (Chapter V).

It is also frequently poor in nutritive salts. In a very dry climate such a soil is particularly unfavourable to vegetation because it cannot form humus. In a climate with a considerable rainfall the surface-layers of soil tend to be very poor in soluble nutritive salts because these salts are washed out by the percolation of rainwater. Owing to the deficiency in lime and other bases such soils tend to form and accumulate acid humus which immediately affects the vegetation, as we shall see in the sequel. Under certain circumstances however, when there is a high level of well aerated ground-water and a fair proportion of soluble salts, mild humus is formed and the soil may be very fertile, as in the case of many alluvial sands, which form good meadowland, and are excellent for certain kinds of cultivation.

Distribution of the newer sands and sandstones.

The more recent (Secondary, Tertiary and post-Tertiary) coarse-grained sands and sandstones occur almost exclusively in the southern and eastern parts of England and in the Midlands. Some of the more extensive and characteristic examples are the Bunter Sandstone of the Trias (Midlands), the Lower Greensand (Cretaceous) in the eastern Midlands and south-east, the Ashdown Sand belonging to the Wealden series, the Bagshot and other Eocene sands of the London and Hampshire basins, the Pliocene "Crag" of East Anglia, and various "valley" and "plateau gravels" deposited at intervals during post-Tertiary times, and resting on different older rocks.

Heath and woodland.

The most generally characteristic feature of the vegetation of these newer sands and sandstones is the prevalence of heathland. On some of them indeed (*e.g.* the East Anglian sands) no woodland which has really good claims to be considered natural has yet been recognised, and it may be that *Calluna*-heath and grass-heath (heath pasture) is the highest type of plant-community these soils have developed, though it is also possible that they have been completely disforested. But in many cases there is a well-marked natural woodland-type characteristic of coarse sandy soils. Such woodland generally alternates with heathland or dry grassland, which often occupies its edges and open places within the wood and bears the same relation to the wood that neutral pasture bears to damp oakwood. In other words much heathland as well as dry heathy grassland is clearly degenerate woodland of a certain type, while other heathland is possibly primitive. It is frequently difficult or impossible to decide whether the heathland is primitive or has been derived from woodland.

Formation of dry peat.

The occupation of a sandy soil by the heath-community is connected with an important modification of the habitat by the accumulation of a comparatively dry acid peaty humus (*Trockentorf*) in or on the surface soil. The formation of this characteristic type of humus excludes many species of plants from the flora, and at the same time introduces a well-defined vegetation of its own. These facts make it necessary to recognise the existence of a new plant-formation, the heath-formation, allied to the moor-formation but characteristic of a drier climate or of drier edaphic conditions. Where woodland degenerates into heath we must therefore recognise a case of succession in which one formation is replaced by another.

The researches of Wollny, P. E. Müller, Ramann, Graebner and others have made it probable that the fundamental cause of the degeneration of woodland into heath in north-west Germany (and it is natural to extend their conclusions to south-east England, which has much the same climate, and often repeats the same edaphic features) is the slow washing out or leaching of the sandy soil.

Degeneration of woodland.

A brief account of the main features of this process may be quoted. According to Graebner, "when there is a rainfall of 70 cm. (28 inches) or more...the surface-layers are being continually impoverished in mineral salts by washing out or 'leaching,' and the typical plants of the forest-floor are thus starved and give way before the invasion of mosses and shade-bearing heath plants. The matting together of the surface-layers of soil by the rhizoids and rootlets of the invaders prevents the access of oxygen to the soil, and leads to the accumulation of 'acid humus' or 'dry peat' (*Rohhumus*) in place of the original mild humus of the woodland soil. Thus we have the formation of a type of wood with a heathy vegetation, poor in species, on a soil composed of a mixture of sand and dry acid humus or peat. Finally, according to Graebner, either by the leaching of the soil to such a depth that the roots of the trees can no longer obtain enough food, or by the formation below the surface of a layer of 'Ortstein' (moor pan), *i.e.* a hard layer of sand bound together by humous compounds, which the roots of the trees cannot penetrate, the rejuvenation of the wood is rendered impossible, the gaps formed by the dying of the old trees are not filled up, and the forest is eventually replaced by heath[1]."

If we assume that this secular process of the impoverishment of sandy soils has been going on since the glacial period, it is easy to see that, acting on a series of soils of varying original constitution, it will produce a vegetation showing many local differences at any given period.

While some very barren sands may have been originally colonised by heath, others which originally bore woodland may have become occupied by heath before the beginning of the historical period; others again have

[1] Moss, Rankin and Tansley, 1910: pp. 132, 133.

certainly degenerated into heathland during the last few centuries, while yet others are only now beginning to show the changes preparatory to this transformation.

It is generally admitted that the process is greatly accelerated by the extensive felling of trees and clearing of woodland on these soils. Clearing leads to the rapid destruction of the mild woodland humus, and thus facilitates the entrance of the pioneers of the heath vegetation. This factor has certainly been fully operative in the conversion of many of the British woodlands to heath.

We will now see how far the plant-associations of our sandy soils can be interpreted in the light of these general considerations.

The Plant-formation of Sandy Soil

Dry sandy oakwood association (*Quercetum arenosum Roburis et sessiliflorae*). This association is developed on many of the sands and sandstones mentioned. The soil is much poorer in humus than that of the damp oakwood on clays and loams, and the association is much poorer floristically. The following is a list of the more characteristic and commoner species.

Tree layer. Dominant.		*Other trees.*	
Quercus Robur (= Q. pedunculata), or		Betula tomentosa	a
		B. alba	f
Q. sessiliflora (mainly shallow soils), or		Populus tremula	o
Q. Robur and Q. sessiliflora co-dominant (and hybrids)		Ilex Aquifolium	a
		Pyrus Aucuparia	f
		Pyrus Aria	o
Fagus sylvatica	ld		
Shrub layer.		Rubus rusticanus, etc.	f
Cratægus monogyna	a	Ulex europæus	o
Prunus spinosa	f to a	Cytisus scoparius	f
Rubus Idæus	f	Lonicera Periclymenum	a

Phot. S. Mangham

a. *Betula tomentosa*, *Pteris aquilina*, *Holcus mollis*, *Anthoxanthum odoratum*, etc.

Phot. R. S. Adamson

b. *Quercus Robur*, *Pteris aquilina*, *Holcus mollis* (May).

Dry oakwood (*Quercetum Roburis arenosum*). Gamlingay Wood, Cambs., on glacial sandy loam.

Ground vegetation.

Anemone nemorosa	lsd
Viola Riviniana	a
Hypericum pulchrum	f
H. perforatum	f
Oxalis Acetosella	f
Sedum Telephium	o
Potentilla erecta	a
P. sterilis	f
Epilobium angustifolium	la
E. montanum	f
Conopodium majus	f
Galium saxatile	f
Solidago Virgaurea	f
Hieracium vulgatum (agg.)	f
H. boreale	f
Centaurium umbellatum (=Erythræa Centaurium)	o
Digitalis purpurea	f
Veronica officinalis	a
Melampyrum pratense	f
Teucrium Scorodonia	a
Rumex Acetosella	f
Convallaria majalis	r
Scilla non-scripta	lsd
Luzula pilosa (=L. vernalis)	f
Holcus mollis	ld
Agrostis vulgaris	a
Anthoxanthum odoratum	f
Pteris aquilina	ld
Blechnum Spicant	f

The species especially associated with the heath-formation which are often found in the dry oakwood (see p. 99) are excluded from this list.

It will be seen that this association, though frequently with the same dominant, differs decidedly from the damp oakwood association. Thus *Quercus Robur* is often associated with *Q. sessiliflora* (which is practically never found in the typical damp oakwood on clays and loams) while the beech often forms pure societies, reducing the undergrowth to a minimum. The birches are typically more abundant (Plate III a), while *Pyrus Aucuparia* is characteristic. On the other hand the association has many fewer species of trees and shrubs. The ash and the hazel (except apparently where planted[1]) are generally scarce or absent. The same may be said of practically all the other damp oakwood shrubs, except the hawthorn (*Cratægus monogyna*) and the sloe or blackthorn (*Prunus spinosa*).

The ground vegetation also is much poorer and is

[1] Oak-hazel copse is found pretty frequently on fairly dry soil accompanied by the typical dry oakwood ground flora. This is probably due to planting.

naturally characterised by absence of the great number of woodland species which frequent damp soils. Societies of *Pteris aquilina*, *Holcus mollis* (Plate III b), and *Scilla non-scripta* are characteristic. *Potentilla erecta*, *Digitalis purpurea*, *Hypericum pulchrum*, *Galium saxatile* and *Solidago Virgaurea* are also specially characteristic.

Changes in the level of the ground-water, owing to the slope of the ground, sometimes lead to the alternation of the damp and dry oakwood associations in the same wood within very short distances.

SUBORDINATE ASSOCIATIONS

Just as in the case of the formation of clays and loams, so here, there are retrogressive associations, consequent on clearing of the woodland, and represented by scrub and grassland. These are largely represented on the sandy "commons" of southern England just as the neutral pasture and corresponding scrub are represented on the commons on clay soil.

Scrub-association. This is dominated by the shrubs of the dry oakwood whose spines protect them from destruction by browsing cattle, *i.e.* *Cratægus monogyna*, *Prunus spinosa*, *Ulex europæus* and *Rubus* spp. The association consequently closely resembles the corresponding member of the plant-formation of clay and loam.

The gorse (*Ulex europæus*) is perhaps the most abundant and characteristic member of the scrub association. It is frequently a pure dominant on the sandy commons referred to, forming a close scrub under the shade of which little can grow. The gorse actively spreads from seed and forms a progressive society colonising the grass heath association, where it flourishes much more vigorously than under the partial shade of the dry oakwood.

Grass heath association (*Graminetum arenosum*). This association generally forms a close short turf, commonly dominated by the so-called "heath grasses," *i.e.* those inhabiting sandy soil. The tracts of country occupied by this association are often called "heaths[1]," although the true heath plants (*Calluna, Erica*, etc.) may be entirely lacking, but all transitions are found between the grass heaths and the *Calluna*-heaths. It seems probable that the difference is determined in the first place by the amount of lime and other soluble salts in the sandy soil.

The association is characterised by a great number of dicotyledons, very largely annual, and often early flowering ephemeral species. Sometimes indeed these are so numerous that the grasses cannot be said to be dominant at all. Many of these species are central European forms which in this country are confined to south-eastern England.

On very dry grass heaths no close turf is formed and much open soil is exposed. Such places are frequently invaded by weeds from neighbouring sandy arable land.

The following is a list of species occurring in the grass heath association:—

Ranunculus bulbosus	f
Corydalis claviculata	o
Cardamine hirsuta	f
Erophila verna (agg.) (= Draba verna)	a
Teesdalia nudicaulis	o
Viola Riviniana	a
V. canina (= V. ericetorum)	o
Polygala vulgaris (agg.)	f
Dianthus Armeria	r
D. deltoides	r
D. prolifer	vr
Cerastium arvense	o
C. triviale	a
C. semidecandrum	la
Moenchia erecta (= Cerastium quaternellum)	l
Spergularia rubra	f
Arenaria serpyllifolia	f
A. tenuifolia	r
Sagina procumbens	f
S. apetala	f
S. ciliata	o
S. subulata	r
Stellaria graminea	f
Hypericum pulchrum	f
H. humifusum	o
H. perforatum	f
Malva rotundifolia	f
Linum catharticum	f
Geranium molle	f
G. pusillum	o
G. dissectum	f
Erodium cicutarium	f
Ulex europæus	la
Cytisus scoparius	la

[1] This term is even used in East Anglia for tracts of chalk pasture, *e.g.* Royston Heath, Newmarket Heath.

Ononis repens	f
Medicago lupulina	f
M. denticulata	r
M. arabica	r
M. minima	r
Trifolium subterraneum	o
T. procumbens	o
T. dubium (=T. minus)	f
T. filiforme	f
T. arvense	f
Lotus corniculatus	a
Ornithopus perpusillus	la
Vicia lathyroides	o
V. angustifolia, *var.* Bobartii	f
Prunus spinosa	f
Rubus *spp.*	a
Potentilla sterilis	f
P. erecta	a
P. argentea	o
Alchemilla arvensis	f
Rosa *spp.*	f
Saxifraga tridactylites	la
Sedum acre	o
S. anglicum	o
Pimpinella Saxifraga	f
Conopodium majus (=C. denudatum)	a
Daucus Carota	f
Galium verum	la
G. saxatile (=G. hercynicum)	la
Achillea Millefolium	a
Filago minima	o
Gnaphalium sylvaticum	o
Bellis perennis	a
Senecio Jacobæa	la
S. sylvaticus	l
Hieracium Pilosella	a
H. boreale (agg.)	f
H. umbellatum (agg.)	o
Hypochæris radicata	a
H. glabra	a
Leontodon hispidum	f
L. nudicaule (hirtum)	o

L. autumnale	a
Taraxacum erythrospermum	f
Carduus nutans	o
Cnicus lanceolatus	f
Carlina vulgaris	la
Jasione montana	la
Campanula rotundifolia	o
Calluna vulgaris	o
Erica cinerea	o
Centaurium umbellatum (=Erythræa Centaurium)	f
Myosotis collina	f
M. versicolor	f
Veronica officinalis	a
V. arvensis	f
V. verna	vr
Euphrasia officinalis (agg.)	a
Teucrium Scorodonia	la
Thymus Serpyllum (agg.)	la
T. ovatus	f
Plantago Coronopus	o
P. lanceolata	a
Scleranthus annuus	la
Rumex Acetosella	a
Allium vineale	o
Scilla autumnalis	vr
Luzula campestris	a
L. multiflora	o
Carex arenaria (rare inland but)	ld
C. divulsa	o
C. binervis	l
C. pilulifera	o
C. verna	f
Anthoxanthum odoratum	la
Agrostis canina	la
A. setacea	la[1]
A. tenuis (=A. vulgaris)	ld
Aira caryophyllea	f
A. præcox	f
Deschampsia flexuosa	la
Sieglingia decumbens	la
Festuca ovina	la
Pteris aquilina	ld

[1] South-west England.

The East Anglian "heaths," particularly those in the neighbourhood of Mildenhall, Brandon and Thetford, include considerable stretches of dry grassland which are not however quite typical grass heaths, such plants as *Agrostis canina* and *Deschampsia flexuosa* and several other typical grass heath species being absent. The soil often contains appreciable quantities of lime. These areas of grassland alternate with true (*Calluna*) heath. The region has a low rainfall (less than 25 inches *per annum*) and exhibits the nearest approach to steppe-conditions to be found within the British Isles. The vegetation tends to be sparse and the tufts of ling or grass are often wide apart, the intervening spaces bearing a scattered vegetation of ephemerals. It is doubtful if this area ever bore natural woodland.

The existence of widespread steppe conditions, subsequently to the glacial period, is now generally recognised. Between the glacial period and this steppe period, a fairly warm epoch intervened, with the coast-line not far from its present position. In the steppe period a considerable elevation occurred, and much of the North Sea was land. Such a period was favourable to the accumulation of loose sands, like those of the Mildenhall district, and there is evidence that the formation of these sands has been going on for some time. A steppe flora must have been widely distributed during the steppe period, and as we find fossil steppe mammalia in England, we should expect that steppe plants also occurred here. If we have survivors of the glacial flora under suitable conditions, we should expect survivors of the more recently arrived steppe flora also. The Mildenhall sandy tract has several plants not found elsewhere in Britain and these may well be the survivors of that flora[1].

The following plants are confined, or nearly confined, within the British Isles, to this region:—

Silene Otites
S. conica
Holosteum umbellatum
Medicago Falcata
M. sylvestris
Artemisia campestris
Veronica verna
V. triphyllos
Muscari racemosum
Ornithogalum umbellatum
Carex ericetorum

[1] The editor is indebted for this paragraph to Dr J. E. Marr.

CHAPTER IV

THE HEATH FORMATION

Distribution of the heath formation.

THE great heath formation of north-west Europe is typically developed on relatively poor sandy and gravelly soils whose climate is wetter than that which gives rise to steppe. The steppe climate is too dry for tree growth, apart from local edaphic conditions, but the heath formation exists side by side with woods and in many cases has arisen as the result of the degeneration of woodland.

Heath is developed in north-west Germany, Jutland and the Baltic islands, Belgium and Holland, northern and western France and southern and eastern England and Scotland. Most of the German, Belgian and southern and eastern English heaths occur in regions with an annual rainfall between 25 inches and 40 inches (*c.* 60 to 100 cm.), but the Cornish heaths and those of the eastern Highlands of Scotland often receive a rainfall of between 40 and 60 inches (*c.* 100 to 150 cm.) in the year. We do not accurately know what is the upper limit of rainfall consistent with the formation of heathland; above the minimum (*c.* 25 inches) the heath formation is mainly determined by the nature of the soil, though the scarcity of heaths on the western side of Scotland, northern England, Wales and Ireland where there are considerable tracts of country with an annual rainfall exceeding 60 inches probably has a real significance[1]. The Scottish heaths develop a deeper layer of relatively pure acid humus, up to 8 or 12 inches (20 to 30 cm.) according to Hardy, and have much in common with the *Calluna* moor of the Pennines (see p. 113).

At the other extreme we have the East Anglian heaths with an annual rainfall of 25 inches or less, and a minimum of dry peat formation. Between these extremes come the heaths of the south-eastern counties and those of the New Forest and Dorsetshire, with a layer of dry peat seldom more than a fraction of an inch in thickness passing down into sand darkened by humus. The surface layer of dry peat is largely formed by lichens and mosses (*e.g.* *Cladonia rangiferina*, *Polytrichum piliferum*, and other species) which are the pioneers in reconstituting the heath association on bared soil, as can be well seen where the vegetation is destroyed by digging for gravel.

It is possible that the heath formation originally colonised many tracts of the poorer English sands, *e.g.* the Pliocene "crag" and some of the more recent Quaternary sands and gravels of East Anglia and the south, perhaps also the Eocene Bagshot sand. If so, these sands, or at any rate parts of them, have never borne natural woodland.

But apart from this possibility there is no doubt that the heath formation is constantly successfully invading and eventually replacing the natural woodland of many sandy soils. The causes and course of this process have already been briefly discussed.

Within the dry oakwood association, more or less isolated patches of heath vegetation are often found, with bilberry (*Vaccinium Myrtillus*), the characteristic heath grass *Deschampsia flexuosa*, and in well-lighted spots the ling (*Calluna vulgaris*). Though sometimes, no doubt, the occurrence of these plants is dependent on local

[1] See Hardy, 1905: p. 103.

conditions, it may often be interpreted as the first step in the degeneration of the dry oakwood association and its replacement by the heath formation. Thus heath plants are frequently found invading the edge of an oakwood whose centre still shows the typical dry oakwood association. When the entire ground vegetation is of the heath type it is doubtful if the oaks can perpetuate themselves indefinitely from seedlings, and the wood is then doomed.

Oak-birch heath association (*Quercetum ericetosum*). The initial association of the heath formation in its transition from dry oakwood is a very characteristic partially open and generally mixed wood of oak, birch, and in south-eastern England frequently beech, in which the oaks are gradually losing their dominance, and typical heath plants form the ground vegetation. Except where the beech is locally dominant, this kind of woodland allows far more light to reach the ground than does the typical oakwood. The foliage of the birches casts a lighter shade, and the trees are often separated by considerable intervals which are occupied by the heath association. Occasionally areas are met with in which the plants of the ground vegetation of the oakwood still maintain themselves, *e.g. Holcus mollis, Scilla non-scripta, Anemone nemorosa, Oxalis Acetosella*, etc. This variation in facies is characteristic of transitional (intermediate) associations in general, and particularly of associations that form the first stage of a succession in which a new formation is supplanting one already occupying the ground.

Another feature of the oak-birch heath association is its frequent invasion by the Scots pine (*Pinus sylvestris*), which may sow itself freely from neighbouring plantations (Plate IV a), and frequently perpetuates itself in the mixed woodland, whose openness is favourable for such colonisation.

Phot. S. Mangham

a. *Quercus Robur*, *Betula tomentosa*, *Fagus sylvatica*, *Pinus sylvestris*, *Pteris aquilina*, *Calluna vulgaris*. Ashdown Forest, Sussex, on Ashdown sand.

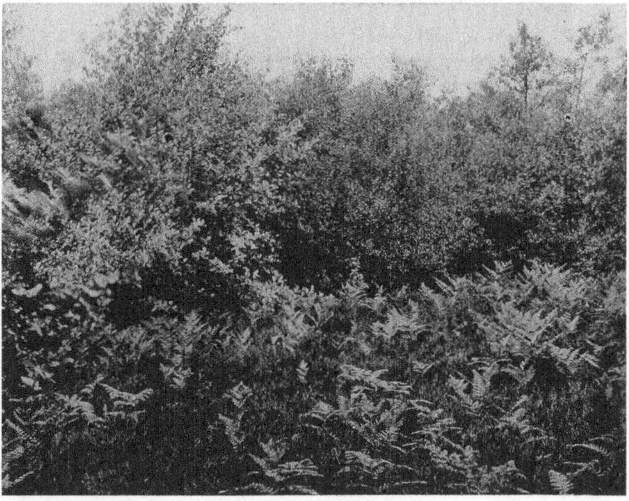

Phot. S. Mangham

b. *Betula tomentosa*, *Pyrus Aria*, *Fagus sylvatica*, *Calluna vulgaris*, *Pteris aquilina*. Toy's Hill, Kent, on Hythe Beds, Lower Greensand.

Oak-birch heath association.

Tree layer (dominance often incomplete).

Quercus Robur	ld	Pyrus Aucuparia	f
Q. sessiliflora (and hybrids)	ld	Castanea sativa (locally planted, rarely subspontaneous)	l
Betula alba (=B. verrucosa)	ld	Quercus Cerris (occasionally planted, very rarely subspontaneous)	r
B. tomentosa (and hybrids)	ld	Prunus Cerasus	o
Fagus sylvatica	ld	Pyrus Malus	o
Pinus sylvestris (local, subspontaneous)	ld	P. Aria	o
Ilex Aquifolium	a		

Shrubs.

Cratægus monogyna	a	Cytisus scoparius	o
Prunus spinosa	a	Rhamnus Frangula	o
Ulex europæus	la	Juniperus communis	l
Rubus Idæus	o	Salix caprea	f
R. cæsius	o	S. cinerea	f
Rubus *spp.*	f	S. aurita	f
		Lonicera Periclymenum	a

Ground vegetation. As we should expect in the case of an association which represents a transition from one formation to another the ground vegetation shows a medley of forms, some belonging to the dry oakwood, many to the heath association, while species of the grass heath also occur, since dry oakwood in process of degeneration, but in which the dry peaty humus characteristic of true heath has not been everywhere established, naturally tends towards the association of grass heath. No attempt will therefore be made to give a complete list of the ground vegetation of oak-birch heath, but a few of the most characteristic and abundant species may be cited.

Calluna vulgaris (where the light is sufficient)	d	Deschampsia flexuosa	a
Vaccinium Myrtillus (in shade)	d	Agrostis canina	o
Erica cinerea	ld	Teucrium Scorodonia	a
Pteris aquilina	a and ld	Potentilla erecta	a
		Polygala serpyllacea	f
		Melampyrum pratense	a

Galium saxatile	a	Scutellaria minor (damp places)	o
Scabiosa succisa	a	Epilobium angustifolium	la
Solidago Virgaurea	f	Orchis ericetorum	f
Veronica officinalis	a	Molinia cærulea	f
Viola Riviniana	a	Wahlenbergia hederacea	o
Blechnum Spicant	f		

While the oak-birch heath type of woodland frequently shows a complete mixture of the various trees (Plate IV), the birches are frequently the only trees over considerable areas (Plate V). They are often hardly dominant in the strict sense because when in open association they scarcely affect the ground vegetation. This local purity of the birch may be due to felling of the oak (many of these woods having been at one time extensively depleted of oak for charcoal and for ship-building) after which the vacant ground is rapidly colonised by birches. The birches also often successfully colonise open heathland. In both of these ways birch heath (*Betuletum ericetosum*) originates.

The beech is also often locally dominant, as in the dry oakwood from which oak-birch heath is derived. When this is the case the ground vegetation is much modified owing to the deep shade cast by the trees. *Vaccinium Myrtillus* (not flowering) and *Leucobryum glaucum* are often the only plants persisting under the shade of the beech, and the ground is frequently quite bare. Both this community (which may be called *Fagetum ericetosum*) and the birch heath are best considered as societies of the oak-birch heath association.

The oak-birch heath is a very picturesque type of woodland and covers wide areas in south-eastern England; it is specially characteristic of much of the Lower Greensand fringing the northern and western edges of the Weald, and of the Ashdown sand in the centre of the Weald (Plate V b). It is also found in Sherwood Forest (Nottinghamshire) and in Delamere Forest (Cheshire),

Phot. S. Mangham

a. Looking west from Press Ridge Warren, Ashdown Forest, Sussex, on Ashdown sand. *Betula tomentosa*, *B. alba*, *Pinus sylvestris*, *Pteris aquilina*.

Phot. S. Mangham

b. Birch wood, Press Ridge Warren. *Betula tomentosa*, *B. alba*, *Pteris aquilina*.

Oak-birch heath association.

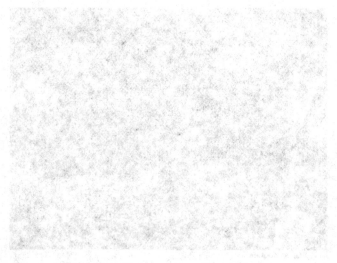

both on the Bunter sandstone. We do not know how long this association can maintain itself unchanged, but there seems good reason to believe that it is in course of degeneration to the heath association.

Heath association (*Callunetum arenosum*). The heath association is a stable association, either resulting, as has been already explained, from the further degeneration of oak-birch heath or developed *de novo* on the poorer sands. The gravelly or relatively coarse sandy soil on which it is nearly always based is characterised by its dark humous surface layers, and generally covered by a thin layer of relatively pure dry peat giving an acid reaction. In some heath soils a hard continuous layer (moorpan, Germ. *Ortstein*, Fr. *alios*) occurs at a distance varying from some inches to a few feet below the surface and often some inches in thickness. This consists of sand or gravel bound into a compact stratum either by humous compounds (*Ortstein* proper of Graebner) or by iron oxide (*Raseneisenstein*). It is formed at the junction of the surface layers of leached sand (often paler in colour) and the unaltered sand below. The shallowing of the soil caused by the formation of moorpan is very inimical to the growth of trees, and heaths and moors in which hard pan is present cannot be afforested unless the pan is broken up or at least holes are made in it for the roots of the young trees to penetrate. Pan is sometimes formed in woodland soils which are undergoing degeneration and may play an important part in the process, but it does not occur in all heaths and there seems no doubt that the process of degeneration described above can take place in its absence. In some of the heaths of southern England, *e.g.* on the Hythe beds of the Lower Greensand, the part of pan is played by a layer of hard chert a few feet below the surface (Plate VII b). This limits the growth of the trees of the oak-birch heath (particularly of the oaks, which are very

Heath soils.

stunted), and this association easily passes over into pure heath[1].

Very occasionally in the south of England the heath association is developed on fine sand or even on clay, but the reasons for this have not been fully investigated[2]. In the north and west of Britain heath is developed over a great variety of siliceous rocks, all of which however give a light sandy or gravelly soil. According to Hardy the association is not developed at an altitude greater than 2000 feet (*c.* 600 m.)[3].

The association, in its pure form, is typically treeless, the dwarf shrub Ericaceæ, generally with linear "ericoid" (*Erica* spp.) or leathery leaves (*Vaccinium* spp., *Arctostaphylos*), being the dominant plant-forms. *Vaccinium Myrtillus* with deciduous leaves and assimilating stems is an exception. The ling (*Calluna vulgaris*) is by far the most widespread and abundant species, frequently covering wide areas as the sole dominant (Plate VI b). With it are often associated other Ericaceous dwarf shrubs, of which *Erica cinerea* (in the drier areas) and *Vaccinium Myrtillus* may be dominant over considerable tracts.

Under the dense shade of *Calluna* very few species can exist. Dwarfed plants of *Vaccinium Myrtillus* and *Viola Riviniana*, which do not flower, mosses such as species of *Polytrichum* and *Hypnum*, and lichens such as *Cladonia*, are commonly found forming a subordinate layer of vegetation, and where the mantle of ling closely covers the ground, very little else occurs. Where the Callunetum is not closed, however, especially in damper

[1] Beech however seems to flourish quite well above chert which is developed near the surface, and this may sometimes account for the local dominance of beech in the oak-birch heath association on these beds.

[2] In some cases it may be due to stagnation of the soil water and the consequent formation of acid humus. In others it appears to be correlated (Rayner, Jones and Tayleur, *New Phyt.* Vol. x., in the press) with excess of magnesium in the soil.

[3] Hardy, 1905, p. 107.

Phot. S. Mangham

a. Heath association invaded by *Pinus sylvestris*. *Calluna vulgaris*, *Ulex europæus*. Pool with *Potamogeton polygonifolius* and *Sphagnum subsecundum* surrounded by *Juncus acutiflorus*, *J. effusus*, etc. Near Wych Cross, Ashdown Forest, on Ashdown sand.

Phot. S. Mangham

b Heath association invaded by subspontaneous *Pinus sylvestris*, same locality.

Heath Formation.

places, several other species are found, but the Callunetum is always poor floristically as compared, for instance, with the grass-heath association.

The following is a general representative list for the southern heaths.

Dominant species.

Calluna vulgaris		E. vagans (Cornwall only)	
Erica cinerea		Vaccinium Myrtillus	
Erica Tetralix (damper peaty heaths)		Campanula rotundifolia	f
		Gentiana Pneumonanthe	r
Viola Riviniana	f	Cuscuta Epithymum (on Ulex and Calluna)	la
V. canina (=V. ericetorum)	o	Veronica officinalis	a
V. lactea	r	Pedicularis sylvatica	la
Polygala serpyllacea	f	Euphrasia officinalis (agg.)	a
Hypericum pulchrum	f	Melampyrum pratense	f
Linum catharticum	f	Teucrium Scorodonia	la
Genista anglica	l	Salix aurita	o
G. pilosa	r	Juniperus communis	la
Ulex europæus	la	Orchis ericetorum	la
U. minor (mainly S. England)	la	Schœnus nigricans	l
		Carex pilulifera	o
U. Gallii (W. and N. England, Wales, Ireland)	la	C. binervis	l
		C. flava	o
Cytisus scoparius	f	C. verna	o
Lotus corniculatus (agg.)	f	Agrostis canina	la
Rubus *spp.*	a	A. setacea (S.W. England)	a
Potentilla erecta	a	Deschampsia flexuosa	a
Galium saxatile	f	Nardus stricta	l
Senecio sylvaticus	o	Molinia cærulea	f
Cnicus pratensis (=Cirsium anglicum)	o	Pteris aquilina	la
		Blechnum Spicant	f
Serratula tinctoria	o	Lastrea montana	r
Wahlenbergia hederacea	o	Lycopodium clavatum	r

The distribution of the common gorse (*Ulex europæus*) on heaths is of some interest. It does not appear to be a typical member of the Callunetum but rather belongs to neutral pasture and grass heath. It is very often locally dominant along roadsides and by villages and cottages and from such situations it invades typical heathland, mainly along the sides of tracks and paths. Professor Weiss' observations[1] have shown that ants carry away the seeds for the sake of the bright orange oily caruncle (*elaiosome*), which they bite and tear as they push the seed along. It is probable in fact that *Ulex europæus*, and also the broom (*Cytisus scoparius*), are typical myrmecochorous plants. Ants often make use of human tracks, and it is likely that the invasion of heaths by the gorse (and less commonly by the broom) is brought about through their agency.

Wet heath sub-association. In constantly wet places, for instance, where owing to local hollows the ground-water reaches the surface and a bog is formed, a characteristic sub-association is found (Plate VI a). In such situations peat is frequently accumulated to a greater depth than on the dry heaths, and owing to its acid character the association contains many species of the moor formation (Chapter XI), though in very different proportions.

Viola palustris	o	J. supinus	la
Hypericum Elodes	o	J. acutiflorus	f
Drosera rotundifolia	la	Eleocharis multicaulis	r
D. intermedia	l	Scirpus cæspitosus	r
D. anglica	r	S. setaceus	r
Oxycoccus quadripetala (= Vaccinium Oxycoccus)	r	Schœnus nigricans	r
		Eriophorum vaginatum	r
Calluna vulgaris	f	E. angustifolium	o
Erica Tetralix	a	Rhynchospora alba	r
E. ciliaris (Dorset and Cornwall)	l	Carex pulicaris	r
		C. ovalis	f
Pinguicula vulgaris	o	C. elata (=C. stricta Good.)	l
Scutellaria minor	o	C. Oederi	r
Gentiana Pneumonanthe	l	Molinia cærulea	a
Salix repens	a	Nardus stricta	l
Myrica Gale	la	Osmunda regalis	r
Malaxis paludosa	r	Lycopodium inundatum	r
Juncus squarrosus	o	Sphagnum *spp.*	a

[1] *New Phytologist*, Vol. VII. (1908), p. 27, and Vol. VIII. (1909), p. 81.

The following are the most noteworthy regions of English heathland.

1. *East Anglian heaths.* These occupy considerable stretches of flat sandy country in north-east Norfolk and south-east Suffolk on the Pliocene "crag," and again in west Norfolk and north-west Suffolk on Quaternary deposits overlying the chalk. On the latter grass heath alternates with Callunetum and this is probably due to the varying proportions of lime in the soil. Some of the grass heaths approach chalk pasture in the character of their vegetation. It is quite possible that the East Anglian heaths represent primitive heathland which has never been colonised by trees.

2. *South-eastern heaths.* The heaths of the south-eastern counties are developed partly on the sandy beds of the Lower London Tertiaries, partly on the Lower Greensand to the north and west of the Weald, and partly on the Ashdown sand in the centre of the Weald. They are, for the most part at any rate, derived from oak-birch woodland, which still exists in considerable quantity. The heaths of Ashdown Forest are the most extensive in this region (Plates IV—VII).

3. *Heaths of the London basin.* The Bagshot sand and the overlying plateau gravels to the south-west of London bear extensive heaths. In this region *Pinus sylvestris* has very extensively colonised the heathland by sowing itself from neighbouring plantations. Little semi-natural woodland exists in the district, and it is impossible to decide, at any rate at present, whether the heaths are primitive or derived from woodland.

4. *Heaths of the Hampshire basin*[1]. The Hampshire basin is a broad syncline of Tertiary rocks, consisting of alternating sands and clays, occupying south Hampshire and extending eastwards into Sussex, westwards into

[1] Mr W. M. Rankin has contributed the substance of these paragraphs.

Dorsetshire and southwards into the Isle of Wight. The plateau formed by these beds was covered in late Tertiary times by extensive sheets of river and estuarine gravels, which have since been dissected by river systems forming shallow valleys. The plateau gravels and the Tertiary sands now bear wide stretches of heath, mainly to the west of Southampton water. The New Forest alone has 30,000 acres (*c.* 12,140 hectares) of heathland, and further to the west, in Dorsetshire, are even greater areas, of which the "Egdon heath" of Thomas Hardy's novels forms part. These Dorsetshire heaths are broken only by occasional plantations of conifers, but in the New Forest region heaths alternate with considerable areas of dry oakwood. On the edges or even throughout some of these woods degeneration through the stage of oak-birch heath to the heath association can be observed, while some of the heaths still bear isolated trees of holly, yew, more rarely of whitebeam (*Pyrus Aria*) and bushes of hawthorn (*Cratægus monogyna*). But much of this heath area is certainly very old. Domesday Book and Leland mention the great "bruaria" of Dorsetshire and the Southampton district.

The general level of the heaths of this region hardly reaches 100 feet (*c.* 30 m.) of altitude, though in places a height of 400 feet and more (*c.* 120 m.) is attained.

Soils. All the heath-bearing soils (Bagshot, Bracklesham and Barton sands, valley and plateau gravels, dunes and shingle banks of coastal origin) of the Hampshire basin are poor in soluble minerals. The typical heath soil is covered by a thin layer of dry peat generally but a fraction of an inch in thickness, though in constantly damp places it may attain sufficient thickness to allow of its being flaked off in "turves" for fuel. Below the layer of dry peat are a few inches of dark peaty soil through which the root systems of the heath plants ramify horizontally. The Bagshot and Bracklesham sands contain appreciable quantities of iron silicate. The water from springs in these beds is frequently red-brown in colour. In places the limit between the six inches of surface soil and the subsoil is marked by a layer of ferruginous "pan."

Grass heath.

The grass-heath association occupies but a small fraction of the area, and is often in close relation with scrub and woodland of the dry oak type, from which it has probably degenerated; in other places grass heath is being invaded by Callunetum. The area of grass heath has been much reduced of recent years by the extensive strawberry cultivation in the neighbourhood of Southampton. Grass heath also occurs as a late stage of the sand dune succession. The composition of the association presents no remarkable features.

Heath association.

The heath association proper is characterised by an overwhelming dominance of *Calluna*. The ling gives fairly close but rather stunted growth, not attaining the luxuriant knee-deep growth of the Pennine and Scottish moors. On the driest spots *Erica cinerea* is freely mixed with the dominant; in moister places *Erica Tetralix* takes its place. In the west of the region, from the borders of Dorset to near Bournemouth, the pretty ciliate heath (*Erica ciliaris*) occurs, mainly in association with *E. Tetralix* in the damper places. This species is a feature of Wareham heath, and occurs again in Cornwall. The general floristic composition is not remarkable. A feature of the Hampshire heaths is the abundance of *Ulex minor*, which creeps on the surface of the soil. *Ulex europæus* is rarely common beyond the immediate vicinity of cottages. *Ulex Gallii*, which is practically absent from the Wealden area to the east, shows a steady increase from east to west of the region, corresponding with its western distribution in the country generally.

The bracken fern (*Pteris aquilina*) is rarely common away from the stream-sides, where it is generally associated with trees, and very likely marks the relics of degenerated woodland.

Pinus sylvestris apparently owes its prevalence on some of the New Forest heaths to its introduction into Ocknell Clump in 1776, though there is some evidence of its much earlier occurrence in the district. It flourishes exceedingly on the poor soils, and sows itself freely on the unenclosed heaths. The natural afforestation of the heathland by the pine is however often checked, just as it is on the Bagshot sand and plateau gravels of the London basin, by the frequent forest fires set alight by careless or mischievous passers (see p. 117).

5. *Western heaths.* On the Devonian sandstones and grits of Exmoor and on the granite of Dartmoor extensive heaths are developed, but in both regions local patches of the true moor formation occur on deep peat.

Of the Cornish heaths those of the Lizard peninsula are interesting because of the occurrence of *Erica vagans*, the so-called "Cornish heath," a plant of south-west Europe which occurs nowhere else in the British Isles. It is dominant over large areas of the Lizard heaths in association with the common species—*Calluna vulgaris*, *Erica cinerea* and occasionally *E. Tetralix*. The vegetation of these heaths is not otherwise remarkable. *Ulex Gallii* is abundant and forms a beautiful feature, as on most of the western heaths. The abundance of *Schœnus nigricans*, which is very local further east, is also striking. Most of the species of south-west European plants for which the Lizard district is famous occur on the grassland of the sea cliffs or in their clefts, and not on the heaths.

Erica ciliaris is also found in several localities in Cornwall, though not in the Lizard peninsula. As in Dorsetshire it generally occurs as a member of the wet-heath association.

6. *Midland heaths.* The heaths occurring on light soils in the Midlands have not been investigated. The natural vegetation of the Lower Greensand of the eastern Midlands appears to be essentially similar to that of the Weald. Dry oakwood, degenerating into heath, and extensive sub-spontaneous pinewoods are present. Some of this country is however properly forested—notably the Duke of Bedford's estate at Woburn.

The vegetation of the Bunter sandstone of Sherwood Forest in Nottinghamshire and Delamere Forest in Cheshire also consists of oak-birch heath, grass heath and Callunetum.

7. *The heaths of north-east Yorkshire.* These have been studied by W. G. Smith, whose results are still unpublished. The following account is based on material supplied by him.

The heaths of this area are situated on the plateau of the Oolite *massif* and have an average elevation of about 1250 feet (*c.* 380 m.) descending to 800 feet (*c.* 244 m.) on the North Sea coast. They are developed on a few inches of humous sand, covered by 2 or 3 inches of sandy peat, and passing down into sand, derived mainly from the Estuarine sandstones, but partly from the "Calcareous grit" and other sandstones of the Oolite series. The greater part of this upland heath is free from glacial deposits.

These heaths, like all the upland areas dominated by *Calluna* in the north of England and Scotland, are called "moors" and are used for the preservation and shooting of grouse (*Lagopus scoticus*), but over the warmer sandy soils which cover most of the area there is rarely any great extent of deep peat, so that the substratum is that of the heath formation.

The mean annual rainfall of these uplands is about 35 inches (*c.* 90 mm.). The climate is fairly dry during the growing season but decidedly damper than that of the East Anglian heaths. The country is subject to sea-mists during spring and early summer.

The vegetation consists almost exclusively of the typical heath association dominated by *Calluna vulgaris*. It is very poor floristically, but contains a few species which are rare or absent on most of the southern heaths, such as *Vaccinium Vitis-Idæa, Juncus squarrosus, Cornus suecica, Nardus stricta, Empetrum nigrum,* and *Listera cordata.*

Five "facies" are recognised.

(i) *Dry series.*

(*a*) Typical Callunetum on shallow sandy humus, with the dwarf form of *Vaccinium Myrtillus* growing beneath the shade of the dominant. *Listera cordata* is locally frequent, and *Agrostis* spp., *Aira* spp., and *Rumex Acetosella* are present.

(*b*) *Calluna-Vaccinium* heath (*Vaccinium Myrtillus* co-dominant), on rocky slopes of the upper valleys, forming a transition to Birch-heath further down the same slopes, and containing *Vaccinium Vitis-Idæa, Empetrum nigrum* and *Blechnum Spicant.*

(*c*) *Calluna-Pteris* heath on shallow to deep sandy soils developed on the edges of the plateau and forming a transition to the dry oakwoods of the valley sides. Scrub of *Cratægus monogyna, Betula tomentosa* and *Pyrus Aucuparia* is present, with the characteristic plants of the dry oakwood association.

(ii) *Wet series.*

(*d*) *Calluna-Nardus* heath on moister humous soils more or less badly drained. *Juncus squarrosus, Molinia cærulea,* forma *depauperata,* and *Erica Tetralix* occur in the damper places. This facies forms a transition to the grass-heath association which occurs on adjacent tracts and frequently coincides with débris of glaciation at the lower levels.

(*e*) *Calluna-Tetralix* heath, a form of the wet heath

sub-association (p. 106) on peaty humus with abundant water. *Erica Tetralix* is co-dominant or locally dominant, and species of *Sphagnum* are always present. *Nardus stricta* and *Juncus squarrosus* are more or less frequent, while *Empetrum nigrum*, *Eriophorum vaginatum*, *Molinia cærulea*, forma *depauperata*, and *Cornus suecica* (very rare) also occur.

This sub-association forms a transition to true *Calluna* moor, which occurs on the plateau of the watershed in the west of the district, and in certain valleys. Peat is regularly exploited on these moors.

SCOTTISH HEATHS, by W. G. SMITH

The heaths of Scotland occupy large areas in the eastern and northern Highlands, rising to an altitude of 2000 feet (*c.* 610 m.) in regions where the mean annual rainfall is from 30—60 inches (*c.* 76 to 152 cm.). It is this heathland which gives the river-basins of the Tay, Spey, and Dee a physiognomy quite distinct from the moorlands of the Western Highlands. The dominance of *Calluna* is best seen during the flowering season (August and September), when the purple heather-clad slopes stand out in striking contrast to any adjoining grassland or moorland. This continuity of *Calluna* is much less common in the Southern Uplands[1] and the Western Highlands. The *Calluna* heaths are always known in common language as "moors," and are the typical "grouse-moors" of the Highlands; as such they are systematically burned in rotation every ten to fifteen years when the *Calluna* has

[1] Thus we have on the north-western edge of the Southern Uplands along the main watershed of Lanarkshire and Ayrshire, an area of 430 sq. miles (now under survey by W. G. Smith), about half of which is uncultivated land, but not more than 10 sq. miles are *Calluna* heath of the above type, and this lies on summits or slopes with little boulder clay; the bulk is grass moor, wet grass pasture, or peat bog, mainly on badly drained boulder clay.

grown high, woody and thinly leaved. They approach in character the true *Calluna* moors (p. 275) of the Pennines, in that they occur on a greater depth of surface peat than the English heaths and contain more species characteristic of the moor formation, especially where drainage becomes impeded in hollows or on the flatter terraces. Nevertheless they are always developed, like the southern heaths, over sand or gravel, and are not so damp (except locally) as typical *Calluna* moor. We may consider that there are two edaphic factors which control the heath formation, the sandy or gravelly soil and the accumulation of acid peaty humus, whereas the moor formation is controlled by the acid peat alone.

The thickness of the superficial peaty layer is commonly 4 to 8 inches (*c.* 10 to 20 cm.), though in extreme cases, according to Hardy[1], it may reach a depth of 12 inches (30 cm.). Below, the peat passes into sand or gravel, and at a depth of 12 to 24 inches (30 to 60 cm.) from the surface a layer of "moor-pan" frequently occurs.

The topography of the *Calluna* heath varies to a considerable extent, as it has a vertical range from the valleys upward of 1500 feet (*c.* 450 m.), and its terrain includes steep slopes, stream valleys, and undulating summits. It is therefore natural to expect variation of facies, and while *Calluna* is the dominant species, the other associates vary considerably. Several of these owe their outstanding position in the association to their quick recovery after the moor-burning, so that they precede *Calluna* during the first year or two of regeneration. So far as records are available, the following are species which thus temporarily occupy the ground. *Cladonia rangiferina*, and other species, often form almost the only covering; *Erica cinerea* recovers quickly on dry soils, *Vaccinium Myrtillus* and *V. Vitis-Idæa* on steep slopes, *Arctostaphylos Uva-ursi* at higher altitudes (towards

[1] Hardy, 1905, p. 106.

2000 feet); heath-grasses (*Festuca ovina, Agrostis* spp., *Deschampsia flexuosa, Anthoxanthum odoratum*) come quickly on sandy humus, *Nardus stricta* on moister humous soils; while *Juncus squarrosus, Scirpus cæspitosus* and *Eriophorum vaginatum* follow the wetter peaty soils, generally accompanied by *Erica Tetralix.*

Several associates of the Highland *Calluna* heath are characteristic, although their occurrence is generally rather sparse; they include *Genista anglica, Lycopodium clavatum, Trientalis europæa, Antennaria dioica, Melampyrum pratense* var. *montanum, Listera cordata,* and *Pyrola media. Juniperus communis* occurs abundantly in some districts, generally along stream valleys, but in other localities it is quite rare. Seedlings of birch and mountain ash (*Pyrus Aucuparia*) have been observed all over the heaths in many districts (R. Smith). The bracken (*Pteris aquilina*) occurs to some extent on steeper slopes, but it rarely assumes the dominance it has on many English heaths. *Ulex europæus* and *Cytisus scoparius* are sometimes plentiful near roads, but they belong more to the grassland type. Various other plants may occur on the Scottish as on the English heaths, but they are neither so characteristic nor so generally distributed as the species mentioned, and are generally to be regarded as relicts of or invaders from some other association.

The following is a generalised list of the more characteristic species from four stations near Blair Atholl in the Tay valley, Perthshire (R. Smith).

Dominant.
Calluna vulgaris

Locally sub-dominant.

Vaccinium Myrtillus	Arctostaphylos Uva-ursi
V. Vitis-Idæa	Erica Tetralix
Empetrum nigrum	

Locally abundant.

Erica cinerea	Hypnum *spp.*
Nardus stricta	

Frequent.

Polygala vulgaris (agg.)	Deschampsia flexuosa
Potentilla erecta	Agrostis vulgaris
Galium saxatile	Anthoxanthum odoratum
Antennaria dioica	Festuca ovina
Luzula multiflora	Blechnum Spicant
Carex Goodenowii	Lycopodium clavatum
C. dioica	Cladonia *spp.*

Sparse or local.

Genista anglica	Trientalis europæa
Melampyrum pratense *var.* montanum	Juniperus communis

A typical list given by Hardy[1] from the Dee valley (Aberdeenshire) is very similar.

Pinewood association (*Pinetum sylvestris*).

It is generally held that within the limits of these islands *Pinus sylvestris* is native only in Scotland. Certainly it was formerly native in many parts of England and Ireland, where its remains are found buried in peat. Clement Reid writes: "*Pinus sylvestris* seems to have been abundant throughout Britain during part of the Neolithic Period, for its cones are abundant at the base of peat-mosses and in 'submerged forests.' It afterwards disappeared from the South of England and only recently has been re-introduced[2]." It is just possible that it may have survived in small numbers in the South of England till the extensive clearing of the lighter soils gave it a fresh opportunity of asserting itself. But it was certainly re-introduced into the south in the eighteenth century, and perhaps earlier, and it quickly spread very rapidly over some of the sandy soils, where it now forms extensive sub-spontaneous woods.

In Scotland *Pinus sylvestris* exists as an apparently endemic variety (var. *scotica* E. and H.) with short grey needles. This tree at one time covered large areas in the

[1] l.c. p. 109.

[2] *The Origin of the British Flora*, 1899, p. 100.

Phot. A. G. Tansley

a. Subspontaneous pinewood and chert diggings. *Pinus sylvestris, Calluna vulgaris, Erica cinerea, Pteris aquilina.* Crockham Hill Common, Kent, on Hythe Beds of Lower Greensand.

Phot. A. G. Tansley

b. Side of chert diggings. *Pteris aquilina, Vaccinium Myrtillus, Calluna vulgaris.* Same locality.

Heath Formation.

Highland straths (broad valleys) and glens, extending for considerable distances up the mountain slopes. The former area is now very greatly restricted, but considerable tracts of forest, largely open owing to heavy exploitation of the timber, still remain in some places.

It is of interest to note that both native and subspontaneous pine is in this country apparently always associated with heathland. The sub-spontaneous pinewoods of southern England occur mainly on the heaths of the Bagshot Sand and of the Lower Greensand, though the pine sows itself fairly freely on some of the East Anglian heaths and also on those of north-east Yorkshire. In the Highlands of Scotland the native pine woods seem to be developed on the same habitats as the heath association, to which they give place when heavily cleared. The pinewood association is therefore naturally included in the heath-formation.

English sub-spontaneous pinewoods.

The sub-spontaneous pinewoods of the southern sands may be very briefly dealt with. In the neighbourhood of mature plantations or of mature self-sown woods thousands of pine seedlings in all stages of growth may frequently be found colonising adjoining heathland, often in company with the birch. The pine seedlings sometimes push up from among fairly thick heather (Plate VI a and b). In this way many of these heaths are being rapidly converted into pinewoods, the pine easily beating the birch in the later stages of competition by reason of the much deeper shade which it casts. The process is only checked by the frequent heath and forest fires which occur in dry summers and autumns.

The close pinewood association is very poor floristically, partly on account of the deep shade and partly because of the thick layer of pine needles which carpets the floor of the wood. The shade forms of *Pteris aquilina* and *Vaccinium Myrtillus* often occur, and the silvery tussocks of the moss *Leucobryum glaucum* are also frequent. Occasionally young beeches form a sparse shrubby undergrowth. In the more open spots various members of the heath association occur, and if the wood is felled, this association quickly colonises the ground.

Scottish pinewoods.

Good examples of native Scottish pinewoods are to be found in the Black Wood of Rannoch (on the southern shores of Loch Rannoch in Perthshire (Plate VIII a)), and in Rothiemurchus Forest in Strathspey (Inverness-shire)[1]. The latter occupies a great basin enclosed by the Cairngorm mountains and formed part of a much larger forest which stretched on both sides of the Spey valley for a considerable distance. Much of this has been replanted, but in Rothiemurchus itself no planting appears to have been done.

Rothiemurchus Forest is situated on gravelly sands of glacial origin forming a gently undulating tract several square miles in extent, having an average elevation of about 1000 feet (*c.* 300 m.), and extending up the lower slopes of the neighbouring mountains. In some places the pines stop short with the lines of moraine on the hill slopes, but in others they extend far up the hill sides, reaching an altitude of 2000 feet (600 m.) or more, probably the highest existing limit of woodland in the British Isles.

The pinewood association of Rothiemurchus Forest is for the most part very open in character, seldom forming a close wood; it is in fact practically a "pine heath," with a very uniform vegetation, poor in species (Plate VIII b).

The peaty character of the surface soil extends to a depth of some inches. Under the pines, below the surface covering of needles, there are about 4 inches

[1] Ballochbuie Forest at the foot of Lochnagar in Aberdeenshire, and Locheil Old Forest on Loch Arkaig in Inverness-shire are also recorded as native pinewoods.

Phot. A. G. Tansley

a. Pine-heath. *Pinus sylvestris*, *Calluna vulgaris*, *Molinia cærulea*. Black Wood of Rannoch, Perthshire.

Phot. A. G. Tansley

b. Pine-heath. *Pinus sylvestris*, *Calluna vulgaris*, *Juniperus communis*. Rothiemurchus Forest, Inverness, on glacial sand and gravel.

Heath Formation.

(10 cm.) of brown humus with interlacing pine-roots and stems and roots of *Vaccinium*, and this is succeeded by about 5 inches (12 to 13 cm.) of black peaty humus with numerous fine rootlets. This layer passes down into peaty sand with numerous stones. On the open heath there is a surface layer of nearly pure peat about 3 inches (7—8 cm.) thick, apparently formed mainly by *Cladonia* and this is succeeded again by several inches of black peaty sand passing down into grey or yellow sand. In places a hard layer of yellow-brown "pan" is met with at the depth of a foot (30 cm.).

The following species occur:

Pinus sylvestris *var.* scotica	a (ld)	Vaccinium Myrtillus	a[1]
Calluna vulgaris	d	V. Vitis-Idæa	a[1]
Hypericum pulchrum	o	Erica Tetralix	o
Oxalis Acetosella	o	Erica cinerea	o
Potentilla erecta	a	Arctostaphylos Uva-ursi	f[2]
Pyrola minor	f	Betula tomentosa	o
P. media	o	Empetrum nigrum	f[2]
P. secunda	o	Deschampsia flexuosa	a
Moneses grandiflora	v.r	Sieglingia decumbens	o
Trientalis europæa	f	Juniperus communis	a
Melampyrum pratense *var.* montanum	f	Blechnum Spicant	o
		Phegopteris Dryopteris	o

[1] Locally co-dominant with *Calluna*.

[2] Towards the upper limits of the pine heath.

Along the paths, even deep in the forest, are many species which apparently do not occur on the heath itself. Among these are *Anemone nemorosa*, *Viola Riviniana*, *Juncus squarrosus*, *Agrostis tenuis*, *Anthoxanthum odoratum*.

When the pinewood is close and casts a deep shade *Calluna* is absent, and the ground vegetation is mainly made up of *Vaccinium Myrtillus*, and *Deschampsia flexuosa*.

The pinewood rejuvenates itself by the self-sowing of seed, freely but not abundantly. One factor which militates against abundant rejuvenation appears to be the thickness and height of the heather, which often reaches 3 feet (*c.* 1 m.) or more, for numerous seedlings occur along the path sides, but none were found in the thick heather.

The juniper (*Juniperus communis*) is abundant and very luxuriant. It sometimes forms tall conical shrubs reaching a height of 10 or 12 feet (*c.* 3 to 3·5 m.) (Plate VIII b).

Diagram showing the probable genetic relations of the plant-communities of the Formation of Sandy Soils and of the Heath Formation.

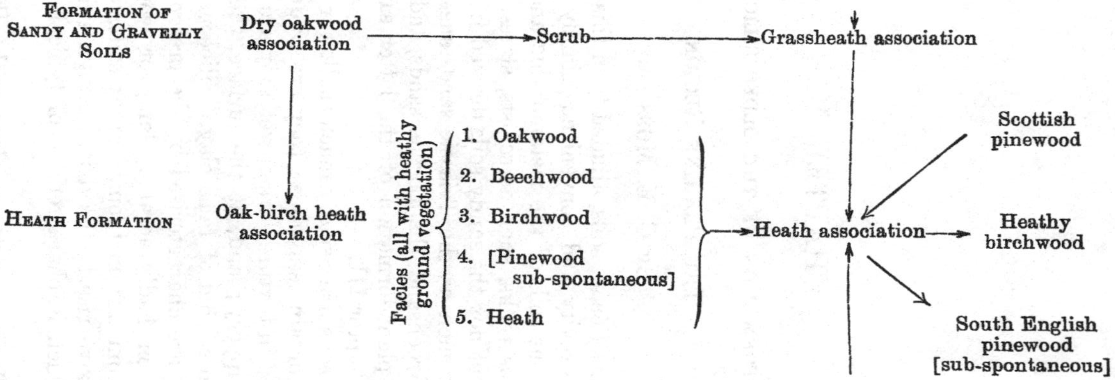

CHAPTER V

THE PLANT-FORMATION OF THE OLDER SILICEOUS SOILS

A. The Pennine Region

By C. E. Moss

The older siliceous soils, alluded to in this chapter for the sake of brevity and convenience simply as siliceous soils, include the soils of the metamorphic and Palæozoic non-calcareous rocks, such as schists, slates, sandstones, and shales, but not the sandy soils derived from the more recent rocks, such as the Triassic sandstones, the Upper and Lower Greensand, the Eocene sands, and the glacial sands. The plant-formation of the latter sandy soils is treated in Chapter III.

Siliceous soils are usually much shallower than sandy soils: they contain less silica: they are frequently more finely grained in texture: when wet they are much more greasy or slippery; and, at the surface, they often degenerate into a kind of false clay. Siliceous soils, when well aerated, are characterised by the presence of "mild humus": when badly aerated, damp or wet, acid peat tends to accumulate on them. Sandy soils, on the other hand, often give rise to the drier, though still acid, peat or humus which is characteristic of typical heaths (see Chapter IV).

These differences, perhaps along with others not yet fully understood or appreciated, between siliceous and sandy soils are related to conspicuous differences in the plant-communities which characterise the two kinds of soil respectively. On clayey soils, the characteristic woodland association is dominated by *Quercus Robur*, or a mixture of this and *Carpinus Betulus*; and such woods on degeneration give rise to neutral grassland. On sandy soils, the woodland associations are more varied in character. *Quercus Robur*, *Q. sessiliflora*, *Fagus sylvatica*, *Betula tomentosa*, and *Pinus sylvestris* may all form more or less definite plant-communities, with many intermediate stages; and these on degeneration give rise to typical heaths often intermingled with grassy patches containing many arenicolous species. On siliceous soils, the typical woodland association is dominated by *Quercus sessiliflora* or, at higher altitudes, by *Betula tomentosa*; and these on degeneration give rise to a type of grassland which we may call siliceous pasture, whose characteristic, wide-spread, and dominant species are *Nardus stricta* and *Deschampsia flexuosa*.

Association of the sessile oak (*Quercetum sessiliflorae*). Over non-calcareous rocks of the Pennine hills in the north of England, oakwoods whose dominant tree is *Quercus sessiliflora* are typically and characteristically developed.

Situation and soil.

The rocks of these hill slopes consist of alternating beds of sandstones and shales belonging to the Pendleside (Yoredale) series, Millstone Grit, and Coal-measures. In all cases, these soils have a very low lime-content. In several analyses, the percentage of lime (calculated as calcium oxide) was so low as 0·02 per cent.; and in no case was it higher than 0·05 per cent. The total soluble mineral-content of all these soils is also very low. The soils often show a marked tendency to allow of the accumulation of acidic[1] humus; and the amount of such

[1] This word is used by Dr Moss for soil of any kind whose waters give an acid reaction [Editor].

humus present in the soil may be roughly gauged by the abundance of such plants as the heather (*Calluna vulgaris*), the bilberry (*Vaccinium Myrtillus*), and the silver hair-grass (*Deschampsia flexuosa*) in the ground flora. The soils are shallow, sometimes very shallow; and on the whole, therefore, the trees are of small dimensions.

Altitudinal range.

The Quercetum sessiliflorae ascends, in general, to about 1000 feet (305 m.); and, above this level, the influence of altitude is seen in the absence of well-developed oakwoods. Derivative scrub occurs, however, as well as small patches of woods of *Betula tomentosa*, above this height. Birch-woods are more extensively and typically developed in the Lake District, and especially in Scotland, than on the southern Pennines.

General character.

The woods of *Quercus sessiliflora* have been described by several writers[1], and variously subdivided into upland and lowland oakwoods, dry oak-woods, and mixed deciduous woods, etc. In the present account, the woods described by these writers are regarded as particular cases of the association of *Quercus sessiliflora*.

In all these woods, this species (*Q. sessiliflora*) is indisputably the dominant tree. On the lower, non-calcareous Pennine slopes, this species grows well and forms fairly large trees on the damper soils up to about 800 feet (244 m.), especially on the shales: above this altitude, especially on soils over the sandstones, the trees are usually of short stature and small girth; and near their upper altitudinal limit, they are little taller than shrubs.

Planted trees.

The pedunculate oak (*Q. Robur*) is totally absent from the great majority of the oakwoods of the Pennine slopes, although it occurs here and

[1] Smith and Moss, 1903; Smith and Rankin, 1903; Crump, *Flora of Halifax*, 1904; Woodhead, 1906; and Moss, 1911.

there in situations where it has obviously been planted; and it sometimes, no doubt, spreads from the plains for some distance up the lower valleys. As a planted tree, mixed with such undisputed aliens as the sycamore (*Acer Pseudo-platanus*) and the larch (*Larix decidua*), it has been observed so high as 1100 feet (*c.* 336 m.).

No conifers are indigenous in these woods; but the larch and the Scots pine (*Pinus sylvestris*) are frequently planted. Other conifers which are occasionally or rarely planted are the black or Austrian pine (*P. austriaca*), the Douglas fir (*Pseudotsuga Douglasii*), and the Redwood (*Sequoia sempervirens*). As sub-fossil timber, the Scots pine is occasionally found buried under the peat of the southern Pennines; and it is rather remarkable therefore that the tree is not indigenous in the Pennine woods at the present time. Even where planted, the Scots pine does not here seem to be very prosperous, as seedling pines are never abundant. It is clear that this species does not thrive on the damp soils of the Pennines nearly so well as on the drier heaths of the south of England.

The beech (*Fagus sylvatica*), although an almost invariable constituent of the larger woods, has little claim to rank as indigenous. In favourable seasons, ripe fruits are formed even at an altitude of 1500 feet (457 m.); and first-year seedlings are frequently seen. Older seedlings, however, have not been observed; and there is no evidence to show that the beech rejuvenates itself in these upland woods.

Associated native trees.

The common birch (*Betula tomentosa*) is, on the whole, the most common and the most constant associate of the sessile oak in these woods. Forms or varieties with glabrous twigs are not uncommon; and the variety *parvifolia* (? = *B. carpatica*) has been observed. Not infrequently the birch becomes locally dominant in places where extensive felling of the oak and no subsequent replanting has taken place. The white birch (*B. alba* = *B. verrucosa*) is probably not indigenous in these woods; and the pendulous form (*B. alba* forma *pendula* = *B. pendula* Roth.) is extremely rare even as a planted tree.

The wych elm (*Ulmus glabra* Huds. *non* Mill.; *U. montana*) is indigenous, and occurs up to about 1000 feet (305 m.). Above this altitude, it occurs in plantations up to 1500 feet (457 m.). It is a constant and sometimes an abundant constituent of the damper woods, but is rare in the drier ones. In favourable localities, seedlings are very common.

The rowan or mountain ash (*Pyrus Aucuparia*) occurs in most of the woods; and, in rocky, upland, and heathery situations, it is often very abundant.

The holly (*Ilex Aquifolium*) also is found in almost every oakwood on the Pennines; but, although it sometimes produces flowers, it rarely fruits.

The ash (*Fraxinus excelsior*) and the alder (*Alnus rotundifolia*) are almost limited to stream sides and marshy places, where, however, they are often very abundant.

Of rarer trees, the aspen (*Populus tremula*) and the small-leaved lime (*Tilia cordata*; *T. ulmifolia*; *T. parvifolia*) are very rarely met with; and possibly they are not indigenous on the siliceous soils of the Pennines. The cherry (*Prunus Cerasus* and *P. Avium*) also occurs in the woods, and is locally not uncommon.

Shrubs.

A number of shrubby species are fairly abundant and characteristic on the damper soils, but there is no species which is generally dominant. Brambles are exceedingly abundant and very many species are recorded by batologists. *Rubus Selmeri* and *R. dasyphyllus* ascend higher than any of the other forms. *R. cæsius* is rather rare and confined to the lower levels. Of the roses, *R. canina* is fairly abundant and *R. arvensis* locally so at the lower altitudes; among the less common

Phot. W. B. Crump

a. *Quercus sessiliflora*, *Holcus mollis*, *Pteris aquilina*.
Heath Moor of Pennine (August).

Phot. W. B. Crump

b. *Quercus sessiliflora*, *Vaccinium Myrtillus*, *Deschampsia flexuosa*.
Hardcastle Crags, Hebden Bridge, Yorkshire, on Millstone Grit.

Sessile oakwood association.

species *Rosa mollis* and *R. tomentosa* occur, but are rather rare.

In the drier parts of the woods the sloe or blackthorn (*Prunus spinosa*) and the hawthorn (*Cratægus monogyna* = *C. Oxyacantha*) are characteristic, and on the outskirts and in open places the broom (*Cytisus scoparius*) and gorse (*Ulex europæus* at lower levels and *U. Gallii* at higher) are locally abundant.

Ground vegetation.

The ground vegetation, as in most types of wood, is very variable from place to place. Three ground sub-associations may, however, be recognised, though transitions between them are numerous.

1. The sub-association of damp situations with well-aerated soil and mild humus. This possesses a fairly rich flora containing many species with more or less conspicuous flowers.

2. The sub-association of drier situations with less humus, characterised by a poorer and more uniform flora (Plate IX a).

3. The sub-association of acid peaty humus characterised by a heathy flora (Plate IX b).

It will be seen that the habitats of these three sub-associations correspond fairly closely with those of the woodland associations of the formation of clays and loams, the formation of sandy soils and the heath formation, *i.e.* to the damp oakwood, the dry oakwood and the oak-birch heath respectively, and the lists of species (pp. 81, 93, 101, 128) emphasise this correspondence. Nevertheless it is held that the common characters of the habitat, which presumably determine the general dominance of *Quercus sessiliflora* throughout the three types of sub-association, are sufficiently marked to consider the whole of this vegetation and the related type of pasture or grassland as belonging to a single formation.

The following is a list of the more characteristic

species occurring in the Quercetum sessiliflorae of the Pennines:—

Trees.

Prunus Cerasus	l	Ulmus glabra (U. montana)	f
Pyrus Aucuparia	f to la	Betula tomentosa	la
Ilex Aquifolium	o	Quercus sessiliflora	d
Fraxinus excelsior	la[2]	Salix fragilis	l[1,2]
Alnus glutinosa	la[2]		

Shrubs.

Ulex Gallii	l	Cratægus monogyna (C. Oxyacantha)	a
U. europæus	f	Acer campestre	f
Prunus spinosa	o	Sambucus nigra	f
P. Padus	la	Viburnum Opulus	f
Rubus spp.	a	Corylus Avellana	f
R. cæsius	r[1]	Salix caprea	f
Rosa canina	o	S. cinerea	f
R. arvensis	la[1]	S. aurita	l[3]
Pyrus Malus	o		

Climbers.

Hedera Helix	a	Lonicera Periclymenum	a

Ground vegetation.

(1) *Sub-association of damp situations.*

Anemone nemorosa	f	V. sylvatica	r
Ranunculus auricomus	l	Geum rivale	la[3]
R. Ficaria	la	G. urbanum	l
Trollius europæus	r[3]	Fragaria vesca	f
Aquilegia vulgaris	r	Spiræa Ulmaria	la[3]
Caltha palustris	f[3]	Chrysosplenium oppositifolium	la[3]
Cardamine amara	l[3]	C. alternifolium	l[3]
C. pratensis	o[3]	Epilobium montanum	f
C. sylvatica (C. flexuosa)	f	E. palustre	l[3]
Viola Riviniana	f	Circæa lutetiana	la
Lychnis dioica (Melandrium rubrum)	f	Sanicula europæa	la
Stellaria Holostea	a	Heracleum Sphondylium	f
S. nemorum	la[3]	Angelica sylvestris	l[3]
Oxalis Acetosella	a	Asperula odorata	l
Vicia sepium	a	Valeriana dioica	r[3]

[1] Chiefly at lower altitudes. [2] Chiefly near streams.
[3] Chiefly in marshy places.

V. sambucifolia	la[3]
Petasites ovatus	la[3]
Cnicus palustris	la[3]
C. heterophyllus	r[3]
Crepis paludosa	l[3]
Lactuca muralis	l
Campanula latifolia	r
Primula vulgaris	r
Lysimachia nemorum	la[3]
Myosotis sylvatica	l
M. scorpioides (palustris)	la[3]
Scrophularia nodosa	a[3]
Veronica montana	la
Lathræa Squamaria	r
Lamium Galeobdolon	a
Ajuga reptans	a
Mercurialis perennis	a
Neottia Nidus-avis	r
Listera ovata	r
Helleborine latifolia	r
Orchis mascula	r
Habenaria virescens (=H. chloroleuca)	r
Iris Pseudacorus	la[1]
Scilla non-scripta	a
Allium ursinum	la
Juncus effusus	la[1]
Luzula pilosa	a
L. maxima	a
Arum maculatum	la
Carex pendula	o[1]
C. helodes (C. lævigata)	o[1]
Carex remota	l
C. sylvatica	v[2]
C. strigosa	v[2]
Milium effusum	la
Deschampsia cæspitosa	o[1]
Holcus lanatus	o[1]
Melica uniflora	o
Festuca gigantea	o
F. sylvatica	r
Bromus ramosus (=B. asper)	o
B. sterilis	la
Athyrium Filix-fœmina	a
Lastrea Filix-mas	a
L. aristata (= L. dilatata)	a
L. montana	la[1]
Polystichum aculeatum	r
Phegopteris Dryopteris	r[1]
P. polypodioides	r[1]
Osmunda regalis	extinct
Equisetum sylvaticum	la[1]
Bryophyta: numerous and abundant	

(2) *Sub-association of drier situations*[4].

Viola Riviniana	f
Polygala serpyllacea	f
Arenaria trinervia	f
Potentilla erecta	f
P. procumbens	l
Conopodium majus	o
Galium saxatile	a
Hieracium boreale	o
Campanula rotundifolia	o
Digitalis purpurea	o
Lamium Galeobdolon	o
Teucrium Scorodonia	f
Rumex Acetosella	o
Scilla non-scripta	ls
Luzula pilosa	o
Carex binervis	o
Anthoxanthum odoratum	o
Holcus mollis	s
Pteris aquilina	ls
Lastrea aristata	o

[1] Chiefly at lower altitudes. [2] Chiefly near streams.
[3] Chiefly in marshy places.
[4] The "mesopteridetum" of Woodhead, 1906.

(3) *Sub-association of more open situations with much acid humus.*

Corydalis claviculata	o	Pyrola media	r
Lathyrus montanus	o	P. minor	r
Ulex Gallii	la	Teucrium Scorodonia	f
Potentilla erecta	f	Luzula multiflora	f
Galium saxatile	a	Carex pilulifera	o
Solidago Virgaurea	o to f	Deschampsia flexuosa	s
Calluna vulgaris	la	Molinia cærulea	l
Vaccinium Myrtillus	ls	Pteris aquilina	ls
V. Vitis-Idæa	l	Blechnum Spicant	o to f
Melampyrum montanum	la	Lastrea aristata	o

Subordinate (Retrogressive) Associations

Scrub associations. Every stage can be traced from typical woodland to open scrub; and there can be little or no doubt that, so far as the Pennines are concerned, the great majority of examples of open scrub are simply retrogressive woodland. In ancient times, when the woods ascended to their climatic limits, there was doubtless a natural region of climatic scrub at a higher level than the woods; but the woodland limit has been depressed some hundreds of feet since that particular period.

There can be no doubt that a great deal of the degeneration of the woodland has been brought about by the indiscriminate felling of trees, the absence of any definite system of re-planting, and the grazing of quadrupeds. It is doubtful, however, if these causes are quite sufficient to account for all the facts. It has to be remembered that the population of the remoter valleys, many of which are now almost or wholly treeless, is very small. A matter which is perhaps not sufficiently emphasised is that, in a closed plant association, seedlings, especially of plants with large seeds like the oak, are rarely found. Now, as time goes on, the ground vegetation of a wood tends to become quite closed; and this simple fact, in itself, is probably one of great importance in the question of the rejuvenation or degeneration of forests. Some foresters make use of their knowledge of the fact, and go to great trouble in keeping the ground vegetation open by removing the woodland "weeds."

On many open hillsides, trees are more or less thinly scattered. In some such cases, the ground vegetation is of a grassy type; in others, heathy undershrubs prevail. It is believed that such examples of scrub are derived from woodland, because every transition occurs between woodland and scrub and because scrub frequently occurs within the woodland limits of altitude. Further, some of these scrubby places still retain the place-name "wood"; and, in one or two cases, it is stated by the local inhabitants that certain examples of scrub were woodland in comparatively recent times.

Whilst the majority of the examples of scrub that one meets in the British Isles are apparently of this retrogressive nature, scrub of a progressive nature, that is, scrub which is developing towards woodland, is sometimes met with, particularly in deserted quarries. Progressive scrub also occurs on fens and moors (see pp. 236, 250).

Siliceous grassland associations. Grassland occurs on the uncultivated slopes of the hills, where every transition can be traced from woodland, through scrub, to a clean turf of *Nardus stricta* and *Deschampsia flexuosa*. So far as one can judge, there are no soil changes corresponding to these different associations.

At the present time, grassland and scrub ascend to higher altitudes than the woods; but it seems probable that almost the whole of the land now occupied by grassland and scrub was once wooded. On some of the highest summits, however, grassland now occurs on ground which has probably never been tree-clad—at least, not in post-Tertiary times.

A certain amount of grazing of sheep and cattle takes place on these uncultivated hill-slopes, but the amount is, on the whole, rather small. The land is not artificially manured or drained.

Two types of siliceous grassland have been described[1] as occurring on the Pennines, and have been distinguished as wet and dry. The most abundant and characteristic grass of the drier siliceous grassland is the mat-grass (*Nardus stricta*) and that of the wetter siliceous grassland is the purple moor-grass (*Molinia cærulea*). The former may be termed *Nardus* grassland and the latter *Molinia* grassland. The two species are respectively dominant in the two associations, since they nearly monopolise the ground and form the great bulk of the turf in which the other plants are rooted; and the associated species are therefore more or less controlled by the dominant plants. To a limited extent, the associations are layered plant-communities; and the smaller plants receive a certain amount of shade and shelter from the dominant ones. As is usual in plant-associations, one or other of the dependent species occasionally becomes social; and thus plant-societies arise.

(i) *Nardetum strictæ.* The typical Nardetum of the Pennines occurs on steep shaley slopes. In summer, this association is characterised by a grassy turf, grey-green in colour, dry and slippery. In late autumn, winter and early spring, the ground is damp and sodden; and the bleached haulms of the mat-grass (*Nardus stricta*) give tone to the landscape, and may be recognised at a considerable distance. The silver hair-grass (*Deschampsia flexuosa*) is, on the southern Pennines, a constant associate;

[1] Smith and Moss (1903); Smith and Rankin (1903); Lewis (1904, *a* and *b*).

and, in early summer, the tall, purple scapes of this grass are very conspicuous. There seems little doubt, however, that the Nardetum of the southern Pennines is ecologically identical with that of the Wicklow Hills[1] and with that of the northern Pennines[2], even though the silver hair-grass is not included in the lists of the association by the authors mentioned.

The two grasses (*Nardus stricta* and *Deschampsia flexuosa*) of the association remain abundant or even dominant up to about 1500 feet (457 m.). Below about 1250 feet (381 m.), the common bent grass (*Agrostis tenuis* = *A. vulgaris*) is often very abundant; and, in the late summer months, its delicate and purple panicles may colour a whole hillside. As lower altitudes are approached, this species becomes increasingly abundant at the expense of the mat-grass, and associated species become more and more abundant. The sheep's fescue-grass (*Festuca ovina*) is also often associated, and this species sometimes forms plant-societies or facies.

The shady hill-slopes which encircle the upper Edale valley afford an extensive and continuous expanse of the typical Nardetum. On the north of this upland valley are the slopes of the Peak, on the south the slopes of the Mam Tor range, and on the west the slopes of the Colborne moors. In the sheltered Grindsbrook clough, the bracken (*Pteris aquilina*) asserts itself very strongly: the small furze (*Ulex Gallii*) occurs in small patches here and there, and the springs of water on the hillsides are marked by clumps of the common rush (*Juncus effusus*).

The last three species give to the association very different aspects or facies. The bracken, where the soil is dry and the locality sheltered, sometimes occurs in extensive sheets, masking the otherwise dominant grasses.

[1] Pethybridge and Praeger, 1905.
[2] Lewis, 1904.

The small gorse (*Ulex Gallii*) is never so prominent on the southern Pennines as, for example, on the Malvern Hills, in Devonshire and Cornwall, on the Wicklow Hills (Plate X), or in the west of Ireland; but it occurs in patches in dry and fairly exposed localities. Rushes (*Juncus effusus*, and sometimes *J. conglomeratus*), in damp places, and independently of conditions of shelter or exposure, are abundant and characteristic associates. The bracken and the rush, in fact, are in many places harvested by the upland farmers.

The relationships of the various facies of the *Nardus* association may be indicated diagrammatically as follows:—

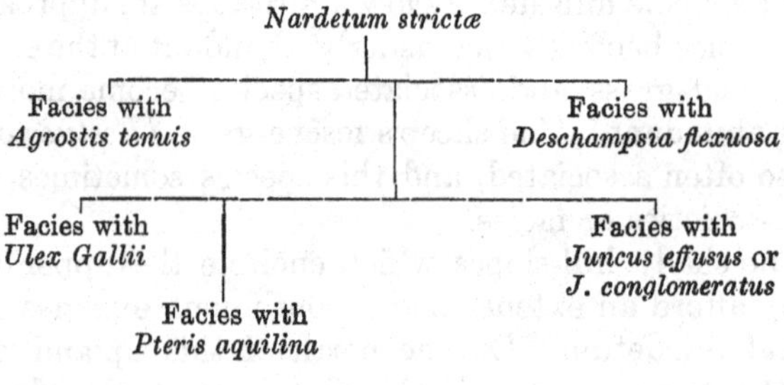

Just as there are many localities which are difficult to describe as wood or scrub, so there are many localities which are difficult to describe as scrub or grassland; and the transitions between these physiographical types are complete. Doubtless, the great bulk of the area now occupied by scrub or grassland was formerly woodland; and, in fact, several localities now occupied by scrub or grassland retain the place-name "wood." Several species of plants which are characteristic of woodland occur here and there; and, whilst some of these are no doubt recent migrants, they may, on the whole, be fairly regarded as relics of pre-existing woodland (see also Moss, 1911).

The following is a list of the more typical and abundant

PLATE X

Phot. R. Welch

Siliceous grassland. Facies of *Ulex Gallii*, Wicklow mountains.

plants which occur in the *Nardus* pasture: full lists are given by Moss (*op. cit.*).

Dominant species.	Nardus stricta
Sub-dominant species.	Deschampsia flexuosa
Locally sub-dominant species.	Pteris aquilina
Locally abundant species.	
Ulex Gallii	Agrostis vulgaris
Juncus effusus	Festuca ovina
Less abundant and rare species.	
Ranunculus acris	Erica cinerea
R. bulbosus	Calluna vulgaris
Polygala serpyllacea	Chrysanthemum Leucanthemum
Viola lutea	Centaurea nigra
V. Riviniana	Crepis virens
Hypericum pulchrum	Leontodon hispidus
Cytisus scoparius	Hieracium Pilosella
Ulex Gallii	Gentiana Amarella
Lotus uliginosus	Veronica officinalis
Vicia sepium	Euphrasia officinalis (agg.)
Lathyrus montanus	Plantago lanceolata
Potentilla erecta	Rumex Acetosella
P. procumbens	Empetrum nigrum
Linum catharticum	Anthoxanthum odoratum
Pimpinella Saxifraga	Briza media
Conopodium majus	Carea binervis
Galium saxatile	C. pilulifera
Scabiosa Succisa	Juncus squarrosus
Campanula rotundifolia	Blechnum Spicant
Jasione montana	Ophioglossum vulgatum

In many places, such dwarf shrubs as *Erica cinerea*, *Calluna vulgaris*, and *Empetrum nigrum* become very abundant; and thus a type of vegetation is produced which is transitional from Nardus grassland to moorland (see diagram on p. 137).

(ii) *Molinietum cæruleæ.* The wetter type of siliceous grassland (see p. 132), dominated by purple moor-grass (*Molinia cærulea*), is of far less extent on the southern Pennines than the *Nardus* grassland, and much more local in its occurrence.

The *Molinia* pasture occurs in the more badly drained places. In a general way, the *Molinia* grassland affects the ground overlying the sandstone rocks, whilst the *Molinia* pasture is more characteristic of the steep slopes of the shales. The soil of the *Molinia* grassland is wet, often very wet, and more or less peaty and acidic. However, as shown in Chapter X, the purple moor-grass is by no means confined to acidic soils. On such soils, the *Molinia* association frequently forms a transition from grassland to moorland. The moorland relations of the habitat are seen in the peaty soil (always acidic on siliceous soils), often supersaturated with moisture, and in the abundance of associated species which characterise certain parts of the moorland.

Molinia is often very abundant in the degenerating oak and birch woods of siliceous soils; and there can be little, if any, doubt that some parts of the *Molinia* grassland have resulted from the degeneration of such woods. Transitional areas occur, for example, near Crowden railway station in north Derbyshire. The following is a list of the species which characterise many of the Molinieta of the Pennines:—

Ranunculus Flammula	la	J. acutiflorus	la
Viola palustris	r to o	Scirpus cæspitosus	r to o
Drosera rotundifolia	r	Eriophorum vaginatum	r to o
Hydrocotyle vulgaris	la	E. angustifolium	la
Erica Tetralix	la	Carex curta	r
Calluna vulgaris	la	C. echinata	la
Oxycoccus quadripetala	l	C. Goodenowii	o to a
Pinguicula vulgaris	r	C. flacca	o
Taraxacum palustre	r	C. panicea	o to a
Empetrum nigrum	la	C. flava (agg.)	la
Orchis ericetorum	r to o	Agrostis canina	l
Narthecium ossifragum	r to la	Deschampsia flexuosa	r to a
Juncus effusus	la	Molinia cærulea	a to d
J. conglomeratus	la	*var.* depauperata	r to d
J. squarrosus	la	Nardus stricta	la

Diagram showing the relationships of the plant-communities of the Formation of Siliceous Soils.

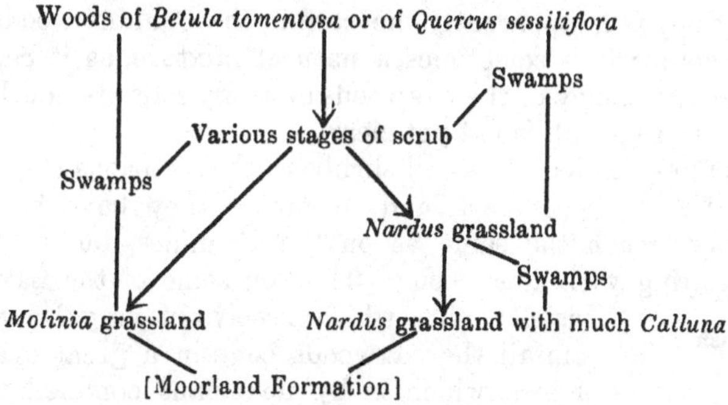

B. Other Regions

The formation of siliceous soils above described appears to be developed on all the siliceous rocks of the Pennine chain; its extension in the British Isles is however far wider than that. So far as they have been investigated the formation appears to be almost co-extensive with the siliceous rocks of Palæozoic age below a certain altitude; these, as was pointed out in Part I, occupy a large part of western England, Wales, Scotland and Ireland. Similar woods also occur on some of the harder secondary rocks, *e.g.* the non-calcareous oolites of north-east Yorkshire.

Distribution of Quercetum sessiliflorae.

Woods of *Quercus sessiliflora* have been recognised in the Lake District and in North and Central Wales on Ordovician rocks, in west-central England on siliceous Silurians, in Devonshire and Cornwall on Upper Greensand and on various Palæozoic strata, in Co. Wicklow on Ordovicians, in Co. Kerry on Old Red Sandstone, and in Co. Galway on metamorphic rocks. In Scotland such woods have also been recognised, but their extension has not been studied. In some of the cases mentioned the woods of

Quercus sessiliflora are strikingly pure; in others the pedunculate oak (*Q. Robur*) is more or less mixed with the sessile-fruited species, but whether this is always due to planting, as is apparently the case in the Pennine woods, or whether it is sometimes a natural mixture, as it certainly is in many of the oakwoods of sandy soils previously described (p. 92), is not yet clear.

The associated trees and shrubs in these various woods are, for the most part, and so far as they have been studied, much the same as on the Pennines, but some interesting variations occur. Thus on some of the Lake District hills which receive a very heavy rainfall the oakwoods contain a great deal of ash, which is by no means confined to streamsides as on the Pennines, but spreads through the woods in all directions, largely replacing the oak, which has probably been extensively removed. The ground vegetation, however, is that of a typical oakwood of siliceous soil. Some of the Devonshire woods of *Quercus sessiliflora* also contain much generally distributed ash.

Oak-ash woods.

The ground vegetation of most of these woods closely corresponds with the three types recognised in the Pennine woods or with transitions between them, but, as in the case of the associated trees, variations occur.

Bryophyte ground vegetation.

Thus in the oakwoods of Co. Wicklow at Glendalough the ground vegetation is largely dominated by mosses, species of *Dicranum*, *Hypnum* and *Polytrichum*, with local patches of *Sphagnum* and also many Liverworts; a similar type of ground vegetation occurs in the woods of *Q. sessiliflora* on the sides of the valleys near Aberystwyth in Central Wales.

Killarney woods.

The woods of *Quercus sessiliflora* at Killarney in Co. Kerry, south-west Ireland, are specially interesting as being the chief station in the British Isles of the strawberry tree (*Arbutus*

PLATE XI

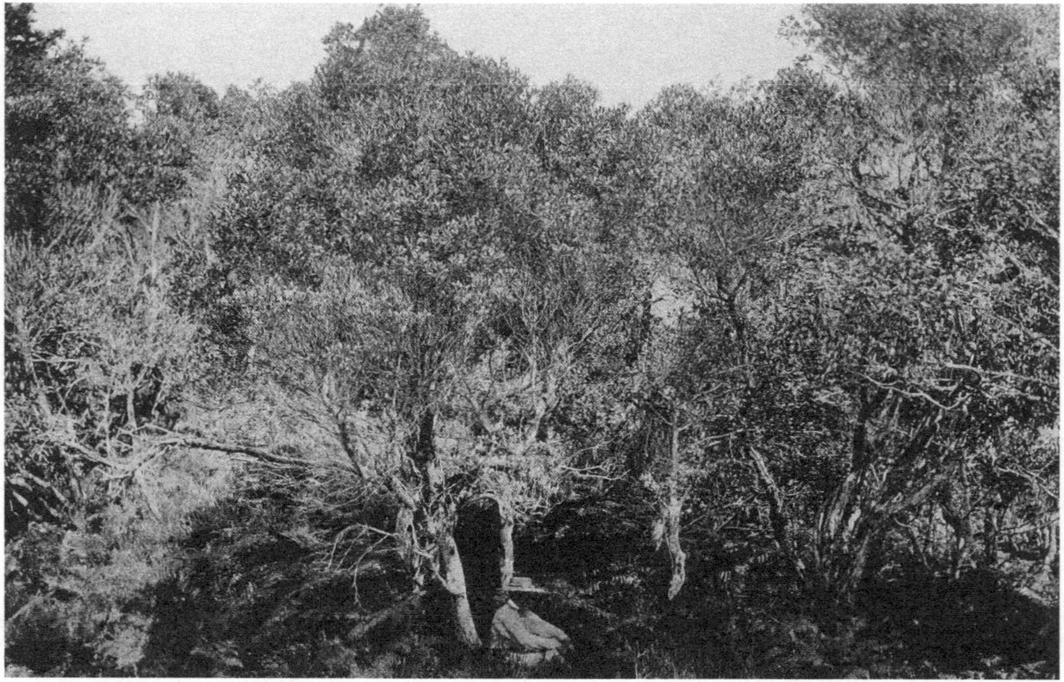

Phot. R. Welch

Society of *Arbutus Unedo* in Quercetum sessilifloræ. *Quercus sessiliflora*, *Pteris aquilina*, *Calluna vulgaris*. Upper lake shore, Killarney, on Old Red Sandstone.

Unedo) which is found besides only in a few isolated stations nearer the Kerry and Cork coasts, to the south and south-west of Killarney[1]. The nearest continental station of the arbutus is in south-western France, about 400 miles (*c.* 640 km.) from its Irish stations.

These woods occur both on Old Red Sandstone and on Carboniferous Limestone, and the arbutus is freely scattered through them; in one small portion of the area it is locally dominant (Plate XI).

The composition of the vegetation is shown in the following lists of the more abundant and the more characteristic but less frequent species.

Quercus sessiliflora	d	Hedera Helix	f
Taxus baccata	a	Scutellaria Nicholsoni	l
Ilex Aquifolium	a	Euphorbia hiberna	f
Arbutus Unedo	a	Allium Scorodoprasum	vl
Fraxinus excelsior	o	Neottia Nidus-Avis	r
Pyrus Aucuparia	o	Cephalanthera longifolia	vr

Old Red Sandstone.		*Carboniferous Limestone.*	
Betula tomentosa	a	Betula tomentosa	o
——		Pyrus rupicola	l
Juniperus sibirica	l	——	
Vaccinium Myrtillus	a	Corylus Avellana	f
Calluna vulgaris	a	Cratægus monogyna	o
Erica cinerea	a	Euonymus europæus	o
Myrica Gale	f	Rhamnus catharticus	l
——		R. Frangula	r
Saxifraga Geum	l	Rubus saxatilis	l
S. umbrosa	f	——	
Pinguicula grandiflora	f	Vicia sylvatica	vr
Melampyrum hians	f	Rubia peregrina	f
Listera cordata	vr	Galium asperum	r
Luzula sylvatica	a	Orobauche Hederæ	l
Molinia cærulea	a	Calamintha montana (= C. officinalis)	l
Hymenophyllum tunbridgense	f	Helleborine atroviridis	l
H. peltatum	l	Lastrea Thelypteris	vl

[1] The editor is indebted to Dr R. W. Scully for very kindly contributing the substance of this account.

Old Red Sandstone (cont.):		Lastrea aristata	f
Trichomanes radicans	vr	L. æmula	f
Asplenium acutum	vr	Osmunda regalis	f

Arbutus Unedo appears to show no special soil proclivities at Killarney, occurring not only on both sandstone and limestone, but also both in the shallow soil of dry rocky woods and in the deeper soil of damp almost marshy hollows. It grows considerably more luxuriantly than in the Mediterranean "maquis," the finest trees reaching a girth of 8 feet (2·4 m.) or in extreme cases 14 feet (4·2 m.). This is probably owing to the fact that its growth is not checked by the hot dry Mediterranean summers. The common character of the two habitats is of course the mild winter, the mean January temperature of Killarney being nearly the same as that of the Riviera.

The present range of the species in the Killarney region extends over an area roughly 6 miles (*c.* 9·7 km.) north and south by about 3 miles (*c.* 4·8 km.) east and west, excluding outlying stations where a few trees still exist as remnants, probably, of woods long since disappeared. There can be no doubt that the arbutus has much decreased both in abundance and in the extent of its Irish area, even in comparatively modern times. Thus Dr Smith, whose *History of Kerry* was published in 1756, makes constant references in that work to the recent destruction of this tree in large quantities for smelting purposes, as well as by a disastrous fire which caused great havoc among them; while if place-names can be relied on as a clue to its former distribution, the arbutus appears to have extended to Co. Clare and even as far north as Co. Mayo, where an island in Clew bay derives its name from the tree.

Of the other plants occurring in the same woods *Saxifraga Geum* and *Pinguicula grandiflora* are confined, within the British Isles, to the south-west of Ireland, *Saxifraga umbrosa* is confined to Ireland and is mainly found in the west and south, *Euphorbia hiberna* is mainly though not quite exclusively Irish. All these belong, outside the British Isles, mainly or exclusively to the Iberian peninsula.

Birchwood association (*Betuletum tomentosæ*).
(Plate XII.)

It has already been shown that on the siliceous rocks of the Pennines a fringe of birchwood frequently occurs immediately above the oakwoods at an altitude of 1000—1500 feet (*c.* 300—450 m.), but the association is, on the whole, poorly developed. In Scotland, however, birchwoods are the characteristic woods of the sides of the Highland valleys, where they are often extensively developed above the upper limit of the oakwoods (*c.* 1000 feet = 300 m.) and sometimes, though rarely, attain a height of about 2000 feet (*c.* 600 m.) on the slopes of the higher mountains (Plate XII a). This increase of altitude is of course in accord with the general rule that the greater the height of a hill-mass the higher any given altitudinal zone of vegetation will ascend.

W. G. Smith writes of the Scottish birchwoods[1] that their occurrence may indicate:—

(1) "An oakwood from which the oak, originally dominant, has been removed, while the less valuable birches have been left, and along with naturally sown seedlings have in time formed a wood.

(2) A wood-clearing or a piece of moorland preserved from excessive grazing by sheep, deer, and rabbits. Under these conditions, birch seedlings and mountain ash frequently develop in such numbers that a thicket or wood results....

(3) The sub-alpine deciduous wood, the highest zone of woodland in the Highland valleys."

The first two cases we have already seen constantly

[1] W. G. Smith, 1905, p. 5.

occur in connexion with the oak-birch heath association in southern England, and are to be regarded merely as stages in the multiform succession exhibited by the heath formation. There is little or no distinction of habitat between such birchwoods and the associations which precede or follow them in succession.

The third kind of birchwood, however, has a different status, for it occupies a distinct climatic zone, above that of the oakwood, but the edaphic factors of its habitat appear to be the same. It might perhaps be regarded as a separate climatic formation, but its characters are too negative, and it is better considered simply as a separate, climatically differentiated, association of the formation of siliceous soils.

Robert Smith, who first described the Highland birchwoods from the ecological standpoint, treated them along with the pinewoods of the same region, and wrote[1]: "The subordinate vegetation in the sub-alpine woods is of a less marked character than that in the temperate woods [*e.g.* the oakwoods], hardly differing, except in the proportion of the individuals present, from the adjoining moorland or hill-pasture....In the pinewoods heather and its associates usually constitute the plant carpet, whilst in...birchwoods the ground is more often covered by grasses and bracken, although this rule is by no means invariable."

The sub-alpine Scottish birchwoods possess, in fact, a ground vegetation made up of the oakwood species which can endure the higher altitude together with any species of siliceous grassland. They appear to have no species peculiar to them. Thus these birchwoods take their place just like those of the Pennines as a climatic zonal association of the formation of siliceous soils; while the pinewoods, as we have already seen (p. 117), appear to be naturally included in the heath formation. The

[1] R. Smith, 1900, p. 451.

Phot. A. G. Tansley

a. *Betula tomentosa*; *Myrica Gale* on the right: 900–1250 feet (*c.* 270–380 m.) alt., near Kinloch Rannoch, Perthshire. Slopes of Schiehallion behind.

Phot. A. G. Tansley

b. *Betula alba* with *Myrica Gale* (wet heath sub-association) in foreground. Wood of *Betula tomentosa* behind. Near Lochan Eilein, Inverness-shire.

Highland birchwoods.

birch, as in the south, may also colonise, and even form woods on, heathland, but it is almost certainly unable to withstand the competition of pine on such soils, because of the deep shade cast by the latter, if the two species meet under equal conditions.

The Bracken Fern in Siliceous grassland.

Pteris aquilina is an abundant constituent of the more open woods on siliceous soils except in the dampest places, and it very frequently persists after the woodland has disappeared. An enormous extent of siliceous grassland in the north and west of England, and in Scotland, is covered with this fern, which rapidly spreads and covers the ground as soon as the shade of the trees is removed. Under its shelter several woodland species often maintain themselves, *e.g. Scilla non-scripta*, *Anemone nemorosa*, *Oxalis Acetosella*. The bracken does not fully expand its fronds till early June in the south and late June in the north of Great Britain, so that during the spring the grasses and other low-growing constituents of the turf can vegetate. In the late summer and autumn however the light is cut off from them by the dense foliage of the bracken. Thus the area occupied by *Pteris* is withdrawn from pasturage during a considerable part of the year and it thus becomes quite a pestilent weed of the semi-natural siliceous pastures. Enormous tracts of land are greatly lowered in value from this cause. Deep ploughing is the only method of eradicating the deeply situated rhizomes of the fern, though repeated mowing at the beginning of the growing season will in time starve the plants. The bracken is often cut in the autumn for use as litter.

CHAPTER VI

THE VEGETATION OF CALCAREOUS SOILS

Calcicole and Calcifuge species.

SOILS containing a comparatively large proportion of lime are always marked by the presence and usually by the abundance of certain species of plants—the so-called "calcicole" species. That this feature is not determined by inability of the species in question to flourish in the absence of a large amount of calcium carbonate is sufficiently shown by the fact that practically all of them occur and flourish on soils poor in lime—if not in the same at least in other regions.

Contrasting with the "calcicole" species there are others, called "calcifuge," which appear to be really intolerant of much lime in the soil. Examples of pronounced calcifuge species usually quoted are the maritime pine (*Pinus Pinaster*), the sweet chestnut (*Castanea sativa*), and among native plants, the broom (*Sarothamnus scoparius*), and the foxglove (*Digitalis purpurea*).

Causes of the distinction.

One explanation which has been given of the occurrence or abundance of calcicole species on calcareous soils is that on such soils they are free from the competition of the calcifuge species, and are thus able to hold their own. The same explanation has been given of the occurrence of halophytes on salt-containing soils, and in that case with considerable plausibility, since the great majority of plants are highly intolerant of salt soils. But in the case of calcicole and calcifuge species it is highly doubtful whether competition has much to do with the matter. In the first place the number of calcifuge species is not so great as to make the absence of their competition a consideration of much importance. And further, this theory does not explain but rather conflicts with the abundance and luxuriance of vegetation, other than that made up of distinctively calcicole species, on suitable calcareous soils.

It seems probable, in fact, that there is some positive feature of calcareous soils favourable to vegetation at large and to calcicole species in particular. Whether it is the chemical characters, as such, of these soils, or correlated physical characters, we do not know. This is one of the problems of plant-ecology, which we may hope will shortly find a satisfactory solution.

The plant-formation of calcareous soils.

Meanwhile we have to recognise that the calcareous soils bear a characteristic vegetation, marked by a number of species which are rarely found away from them, at least in the same region (or if found away from them, occurring only on special types of non-calcareous soil), but generally containing also numerous "indifferent" species which are abundant alike on calcareous and on non-calcareous soils. The correlation of calcareous soils with characteristic plant-communities makes it necessary to recognise in these soils a distinct type of habitat bearing distinct plant-formations.

The distinctive calcareous soils of this country are in the first place the limestones in the wide sense, including the chalk, which is a relatively pure and very soft limestone, and secondly the calcareous sandstones and the marls, which are mixtures of lime and sand and of lime and clay respectively.

In England and Wales the most extensively developed

limestones are the chalk (southern, eastern and north-eastern England), the Oolitic limestones of the Jurassic series (west midlands), the Magnesian Limestone of Permian age (northern England), the Carboniferous or Mountain Limestone (Pennines, Wales and Somerset), and various Devonian limestones in Devonshire. Limestone strata of minor importance occur in rocks of all ages. In Ireland the Carboniferous Limestone has a very wide extension, and though it is frequently covered with non-calcareous deposits, it is exposed over considerable areas in the west; but in Scotland there are no extensively developed rocks of this class, though minor limestone bands occur. Some of the soils derived from igneous rocks rich in lime-containing minerals, such as some of the felspars, are also calcareous.

Sub-formations. Though there are more or less important floristic and ecological differences between the vegetation of the chalk, which is mainly developed in the drier regions of the east and south, and that of the older limestones, which are mainly developed in the wetter regions of the west and north, yet these do not seem of sufficient importance to override the essential unity of the limestone vegetation as a whole. We shall therefore consider the whole of this vegetation as constituting a single formation, regarding the vegetation of the older limestones as one sub-formation, that of the chalk as a second and that of the other calcareous soils as a third.

The Sub-formation of the older Limestones

By C. E. Moss

By far the most extensive of the older limestones is the so-called Mountain Limestone of carboniferous age, which forms the centre of the Pennine Range, occurs in

Phot. W. B. Crump

a. Ash scrub. *Fraxinus excelsior*, *Cratægus monogyna*, etc. Ebbor Gorge, Somerset.

Phot. W. B. Crump

b. *Fraxinus excelsior* (coppiced). Damp ground sub-association. *Allium ursinum*, *Lamium Galeobdolon*. Long Wood, Cheddar, Somerset.

Ashwood and scrub on Mountain Limestone.

Wales and in the Mendip Hills in Somerset, and is exposed over considerable areas in Ireland. The vegetation of this rock has been principally studied in the Pennine area and on the Mendips, and the following account is mainly based on these regions.

As in most other cases, the characteristic vegetation of the older limestones is naturally considered as falling into three associations corresponding to woodland, scrub and grassland.

Ashwood association *(Fraxinetum excelsioris)* (Plates XIII and XIV).

Ashwoods are characteristic of the limestone hills of the north and west of England. Woods of this type occur on the limestone hills of Yorkshire[1] and Derbyshire[2], Westmorland[3] and Somerset[4].

Distribution. Those described by Smith and Rankin are in many cases of the nature of scrub (see p. 153), and those described by Lewis are of the ash-birch type (see p. 149). In Somerset, ashwoods are well developed on the slopes of hills of the Carboniferous or Mountain Limestone, of the Dolomitic Conglomerate, and of the Jurassic limestones.

In Derbyshire, they are well developed on the slopes of hills of the Mountain Limestone. The summits of the limestone hills of north Derbyshire attain a height of about 1550 feet (*c.* 470 m.), and their average height is about 1200 feet (*c.* 370 m.). The plateau is dissected by numerous valleys or "dales," most of which are streamless. The dales descend from the plateaux, and the ash woods begin to appear at an altitude of about 1000 feet (305 m.), above which scrub occurs but no genuine woods, and continue to the bottoms of the dales.

[1] Smith and Rankin, 1903. [2] Moss, 1911.
[3] Lewis, 1904 *a*. [4] Moss, 1907.

It is probable that at some past time, the whole of the limestone areas of England below about 1000 feet (*c.* 300 m.), or perhaps even 1250 feet (*c.* 380 m.), were covered by a primæval ash forest, just as similar places on the older siliceous hills were once covered by forests of oak (*Quercus sessiliflora*) and birch (*Betula tomentosa*). The numerous place-names including the word "ash" indicate that the abundance of *Fraxinus excelsior* is of long standing. In north Derbyshire, for example, there are Ashwood dale, Ashford, Money Ash (= "many ash"), and, on the edge of the plateau at the upper limit of woodland, One Ash.

On the Chalk of the south and east of England, the ash is a very abundant and characteristic plant, though its dominance in woods is apparently confined to the south-western region of the chalk outcrop.

Relation to water and to lime.

In the limestone districts above indicated, the ash (*Fraxinus excelsior*) is the dominant tree of the natural and semi-natural woods, whether the soil be damp or dry[1]. In non-calcareous areas, the ash only becomes dominant or abundant where the soil is wet or decidedly damp; and it seems clear that, in any given natural station, the abundance of the ash is due to one of two causes, either to a high lime-content of the soil or to a high water-content. Some of the local foresters are of opinion that the timber of the ash grown on the limestone soils is harder and more durable than that grown on the wet, non-calcareous soils.

Associated trees and shrubs.

The two most frequent woody associates of the ash are the wych elm (*Ulmus glabra* Huds. = *U. montana*) and the hawthorn (*Cratægus monogyna*), both of which are more abundant in the ashwoods than in typical oak

[1] These ashwoods cannot be separated from oakwoods on the basis of a difference in the water-content of the soil, for in each case there is a range from very wet to very dry soils.

and birchwoods. The elm is more abundant at the lower altitudes and in the damper situations, the hawthorn in the drier situations and at the higher altitudes. When the ash is removed or dies out naturally, the elm or the hawthorn becomes locally sub-dominant, and societies of elm and hawthorn are as characteristic of ashwoods as societies of birch and alder are of oakwoods.

Two conifers are native in the ashwoods, namely, the juniper (*Juniperus communis*) and the yew (*Taxus baccata*); but they are somewhat local in their distribution. In Derbyshire, for example, the yew is rather rare, whilst in north Lancashire it is abundant.

The aspen (*Populus tremula*) is the only indigenous species of the genus *Populus*; but sometimes *P. serotina* is planted at the edges of the woods. An interesting society of *P. tremula* occurs in upper Cressbrook Dale.

Of willows which are certainly indigenous, the crack willow (*S. fragilis*) and the osier willow (*S. viminalis*) occur by stream sides, and the sallows (*S. caprea* and *S. cinerea*) occur in the woods themselves. *S. aurita* appears to be absent from the ashwoods of limestone soils.

In Derbyshire, oaks appear to be absent from the ashwoods, except locally as introduced trees; but in other localities *Quercus Robur* and *Q. sessiliflora* may occur.

The beech (*Fagus sylvatica*) is frequently planted on all kinds of soil in this country, but is probably not indigenous in the north of England.

The hazel (*Corylus Avellana*) is a very abundant and characteristic shrub of the ashwoods, where dense thickets of this plant frequently occur.

The alder (*Alnus rotundifolia*) is much less abundant than in the oakwoods.

In the ashwoods of Somerset and Derbyshire, birches (*Betula* spp.) are very rare and local; but on the limestone hills of the northern Pennines, particularly at the higher altitudes, the birch locally becomes subdominant or dominant (Moss, Rankin and Tansley, 1910, p. 142).

One of the most noticeable features of the English ashwoods of calcareous soils is the large number of associated species of trees and shrubs. For example, the following plants are characteristic of ashwoods but are absent or nearly absent from oak and birchwoods on the older siliceous soils.

Tilia cordata	la	Cornus sanguinea	a
T. platyphyllos	r	Viburnum Lantana	la
Rhamnus catharticus	la	Ligustrum vulgare	a
Euonymus europæus	la	Daphne Mezereum	r
Rosa spinosissima	l	D. Laureola	o
R. micrantha	o	Juniperus communis	l
Pyrus Aria	lf	Taxus baccata	l

Ericaceous undershrubs are totally absent from the ashwoods.

Sub-associations of the ground vegetation.

Two species which characterise sub-associations of the oakwoods, namely *Deschampsia flexuosa* and *Holcus mollis*, do not occur in the ashwoods. The following divisions (sub-associations) of the ground vegetation will illustrate the range in habitat and in floristic composition which occurs within the ashwoods.

(1) In marshy places, which occur in ashwoods by stream sides, at the bottoms of the streamless dales, and on slopes where springs arise, such moisture-loving plants as the following occur:

Trollius europæus	l	Polemonium cæruleum	l
Caltha palustris	a	Myosotis Scorpioides (palustris)	lf
Spiræa Ulmaria	a	Mentha aquatica	a
Geum rivale	l	Sparganium ramosum	l
G. rivale × urbanum	r	Scirpus compressus (= Blysmus Carici)	l
Epilobium hirsutum	a	Phalaris arundinacea	l
Valeriana officinalis (Mikanii)	a	Phragmites vulgaris (= P. communis)	l
Petasites ovatus	va		
Cirsium heterophyllum	l		

Phot. W. M. Rankin

a. Wooded hill. Anstwick, Yorkshire.

Phot. W. M. Rankin

b. Ashwood on limestone pavement. *Fraxinus excelsior.* Chapel-le-Dale, Ingleton, Yorkshire.

Ashwood on Mountain Limestone.

(2) Other parts of the woods, which, though not marshy, are nearly always very moist, have a rich and varied ground flora. Sheets of wood-garlic (*Allium ursinum*) (Plate XIII b) and of the lesser celandine (*Ranunculus Ficaria*) are characteristic. The following is a selected list of the species of such damp parts of the ashwoods:

Anemone nemorosa	a	Campanula latifolia	o
Ranunculus Ficaria	a	Myosotis sylvatica	lf
Trollius europæus	l	Lamium Galeobdolon	a
Aquilegia vulgaris	l	Orchis mascula	f
Lychnis dioica	a	Habenaria virescens	
Fragaria vesca	a	(chloroleuca)	o
Oxalis Acetosella	f	Allium ursinum	la
Asperula odorata	la	Arum maculatum	f
Valeriana officinalis		Carex sylvatica	f
(Mikanii)	a	Bromus ramosus	f
V. dioica	l	Triticum caninum	l
Cnicus palustris	a	Hordeum sylvaticum	l
C. heterophyllus	l		

(3) On soils which are drier than the preceding, and which, during the summer months, may become temporarily very dry, sheets of dog's mercury (*Mercurialis perennis*) often occur; and this plant is in Derbyshire very frequently associated with the moschatel (*Adoxa Moschatellina*). At the beginning of April, in the Derbyshire dales, the dog's mercury is about three inches high, its leaves are beginning to unfold, and a few stamens are ripe. At this time of the year, the moschatel is here flowering abundantly, and is almost hidden by the young shoots of the dog's mercury. In the fairly dry portions of the ashwoods of the Peak district, this plant society of dog's mercury and moschatel is a characteristic feature. The society is an excellent example of what Woodhead (1906, p. 345) terms a "complementary" society, as the roots of the dog's mercury reach down to lower layers of soil than the roots of the moschatel, whilst the small

and delicate shoots of the *Adoxa* receive their necessary shade from the larger and more vigorous shoots of *Mercurialis*. Before the end of June, *Adoxa* has entered on its long period of dormancy; and the dull green leaves of the dog's mercury, hiding its ripening berries, occur in extensive and monotonous stretches. It may, therefore, be said that the roots of the two species are edaphically complementary and the shoots seasonally complementary. In the oak and birchwoods, the dog's mercury occurs in more or less local patches, and *Adoxa* is extremely rare; whilst the *Mercurialis-Adoxa* society does not occur.

The dog's mercury is much more abundant in English woods on calcareous soils than in those on non-calcareous soils; and this is a partial confirmation of an observation made by Thurmann[1], who mentions the plant as one of fifty "xerophilous" plants typical of "dysgeogenous" or calcareous soils.

(4) The dry parts of the ashwoods are characterised by stretches of ground ivy (*Nepeta hederacea*), which remains green throughout the whole year and flowers from early spring to late summer. If the ground is stony and composed of old screes, taller herbs occur, such as the hairy St John's wort (*Hypericum hirsutum*), the nettle (*Urtica dioica*), and the wood sage (*Teucrium Scorodonia*). These plants form close herbaceous thickets in summer; and their dead stalks remain upright and rigid throughout the succeeding winter and spring. Locally, the lily-of-the-valley (*Convallaria majalis*) and the stone-bramble (*Rubus saxatilis*) form fairly extensive plant societies, and in these the nodding melic-grass (*Melica nutans*) sometimes occurs.

(5) The very driest parts of the ashwoods occur on the rocky knolls. Here the soil is extremely shallow, and in places the bare rock protrudes. Trees and shrubs are absent, and the absence of shade allows of the growth of

[1] Thurmann, 1849.

saxicolous lichens and bryophytes, of ephemeral species as *Arenaria serpyllifolia, Draba verna,* and *Saxifraga tridactylites*, and of dwarf perennials like *Sedum acre* and *Thymus Serpyllum*. Such a community does not, except in a topographical sense, belong to a woodland association at all, and is to be regarded as an outlier of another association.

Limestone screes and cliffs also occur in the midst of the ashwoods. These, if damp, become in time clothed with the vegetation of the ashwoods; and, by comparing several such localities, it is possible to gain some idea of a progressive succession from bare screes and cliffs to a closed ashwood association. Such a succession supplies the reason, an historical one, why plants like the mossy saxifrage (*Saxifraga hypnoides*) and the limestone polypody (*Phegopteris Robertiana*) are sometimes found on old screes in the midst of existing ashwoods.

Limestone scrub association. The ashwoods merge imperceptibly into calcareous scrub, and many areas occur which it is difficult to classify either as woodland or scrub. An excellent series of such transitional areas may be seen in Monsal Dale and Cressbrook Dale in Derbyshire. As on the siliceous soils, the scrub assumes many different facies. In some cases, the taller woody plants are more or less isolated, and the ground vegetation quite grassy. In other cases, dense thickets of shrubs occur, and the ground flora contains many shade-loving species. In the great majority of cases, however, the scrub appears to be of a retrogressive nature, though progressive scrub communities occur here and there, as at the foot of cliffs and on screes. As the scrub contains many woodland species as well as many grassland species, it is often extremely rich floristically, as in the case of the calcareous scrub in upper Cressbrook Dale above mentioned; but very nearly all the plants of scrub occur in the related woodland or grassland (Plate XIII a).

The species of shrubs of the limestone scrub association are as follows:

Prunus spinosa	o	Rhamnus catharticus	r
Rubus *spp.*	o	Hedera Helix	f
Rosa canina	o	Cornus sanguinea	r
R. mollis	r	Lonicera Periclymenum	o
R. spinosissima	la	Fraxinus excelsior	o
Cratægus monogyna	la	Corylus Avellana	la
Euonymus europæus	r	Salix *spp.*	r

The following are some of the more characteristic species of the ground vegetation:

Thalictrum minus	r	Chrysanthemum Leucanthemum	f
Anemone nemorosa	o	Cnicus palustris	o
Cardamine impatiens	l	C. eriophorus	o
Helianthemum Chamæcistus	a	Picris hieracioides	r
Stellaria Holostea	r to o	Primula veris	f
Geranium sanguineum	l	Teucrium Scorodonia	r
Rubus saxatilis	l	Nepeta hederacea	la
Fragaria vesca	o	Origanum vulgare	f
Poterium Sanguisorba	a	Cypripedium Calceolus	vr
Saxifraga hypnoides	l	Listera ovata	o
Hypericum hirsutum	la	Convallaria majalis	l
H. montanum	l	Luzula pilosa	o
Galium Cruciata	r to o	Avena *spp.*	o
Solidago Virgaurea	o	Koeleria gracilis	o
Inula squarrosa	r	Melica nutans	r
		Brachypodium gracile	f

Limestone grassland association (*Festucetum ovinæ*).

From calcareous scrub, every transition may be traced to pure grassland. Typical calcareous grassland consists of short, grassy turf, largely composed of the sub-aërial parts of the sheep's fescue-grass (*Festuca ovina*). Such grassland is characteristic and widespread on the slopes of the limestone hills of the west and north of England and equally so on the chalk downs of the south (see p. 173).

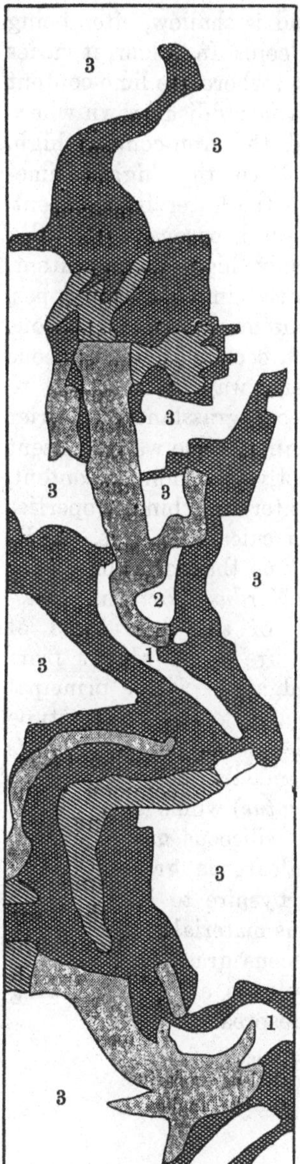

FIG. 5. Chart of associations of the Limestone Formation of Cressbrook Dale and Monsal Dale, Derbyshire. The natural and semi-natural associations are mainly confined to the valley sides. The lower half of the figure is occupied by Monsal Dale which has a winding course and a mainly southerly direction; the upper half by the dry tributary valley of Cressbrook Dale, which runs southwards and joins Monsal Dale about the middle of the figure.

I. Cultivation: (1) of valley bottom alluvium, (2) intakes from limestone slopes, (3) of plateau.

II. Limestone grassland on valley sides and extending on to plateau.

III. Limestone scrub; on valley sides and extending slightly on to plateau.

IV. Ashwood: on valley slopes.

Soil. The soil of limestone grassland is shallow, often being only a few inches deep. In colour, it varies from a whitish grey, where the lime-content is very high, to a brownish or even reddish-brown where the lime-content is lower and the iron-content high. The more newly-formed soils have the higher lime-content, and the older soils have the lower lime-content. The newer soils are also dry and porous: the older ones have often a comparatively high water-content. Marshes occur here and there on the limestone hill-slopes, so that the range of water-content on the calcareous grassland is similar to that which occurs on the siliceous grassland. However, in a general way, it is correct to say that the soils of the calcareous grasslands are drier than those of the siliceous grasslands. The water-content of non-calcareous soils varies directly as the humus-content, a result doubtless due to the water-absorbing properties of humus. The water-content of calcareous soils, on the other hand, is greater the less lime they contain.

Differences from siliceous grassland. The two principal grasses (*Nardus stricta* and *Deschampsia flexuosa*) of the drier type of siliceous grassland are quite absent from calcareous grassland, and so is the principal grass (*Molinia cærulea*) of the wetter type of siliceous grassland (cf. pp. 132–6). The bracken (*Pteris aquilina*), gorse (*Ulex* spp.), and the common rushes (*Juncus effusus* and *J. conglomeratus*) which bring about such characteristic facies of the siliceous grassland, are only of local occurrence on calcareous grassland, and perhaps mainly confined in Derbyshire to leached soils or soils mixed with non-calcareous material. Many other plants characteristic of the siliceous grasslands are also absent from the calcareous grasslands, and the following list is a selection of the more widespread of these:

Spergularia rubra	Cytisus scoparius
Polygala serpyllacea	Genista anglica

Potentilla procumbens	Melampyrum pratense *var.* montanum
Galium saxatile	Empetrum nigrum
Gnaphalium sylvaticum	Salix repens
Jasione montana	Orchis ericetorum
Erica cinerea	Luzula multiflora
Calluna vulgaris	Carex binervis
Vaccinium Myrtillus	Holcus mollis
Digitalis purpurea	Blechnum Spicant

On the other hand, the following widespread plants are a few selected from those which are characteristic of calcareous grassland but absent from or rare in siliceous grassland:

Thalictrum minus	Campanula glomerata
Viola hirta	Origanum vulgare
Polygala calcarea	Plantago media
Anthyllus Vulneraria	Ophrys apifera
Hippocrepis comosa	O. muscifera
Poterium Sanguisorba	Orchis ustulata
Potentilla verna	O. pyramidalis
Helianthemum Chamæcistus	Avena pubescens
Galium asperum	A. pratensis
Asperula cynanchica	Koeleria gracilis (cristata)
Scabiosa Columbaria	Bromus erectus
Leontodon nudicaule	Brachypodium pinnatum

Local variations in floristic composition.

Interesting local differences are seen in the floristic character of calcareous pasture in different parts of the country. For instance, in the north of England, *Sesleria cærulea* is a very abundant and characteristic member of calcareous grassland. This grass appears to reach its southern British limit on the Pennines of mid-Yorkshire, and is absent from the limestone of Derbyshire. *Actæa spicata* and *Primula farinosa* also seem to have found the siliceous soils of the Leeds and Halifax district an effectual barrier against a southern extension of their range. *Carex ornithopoda* is a very local plant in England, occurring on the calcareous pasture, and also on limestone cliffs in Derbyshire.

In the west of England *Helianthemum polifolium*, *Trinia glauca*, and *Koeleria vallesiana* occur locally on the calcareous pasture of the limestone hills.

A larger number of species characterise the chalk of south-eastern England and have failed to reach the limestones of the west and north.

Limestone swamps.

Although swamps are less characteristic of calcareous than of siliceous pasture, they do occur here and there on the slopes of the fissured limestones, and they are characterised by different plant communities from the swamps of siliceous soils. On the Pennines, for example, the following plants occur on the swamps of the limestone hills:

Thalictrum flavum	r	Petasites ovatus	la
Trollius europæus	la	Carduus heterophyllus	la
Caltha palustris	a	Cnicus palustris	a
Spiræa Ulmaria	a	Primula farinosa	la
Potentilla palustris	l	Polemonium cæruleum	la
Geum rivale	la	Myosotis palustris	la
G. rivale × urbanum	o	Mentha *spp.*	a
Chrysosplenium oppositifolium	la	Pedicularis palustris	l
C. alternifolium	la	Orchis latifolia	l
Parnassia palustris	la	Epipactis palustris	r
Epilobium hirsutum	a	Juncus glaucus	la
Valeriana officinalis (V. Mikanii)	a	Scirpus compressus	l
V. dioica	la	Carex disticha	l
Eupatorium cannabinum	la	Festuca elatior	l
		Selaginella selaginoides	l

The "limestone heath."

In exposed situations, especially on plateaux and the higher hill-slopes, the soil of the calcareous pasture tends to lose its high lime-content owing to the leaching action of rain. As this process continues, some of the plants of siliceous grassland may invade the leached area, and in time, a *limestone heath* may be produced. A limestone or calcareous heath is characterised by the occurrence side by side, on soils unmixed with non-calcareous débris, of calcicole plants and calcifuge plants. The former are, in the majority of cases, deep tap-rooted plants; and the roots reach down to the lower layers of soil where the lime-content is high. The latter are nearly always plants with a superficial root-system whose activities are confined to the leached upper layers where the lime-content is low. Thus, such calcifuge plants as *Calluna vulgaris* may be found growing side by side and completely mixed up with such lime-loving plants as *Poterium Sanguisorba*[1].

The following is a list of plants occurring in the limestone grassland association of the Pennines.

Thalictrum minus	l	Pimpinella Saxifraga	o
Ranunculus bulbosus	a	P. magna	r
Cochlearia alpina	l	Conopodium majus	o
Erophila verna	la	Asperula cynanchica	la
E. præcox	la	Galium verum	a
Arenaria verna	l	G. Cruciata	la
Geranium lucidum	la	G. asperum	la
G. sanguineum	r	Scabiosa Columbaria	o
Polygala vulgaris	a	Bellis perennis	o
Viola hirta	o	Antennaria dioica	la
V. lutea	la	Achillea Millefolium	a
Anthyllis Vulneraria	o	Chrysanthemum Leucanthemum	a
Lotus uliginosus	a	Carlina vulgaris	o
Hippocrepis comosa	l	Carduus nutans	o
Potentilla verna	l	Cnicus eriophorus	l
P. Crantzii	l	C. palustris	l
Alchemilla vulgaris *var.* filicaulis	o	Centaurea nigra	a
Poterium Sanguisorba	a	C. Scabiosa	o
Saxifraga hypnoides	l	Picris hieracioides	o
S. tridactylites	la	Crepis virens	a
Sedum acre	a	Leontodon hispidus	a
Hypericum hirsutum	la	L. nudicaule	l
Helianthemum Chamæcistus	a	L. autumnalis	la
		Hieracium Pilosella	a

[1] Moss, 1907 and 1911.

Hieracium *spp.*	la	H. conopsea	l
Campanula glomerata	o	Listera ovata	lo
C. rotundifolia	o	Spiranthes autumnalis	r
Primula veris	a	Allium vineale	r
Centaurium umbellatum	r	A. oleraceum	r
Gentiana Amarella	a	Luzula campestris	a
G. baltica	r	L. pilosa	l
Stachys Betonica	o	Carex caryophyllea	la
Origanum vulgare	la	C. flacca	o
Thymus Serpyllum	a	C. ornithopoda	r
Veronica Chamædrys	o	C. pilulifera	l
Euphrasia officinalis (agg.)	a	Anthoxanthum odoratum	o
		Trisetum flavescens	l
Plantago media	la	Avena pratensis	o
P. lanceolata	a	A. pubescens	l
Ophrys apifera	r	Arrhenatherum elatius	la
O. muscifera	r	Koeleria gracilis (cristata)	o
Orchis Morio	la	Briza media	o
O. mascula	la	Festuca ovina	d
O. maculata	l	Bromus erectus	l
O. ustulata	r	Brachypodium gracile	a
O. pyramidalis	r	B. pinnatum	r
Habenaria viridis	la	Asplenium *spp.*	l

Associations of limestone pavements.

A very striking feature of some of the summits of the Carboniferous Limestone, as in the mid-Pennines[1] and the north Pennines, is the intricate maze of deep and narrow clefts of the limestone rocks. Limestone rocks weathered in this way are spoken of as "pavements." They are also of great extent at low altitudes on the limestone plain of Co. Clare, in west Ireland[2] (Plate XV).

The exposed surface of the rock is very bare, and, although rather rich in lichens and mosses, is extremely poor as regards its Phanerogamic flora. Within the clefts, however, ferns and flowering plants are abundant. The clefts are sheltered from wind and sun; and, there

[1] Smith and Rankin, 1903, p. 19 of separate.

[2] R. Lloyd Praeger, 1909, p. 50 et seq.

PLATE XV

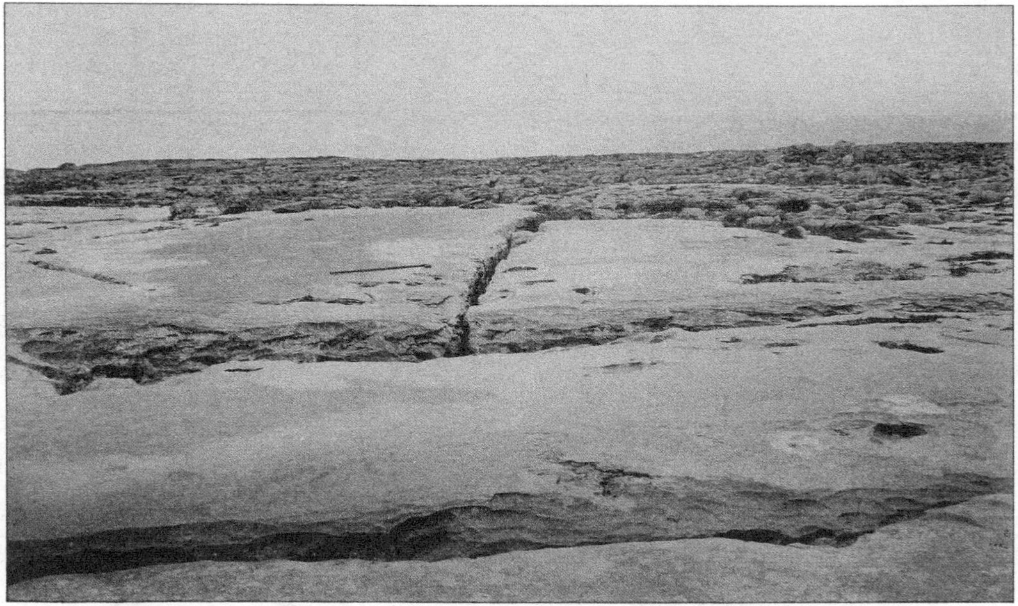

Phot. W. Welch

Mountain Limestone pavement. Black Head, Co. Clare, West Ireland.

being little evaporation, the soil is almost always moist. Many of the plants found are those of the ash-woods, such as the dog's mercury (*Mercurialis perennis*), the wood garlic (*Allium ursinum*), and the hartstongue fern (*Phyllitis Scolopendrium* = *Scolopendrium vulgare*); but in addition to the woodland species, there are *Ribes petræum* Sm., the green spleenwort (*Asplenium viride*), *Lastrea rigida*, and the holly fern (*Polystichum Lonchitis*). The following is a more complete list:

Thalictrum minus	f	Hedera Helix	la
Anemone nemorosa	f	Lactuca muralis	o
Actæa spicata	l	Mercurialis perennis	f
Viola Riviniana	o	Corylus Avellana, dwarfed	la
Geranium Robertianum	o	Listera ovata	o
Oxalis Acetosella	a	Allium ursinum	f
Rubus Idæus	la	Asplenium viride	la
Cratægus monogyna (= Oxyacantha), dwarfed	la	A. Trichomanes	a
Ribes petræum	la	Phyllitis Scolopendrium	f
Heracleum Spondylium	o	Polystichum lobatum	r
		Lastrea rigida	la

Sometimes, in more sheltered places, the limestone pavement becomes completely overgrown with vegetation and an ashwood may then take possession of the habitat (Plate XIV b).

THE SUB-FORMATION OF THE CHALK

By A. G. TANSLEY and W. M. RANKIN

Characters of the chalk. The English chalk uplands, called "downs" in the south and "wolds" in the north, have a wide extension in the south and east of the country. Their well-known soft and rolling contours are very familiar to dwellers in and travellers through the country lying east and south-east of a line drawn from north Yorkshire southward to Buckinghamshire and thence south-westward to the Devonshire coast of the Channel.

The distribution of the chalk and some of its characteristic agricultural features have already been described in Part I (pp. 53–5). Stratigraphically and lithologically the rock is divisible into three, the Upper, Middle and Lower Chalk. The Upper Chalk is the thickest division and also the purest, frequently containing as much as 98 °/ₒ of calcium carbonate. The Lower Chalk is much less pure, containing considerable percentages of clay and siliceous material generally. The Lower and Middle Chalk are generally under cultivation, and it is mainly the Upper Chalk which bears the characteristic woodland, scrub and grassland to be here described. A large proportion of its surface is, however, covered by various later deposits of no great depth, such as clay-with-flints, plateau gravels, and in the regions near the North Sea, chalky boulder clay, together with rainwash from these accumulations. The typical chalk vegetation is thus generally confined to the escarpments and valley sides where the actual chalk comes to the surface.

Vegetation. The vegetation of the chalk proper, as in the case of other plant-formations, shows the obvious division into woodland, scrub and grassland associations with which we have now become familiar. In many respects these associations are distinctive as compared with the corresponding associations of the older limestones. Thus the beechwoods of the escarpments and valley sides of the chalk contrast with the ashwoods of corresponding situations on the older limestones. To this generalisation there are indeed exceptions. The extreme limits of the chalk in Devon, Dorset and the Isle of Wight bear ashwoods in place of beechwoods, and on the northern wolds of Yorkshire and Lincolnshire natural beechwoods are also absent; while the oolitic limestones of the Cotswold Hills in Gloucestershire bear well-developed natural beechwoods.

Phot. S. Mangham

a. Beechwood and chalk pasture on sides of dry valley in chalk escarpment. Chevening Park, Kent.

Phot. S. Mangham

b. Interior of same beechwood. Communities of *Mercurialis perennis* in lighter spaces. [*Cephalanthera grandiflora* and *Neottia Nidus-Avis* occur in the darker shade.]

Beechwood association on Chalk.

The grasslands, as developed in the smooth down pastures, have a more varied flora than, for instance, the pastures of the mountain limestone, containing several species, notably certain orchids, which are confined in England to the south-east. Finally, the chalk scrub also possesses some distinctive features.

Beechwood association of the chalk (*Fagetum sylvaticæ calcareum*) Plates XVI—XVIII.

Distribution of beech-woods.

Natural beechwoods occur on the escarpments and valley sides of the North and South Downs fringing the Weald; they are sparsely represented towards the eastern limits of both escarpments, but attain to very fine development at the western end of the Weald, especially between Arundel and Selborne. A second important beechwood area is that of the Chiltern Hills in Buckinghamshire and south Oxfordshire. From this centre beechwoods extend in one direction across the Thames valley, dying out in Wiltshire, and in the other, north-eastward along the main chalk escarpment into Hertfordshire and Cambridgeshire, where, however, they are very sparsely developed, while they are apparently quite absent from the chalk of Norfolk, Lincolnshire and Yorkshire. A third distinct area is that of the oolitic limestones of the Cotswolds. The western chalk escarpment of Salisbury Plain and the Jurassic limestones in the neighbourhood of Bath, which seem to offer suitable situations, and might serve to connect the Cotswold and Chiltern areas, are, however, apparently quite destitute of natural beechwoods.

"It seems likely that the beech is, speaking geologically, a comparatively recent migrant into this country from the Continent. Such a view is supported by its sub-fossil occurrence in Denmark in the recent, but not in the older peat. Whether the beech be a comparatively recent immigrant or not, it seems that only in the south-east of England has it found climatic conditions enabling it to become dominant and to form woods, beating the ash, its light-demanding competitor on calcareous soils, by reason of the deep shade which it casts. While the beech has every appearance of being native so far west as Cornwall, and while it flourishes and ripens seed so far north as Scotland, where it is often successfully planted on very various soils, it appears quite unable to form natural woods outside the area indicated[1]. It may, however, have had its distribution as a wood-former somewhat curtailed by clearance of forest on the outskirts of its present area, *i.e.* on the chalk escarpment on either side of the Chiltern region, for in Cambridgeshire and Wiltshire indications of natural beechwoods are not wanting[2]."

We may therefore conclude that the beechwoods on the chalk of southern England represent an extension of the continental beechwoods on calcareous soils[3]. In this country they occupy the drier chalk rather than the damper marls, a fact which may be connected with the damper climate of England. The factors which have arrested the further progress north-westward of the beech association are not at all clear, since the beech itself, as we have seen, flourishes and sets seed perfectly well far beyond its limits as a wood-former.

[1] "It is interesting to notice that the distribution of certain associations of land molluscs which show continental affinities, is limited in a similar way. *Helix obvoluta*, the most typical species of these associations, is found nowhere beyond the limits of the beech association."

[2] Moss, Rankin and Tansley, 1910, pp. 143—4.

[3] Graebner (*Die Pflanzenwelt Deutschlands*, 1909) describes the German beechwoods as occupying mostly marly soils. The English beechwoods on chalk must be carefully distinguished from the local beechwoods on sand which belong to the heath formation, and have already been mentioned (p. 102). Both types of beechwood are apparently confined to the south of England.

PLATE XVII

Phot. C. J. P. Cave

Beechwood in winter, with young trees. Ditcham Park, Hampshire.

Beechwood on Chalk.

Habitat. The chalk beechwood association occurs typically on the steep slopes of the escarpments and valleys of the downs—a characteristic position that doubtless gave rise to the local name "hanger" for this type of wood. Sometimes beechwood is found on the more level ground of the chalk plateaux, but in these situations the chalk is generally covered by superficial non-calcareous deposits, and where such is the case the oak (*Q. Robur*) always occurs, either mixed with beech, or forming pure oakwoods of the damp or dry type according to the nature of the soil overlying the chalk. The summits of the downs, where they are bare of superficial deposits, are largely covered with calcareous pasture[1], which has from time immemorial been used as "sheep-walk"; from the constant occurrence of neolithic barrows and ancient camps and track-ways on the down summits it is very doubtful indeed if much of this grassland was ever covered with wood. It has been suggested that the ground water level is too far below the surface for the beech to flourish on these chalk summits and plateaux, and the general restriction of the beechwoods to the steeper slopes may be due to this cause. This restriction may also be partly due to the tendency to clear the gentler slopes first for agricultural or pastoral purposes.

The typical beech hangers are developed on very shallow soil, the rock being generally covered only by a few inches of mild humus; the roots of the trees are largely embedded in the chalk itself.

Composition. The beech is not only the dominant tree in the beechwood association; it typically forms a practically pure high forest in close canopy, and is very rarely coppiced (Plates XVI, XVII). Occasionally an ash is met with, occasionally a whitebeam (*Pyrus Aria*), but these trees cannot compete successfully with

[1] This is especially the case on the South Downs of Sussex and Hampshire, and on Salisbury Plain.

the beech if the latter is growing in close canopy, for their light demand is higher. The oak is extremely rare, and is entirely absent from many beechwoods. *Prunus Avium* is occasionally abundant. The yew (*Taxus baccata*) is abundant in many beechwoods, this species forming a distinct "layer" below the crowns of the dominant trees.

Shrubs are practically absent in a normal beechwood, for the light penetrating the canopy is insufficient to permit of their growth.

The ground vegetation is, for the same reason, very scanty—often indeed quite absent. However, in places where the foliage canopy is comparatively thin, *Mercurialis perennis* is the typical dominant, often covering the ground in continuous sheets (Plate XVI b), while *Sanicula europæa, Viola Riviniana, V. sylvestris, Fragaria vesca* and *Circæa lutetiana* are generally abundant and locally dominant (Plate XVIII a). *Viola hirta*, a pronounced calcicole, is also abundant and characteristic. The orchids *Cephalanthera grandiflora*, and the rarer *Helleborine violacea* and *H. atrorubens*, are also characteristic, and *Habenaria virescens* (*chloroleuca*) is frequent. In the humus the colourless saprophytes *Neottia Nidus-avis* and *Monotropa Hypopitys* are frequent. *Helleborus viridis, H. fœtidus, Atropa Belladonna, Daphne Laureola*, and *Ruscus aculeatus* are characteristic species of the more open spots in or on the outskirts of beechwoods.

Ashwood association (*Fraxinetum excelsioris calcareum*).

The composition of this association does not differ essentially from that already described (p. 147) as the chief association of the formation of the older limestones, though minor floristic differences of course exist. The main interest of the association lies rather in its relation to beechwood—the more typical woodland association of the chalk.

Phot. A. G. Tansley

a. Typical ground vegetation of beechwood in better lighted spots. *Mercurialis perennis, Fragaria vesca, Viola hirta, Viola Riviniana.*

Phot. W. M. Rankin

b. Degenerating beechwood on chalk. *Fagus sylvatica, Pyrus Aria, Taxus baccata, Cratægus monogyna.* Chalk grassland and bases of foliage of trees and shrubs nibbled close by rabbits. Holt Down, near Butser Hill, Hampshire.

Beechwood on Chalk.

Distribution.

Ashwoods characterise the south-western extremities of the chalk outcrop, lying in the Isle of Wight, Dorset and East Devon. The line of contact between the ashwood and the beechwood regions occurs near Butser Hill north of Portsmouth, where the South Downs attain their greatest elevation (889 feet = c. 270 m.). Ashwoods are also developed on those portions of the Upper Greensand (which often forms a low terrace at the foot of the chalk scarp) that are highly calcareous, as in parts of the western Weald, in the Isle of Wight, Dorset and East Devon. Within the beechwood regions, ashwoods scarcely occur on the chalk.

The factors determining the distribution of beechwood and ashwood on the chalk are by no means clear. One factor may be atmospheric and soil moisture. The escarpments of the south-western chalk, such as the Isle of Wight and Purbeck ridges, as well as the heights of Butser Hill, receive a higher rainfall than any other part of the English chalk, while the more clayey and therefore moister nature of the Lower Chalk and somewhat similar Upper Greensand, which carry most of the ashwoods of the Isle of Wight, is another difference which may favour ashwoods rather than beechwoods.

Competition between beech and ash.

The competition between beech and ash is an important factor in this question. Where the two trees are placed in direct competition on equally suitable soils the beech must eventually suppress the ash because of the deeper shade which it casts and its correlated lower demand for light. It may be that the beech, advancing westwards from the continent, across what are now the Straits of Dover, drove the ash, so to speak, before it and came to occupy those soils of south-eastern England previously in possession of the ash; and that this tide of invasion was checked, owing to climatic or other factors somewhere about the line marking the present western and northern limits of the chalk beechwoods.

The equipment of the two trees for such a struggle is very different. The beech, like its relative the oak, is a heavy-seeded tree, and its wide dispersal must depend on the activity of birds and other animals. Many existing beechwoods show signs of failure to regenerate themselves, and this, as in the case of oakwoods, is often attributed to the disappearance from the woods of the herds of swine which used to feed on the mast and acorns, and incidentally keep the soil open and sow the seeds by trampling them into the ground. The greater production of seed by the ash and its dispersal by the wind clearly gives this species an advantage over the beech in colonisation, and *pro tanto* will favour the production of ashwood rather than beechwood on soils equally favourable for both species. On the other hand, given successful colonisation of an ashwood by the beech, the latter will inevitably beat the former in competition because of the deep shade cast by the beech foliage, so that eventually most of the ashes will be starved and the wood converted into a beechwood. On the borders of the beechwood region in East Hampshire this process may in places be observed. Beechwood has been extensively felled, but enough parent trees were left to furnish a supply of self-sown seedlings, and the ash which has sprung up since the beech was felled is here and there being overshadowed and killed by these self-sown beech seedlings as they grow up. In other places, however, in the same district, the beech is not coming back where it has been cleared, and the ground is occupied either by mixed chalk scrub or mainly by ash. The failure of the beech to regenerate and its replacement by the ash is a parallel phenomenon to the replacement of oak by birch (see pp. 102, 141); and it may be stated as a generalisation that extensive clearing tends to handicap the heavy-seeded trees and to lead to their replacement by light-seeded ones which succeed on the same type of soil.

In many cases, also, the abundance of rabbits on the light chalk soil must tell more heavily against the species which produces fewer and less widely distributed seedlings of slower growth. The destructive effect of rabbits is sometimes extreme, as may be well observed on many chalk downs, where the trees and bushes are uniformly eaten close, up to a height that a rabbit can reach, while no tree seedling and scarcely a herbaceous plant can escape destruction, except the perennial constituents of the turf, which are nibbled close to the soil (Plate XVIII b). No woodland has the slightest chance of regenerating itself under such conditions, but where the incidence of this factor is rather less severe the chances will probably be in favour of the ash as opposed to the beech. The more abundant and lighter seeds of the ash will produce a more numerous and widespread crop of seedlings than those of the beech, which will largely germinate where they fall, under the edges of the tree canopy. In this position, where the ground is relatively bare owing to the deep shade of the parent tree, they will easily be seen by rabbits. The ash seedlings, on the other hand, are carried far and wide and frequently germinate in scrub or among tall grass, etc., where their earlier life is to some extent protected.

Thus with a disturbance of primitive conditions a number of factors are seen to tell against the heavy-seeded tree, in spite of its primary advantage of being able to bear shade, to which no doubt it owed its original success.

Yew-woods. The yew (*Taxus baccata*), as already mentioned, is often abundant in beechwoods, where it grows entirely shaded by the beech canopy; it is also found very commonly in the chalk scrub, and sometimes isolated yew shrubs occur in the chalk pasture, where seedlings in various stages of growth are often met with. In places, notably on the borders of Sussex and Hampshire, considerable aggregations of yews are found forming small groves, or in some cases yew-woods of large extent (Plate XIX). These yew-woods are nearly pure, the whitebeam (*Pyrus Aria*) being the only tree at all commonly associated. The ground vegetation is extremely poor, much poorer even than that of the beechwoods, and the soil beneath the shade of the trees is typically bare.

Origin of yew-woods.

The origin of these yew-woods is not clear. They occupy situations quite similar to those of the typical beechwoods, *i.e.* the fairly steep valley sides of the chalk downs, and in some cases the bottoms of dry chalk valleys. It is suggested that they may originate in the following way. If a beechwood possessing numerous yews is destroyed by total felling of the dominant trees, or if it degenerates owing to various causes, chalk scrub and chalk grassland in which yews are prominent will take its place. The yews can and do sow their seeds and produce seedlings freely under such conditions, and the young trees, probably owing to their poisonous foliage, are largely, if not wholly, immune from the attacks of rabbits and other animals. As soon as a locally closed community of yews is formed, competitors are almost entirely excluded, owing to the very deep shade cast by the trees. In this way local yew groves might be formed and an extension of the process might lead to the formation of extensive, nearly pure, yew-woods, since the intervening scrub and grass association would be gradually destroyed by the extension of the yew. The process would no doubt be slow, owing to the slow growth of the trees. On the other hand it would be very sure, because the equipment of the yew for this kind of competition is overwhelming. Nothing but the formation of a very dense scrub of other species or

Phot. A. G. Tansley

a. Yew wood on chalk escarpment. *Taxus baccata, Pyrus Aria, Cratægus monogyna* in foreground.

Phot. A. G. Tansley

b. The same from above. *Juniperus communis* in foreground. Heath on "clay-with-flints" on left with *Calluna vulgaris* and *Ulex nanus.*

Yew wood on Chalk. Kingley Vale, Sussex.

Phot. W. M. Rankin

a. *Taxus baccata*, *Cratægus monogyna*, *Cornus sanguinea*, *Prunus spinosa*, *Rhamnus catharticus*, etc. Chalk pasture with *Senecio Jacobæa* in the foreground. Butser Hill, Hampshire.

Phot. S. Mangham

b. *Fagus sylvatica*, *Pyrus Aria*, *Taxus baccata*, *Cratægus monogyna*, etc. Chalk escarpment near Kemsing, Kent.

Chalk scrub association.

the felling of the yew trees would be likely to arrest it.

The explanation given is in accordance with the observed facts, but at present it lacks corroboration or direct verification. It is possible that other factors are operative, and it may be that the habitat of the yew-wood really differs from that of the beechwood, though we have no clue to such a difference.

Scrub association. On the outskirts of beechwoods and scattered on the chalk generally, but more especially occurring in old chalk pits, by the sides of trackways, and often left to form natural hedges, *e.g.* between the cultivated ground and the open chalk pasture, a very characteristic scrub occurs, consisting of a considerable number of species of shrubs, many of which are found chiefly or almost exclusively on calcareous soils (Plate XX).

It is noteworthy that some of the most abundant species of this scrub are also the dominant shrubs of the clays and loams and of the sands, sandstones and shales, *e.g. Cratægus monogyna* and *Prunus spinosa.* On calcareous soils, however, many other species, which are rare or almost absent on the non-calcareous soils, are abundantly associated.

A characteristic feature of the chalk is the abundance of species of *Rosa.* The fruticose Rubi, on the other hand, are neither so abundant nor so rich in species as on the sand. The Juniper (*J. communis*) is abundant in places and forms a most characteristic open scrub of stiff erect bushes almost black in colour and often several feet high.

The following shrubs and climbers occur:—

ca Clematis Vitalba	a	*Acer campestre	a
Ilex Aquifolium	la	Prunus spinosa	a
ca Euonymus europæus	la	P. insititia	o
ca Rhamnus catharticus	la	P. Cerasus	o

Rubus *spp.* (fruticosus)		Viburnum Opulus	f
R. cæsius	la	ca V. Lantana	a
* Rosa pimpinellifolia	l	* Bryonia dioica	la
* R. Eglanteria		Lonicera Periclymenum	f
(rubiginosa)	la	* Ligustrum vulgare	la
R. micrantha	f	Solanum Dulcamara	f
ca R. agrestis (sepium)	r	S. nigrum	o
R. stylosa *var.* systyla	o	ca Atropa Belladonna	la
R. canina	a	Daphne Mezereum	vr
R. arvensis	a	ca D. Laureola	la
ca Pyrus Aria	a	ca Buxus sempervirens	o
Cratægus monogyna	a to d	Corylus Avellana	la
Hedera Helix	f	Salix caprea	f
* Cornus sanguinea	a	* Ruscus aculeatus	f
* Sambucus nigra	a	* Juniperus communis	la

In the above list *ca* is prefixed to the species which may fairly be described as "calcicole," at any rate in the chalk districts, though none is absolutely confined to the chalk. An asterisk is prefixed to other species which are specially abundant on the chalk.

The chalk scrub often grows exceedingly thick and then excludes almost all ground species; commonly, however, it shelters many woodland species. Thus *Mercurialis perennis* often occurs in sheets under its shade. Among these woodland species are several of the characteristic chalk orchids, such as *Cephalanthera longifolia*, *C. grandiflora*, *Helleborine atrorubens*, *H. violacea*, *Ophrys muscifera*, *Orchis purpurea*, *O. militaris*, *O. Simia*, *O. hircina*, the last three of which are extremely rare. The orchids of chalk grassland are given on p. 178.

Trees of the beech, the whitebeam, the yew and the ash frequently occur in this scrub; the relation to beechwood is often clear, though it differs somewhat from the relation of scrub and woodland in the case of the formations of clays, loams and sands. This is owing to the fact that while the shrub flora of calcareous soils is richer than that of siliceous soils, the shade of the beechwood is so deep as to exclude almost all shrubby undergrowth. Hence while the scrub association of clay or sand is commonly poorer in species than the shrub layer of the corresponding woodland, the scrub association of calcareous soils is much richer and develops largely on its own account in suitable situations. Sometimes this scrub can be clearly recognised as a progressive association which has colonised open chalky soil and in which such trees as the ash may spring up freely. In such cases an ashwood may develop, with its undergrowth consisting of the shrubs of the chalk scrub and a fairly rich ground vegetation. Whether in these cases the beech normally replaces the ash in succession is doubtful. It is probably generally prevented from doing so by the want of enough adult trees to furnish an adequate supply of seed.

Chalk grassland association [Chalk pasture] (*Festucetum ovinæ calcareum*).

This association, which is the typical association of the South Downs[1], and has a very wide extension in the southern counties of England, is closely similar to the limestone grassland previously described. It has the same general dominant, the sheep's fescue (*Festuca ovina*), and a large proportion of the associated species are also the same, but a number of species confined to the south (and particularly the south-east) of England occur in the chalk pasture only. Swamps appear to be entirely absent.

History of chalk pasture.

There is good reason to suppose, as we have already pointed out, that much of the chalk pasture is extremely old, and much of its area has possibly never been occupied by a tree association—perhaps because of an inadequate supply of underground water. The chalk grassland, which forms a very excellent light crisp pasture, has from time immemorial supported considerable flocks of sheep. The smooth curves of the chalk downs are occasionally broken by ancient trackways, camps and

[1] Sussex, Hampshire, Dorset and Wiltshire. The North Downs of Surrey and Kent have much less.

other earthworks of many periods from the Neolithic onwards. It has been suggested[1] that the original purpose of many of these works was to shelter and defend the flocks from the attacks of predatory animals such as wolves, coming from the forests of the lower country. Be that as it may, it seems unlikely that primitive man was responsible for the disforestation of such great areas of the chalk upland as are marked by traces of his existence, and the conclusion is therefore indicated that much of this grassland is primitive, or at least has existed since the conditions of climate resembled at all closely those at present obtaining. There may well have been originally more scrub than there is now.

In addition to this possibly primeval grassland, occupying much of the rolling summits of the downs, there is much pasture on the slopes of the escarpments and valleys, and this has probably taken the place of beechwood and ashwood destroyed by man.

Characters of chalk pasture.

The chalk grassland is almost everywhere used for sheep pasture, and where the grazing is heavy the herbage is eaten very close. When rabbits are abundant the turf is even more closely nibbled, and scarcely a herbaceous plant is able to rise more than an inch or so above the surface of the soil (Plate XVIII b). In places freer from this constant grazing and nibbling the vegetation is able to develop much more vigorously, and a rich and varied flora occurs (Plate XX b).

The turf is springy, being formed of a close mat of wiry herbage, often easily separable from the substratum. It is made up of a multitude of species, among which *Festuca ovina* is typically, but by no means invariably, dominant. Besides the grasses and less abundant sedges there are many dicotyledonous species, and the association

[1] By the Messrs Hubbard.

varies from place to place, partly owing to minor variations of the depth and nature of the soil, partly to the incidence of exposure and of grazing.

On the North Downs of Surrey and Kent the grassland is more frequently mown, and in place of the sheep pasture dominated by *Festuca ovina*, a mixed grass association is present in which *Bromus erectus*, *Trisetum flavescens* or *Brachypodium gracile* (*sylvaticum*) are frequently dominant.

Soil.

The soil of the chalk grassland is typically very shallow, often not exceeding an inch (2·5 cm.) in depth. If the whole depth of soil, down to the weathering rock, be taken together, estimations show a very high percentage of calcium carbonate; but the upper layers which accompany the greater part of the intricate network of roots and rhizomes of which the turf is composed, are typically extremely poor in lime. Corresponding with this very different nature of the surface layer from that of the underlying soil, two groups of plants, whose root systems are hardly competitive, may be distinguished [1].

The lower, calcareous, soil is tenanted only by roots of plants such as the rock-rose (*Helianthemum Chamæcistus*), the salad-burnet (*Poterium Sanguisorba*), the squinancy-wort (*Asperula cynanchica*), by the tubers of orchids, and generally by the underground systems of most of the plants which may be called calcicole. The upper non-calcareous layer, on the other hand, which is richer in humus but more exposed to drought, is occupied by the underground parts of shallow-rooted plants such as *Festuca ovina*, which occur in all dry grassland.

The "chalk heath."

In the non-calcareous upper layer actual heath plants such as *Calluna vulgaris* and *Potentilla erecta* sometimes occur. The mixture of these with deeper rooting calcicoles gives

[1] This is an example of a type of association which has been distinguished by Woodhead as *complementary* (Woodhead, 1905, p. 345).

rise to a curious sub-association which may be called *chalk-heath*, parallel with the limestone heath (p. 158).

Very characteristic of chalk pasture are numerous species of Orchidaceæ, some of which are extremely rare.

The exposure of the chalk pasture association to the atmosphere and to radiation is extreme. There is little or no shelter from the full effect of sunlight, wind, rain and radiation. The very thin soil is able to retain very little water, such aerial water as is not quickly evaporated rapidly passing down into the chalk below. On some chalk summits and steep slopes the soil almost disappears owing to its "creep" to lower levels, and the vegetation becomes open. During protracted summer droughts much of the herbage commonly withers, but the deeper rooting plants are to some extent protected by thick underground organs and sometimes by xerophilous adaptations.

Climatic conditions.

The following is a list of species occurring in chalk pasture. The species clearly belonging to the heath association, sometimes developed on the non-calcareous soil overlying the chalk, are excluded :—

Dominant—Festuca ovina.

	Thalictrum minus	o	ca	S. nutans	la
ca	Anemone Pulsatilla	l		Cerastium arvense	lf
	Ranunculus acris	f		C. triviale	a
	R. bulbosus	f		Arenaria serpyllifolia	f
ca	Reseda lutea	f		A. leptoclados	f
	R. Luteola	f		Linum catharticum	a
ca	Helianthemum Chamæcistus	f to a	ca	L. perenne	vl
				L. angustifolium	o
ca	Viola hirta	a		Ononis spinosa	o
ca	V. calcarea	r		Medicago lupulina	f
	Polygala vulgaris	a		Trifolium procumbens	f
	P. oxyptera	l		T. minus	f
ca	P. calcarea	la		T. filiforme	f
ca	P. austriaca	vr		T. pratense	a
	Dianthus deltoides	r	ca	Anthyllis Vulneraria	a
ca	Silene latifolia (= Cucubalus)	f		Lotus corniculatus	a
			ca	Astragalus danicus	l

ca	Hippocrepis comosa	a
ca	Onobrychis viciæfolia	f
	Vicia Cracca	f
ca	Spiræa Filipendula	lf
	Fragaria vesca	f
ca	Poterium Sanguisorba	a
	Sedum acre	f
	Pimpinella Saxifraga	a
ca	Seseli Libanotis	r
	Daucus Carota	a
	Galium verum	f
ca	Asperula cynanchica	f
ca	Scabiosa Columbaria	f
	S. arvensis	f
	Bellis perennis	a
	Erigeron acre	l
	Filago germanica	f
ca	Inula Conyza	a
	Achillea millifolium	a
	Chrysanthemum Leucanthemum	f
ca	Senecio integrifolius (campestris)	r
	S. erucifolius	f
	S. Jacobæa	a
*	Carlina vulgaris	a
*	Carduus nutans	la
*	C. crispus	la
	Cnicus lanceolatus	a
ca	C. acaulis	a
	Centaurea nigra	a
ca	C. Scabiosa	a
*	Cichorium Intybus	o
	Picris hieracioides	f
	P. echioides	f
*	Crepis taraxacifolia	f
	C. fœtida	r
	C. capillaris (virens)	f
ca	C. biennis	o
	Hieracium Pilosella	a
	Hypochæris radicata	a
	Leontodon hispidum	f
	L. nudicaule	f

	L. autumnale	a
	Taraxacum officinale	a
	Tragopogon pratense (agg.)	f
ca	Phyteuma orbiculare	vl
ca	Campanula glomerata	la
	C. rotundifolia	f
	Primula veris	la
ca	Blackstonia perfoliata	a
	Centaurium umbellatum (=Erythræa Centaurium)	a
	C. capitatum	r
*	Gentiana Amarella	f
	Myosotis arvensis	f
	M. collina	f
*	Echium vulgare	la
*	Verbascum Thapsus	f
*	V. Lychnitis	o
*	V. nigrum	o
	Linaria vulgaris	f
	Euphrasia officinalis (agg.)	a
	Orobanche elatior (on Centaurea Scabiosa)	l
	O. Picridis	r
ca	Origanum vulgare	f
*	Thymus Serpyllum	a
*	T. Chamædrys	? f
*	Clinopodium vulgare	f
*	Calamintha Acinos	f
*	C. montana (officinalis)	f
	Salvia Verbenaca	o
	Nepeta hederacea	f
	Prunella vulgaris	a
ca	Teucrium Botrys	vr
	T. Scorodonia	la
ca	Ajuga Chamæpitys	r
ca	Plantago media	a
	P. lanceolata	f
ca	Thesium humifusum	o
	Listera ovata	o
	Spiranthes spiralis (autumnalis)	f

ca Orchis pyramidalis	la	ca Trisetum flavescens	lsd
ca O. ustulata	l	ca Avena pubescens	f
* O. Morio	f	A. pratensis	o
O. maculata	a	Arrhenatherum elatius	f
ca Aceras anthropophora	la	Cynosurus cristatus	f
ca Ophrys apifera	la	ca Koeleria gracilis (cristata)	f
ca O. fuciflora (arachnites)	vr	Dactylis glomerata	f
ca O. sphegodes (aranifera)	lf	* Briza media	la
ca Herminium Monorchis	o	* Festuca rigida	f
* Habenaria conopsea	la	F. elatior	f
H. albida	vr	ca Bromus erectus	la
* H. viridis	r	B. mollis	o
ca Iris fœtidissima	l	Brachypodium gracile (sylvaticum)	ld
Luzula campestris	a	B. pinnatum	o
Carex flacca (glauca)	a	Lolium perenne	f
C. verna (præcox)	f	Ophioglossum vulgatum	o
C. humilis (clandestina)	r	Phegopteris Robertiana (calcarea)	vr
Phleum pratense	f		
Agrostis vulgaris	o		

ca = calcicole * specially abundant on chalk

The following is a hypothetical scheme of the relationships of the associations of the chalk sub-formation :—

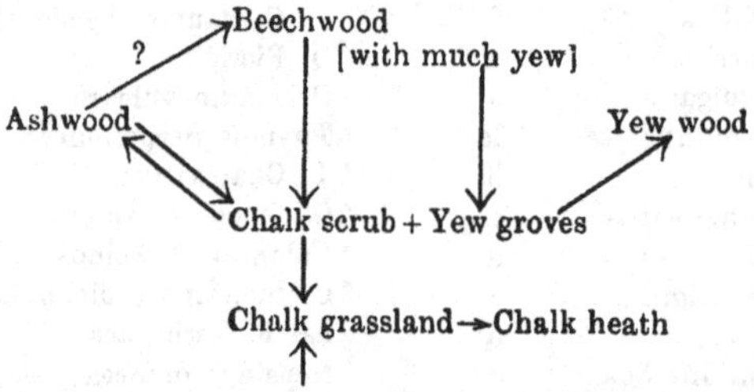

It will be useful to add some results of a detailed study of the composition of a small area of the chalk grassland association on Fleam Dyke, Cambridgeshire[1]

Fleam Dyke is one of a series of very ancient military earthworks occurring in south-east Cambridgeshire and extending in a south-easterly direction from the Fenland to the chalk uplands, which are largely

[1] The results of this study, which was carried out by the Marshall Ward Society, of the University of Cambridge, were very kindly placed at the Editor's disposal by Mr R. H. Compton, late Secretary of the Society.

covered by chalky boulder clay. These earthworks cross, approximately at right angles, the line of the early pre-Roman road called Icknield Way, which follows the line of the chalk outcrop between the boulder clay country (which was no doubt once covered with thick forest) to the south-east, and the (originally) impassable Fenland to the north-west; and they are supposed to have been thrown up to facilitate the obstruction of the road to the passage of hostile forces.

Fleam Dyke, in its least altered portions, is a narrow and steep-sided bank of solid chalk, its summit lying about 12 feet above the general level of the surrounding country. On its south-west side is the ditch from which the chalk was dug. The surrounding country is almost entirely arable, but for a length of about two miles the dyke has been left almost untouched, though here and there pines and birches have been planted. The south-west side of the dyke bears a typical chalk grassland association; on the north-west side a good deal of bare chalk is exposed and the vegetation is taller though composed of few individual plants. At the bottom of the ditch are numerous rabbit burrows.

The following is a list of the species which occur in the chalk grassland association of the dyke.

Festuca ovina d.

Thalictrum minus	r	Anthyllis Vulneraria	a
[Mahonia vulgaris]	r	Lotus corniculatus	a
[Papaver Rhæas]	r	Astragalus danicus	la
[P. hybridum]	vr	Hippocrepis comosa	va
[Fumaria officinalis]	r	Onobrychis viciæfolia	o
[Brassica nigra]	r	Vicia Cracca	r
Reseda lutea	o	Prunus spinosa	vr
Helianthemum Chamæcistus	va	*Spiræa Filipendula	la
		Poterium Sanguisorba	ls
Viola hirta	o	Rosa canina (agg.)	o
Polygala vulgaris	a	R. Eglanteria (rubiginosa)	r
[Lychnis alba]	r	Pyrus Malus	vr
Silene latifolia (Cucubalus)	o	Cratægus monogyna	la
Cerastium vulgatum	o	Pimpinella Saxifraga	la
[Stellaria media]	o	Peucedanum sativum	la
Linum catharticum	la	Daucus Carota	la
Rhamnus catharticus	vr	Viburnum Lantana	vr
[Acer Pseudoplatanus]	vr	Galium verum	a
Ulex europæus	r	Asperula cynanchica	a
Ononis spinosa	r	Scabiosa Columbaria	o
Medicago lupulina	a	S. arvensis	l
Trifolium procumbens	o	Erigeron acre	r
T. pratense	o	Achillea Millefolium	o

*Chrysanthemum Leucanthemum	o	Thesium humifusum	vr
Senecio Jacobæa	o	[Urtica dioica]	l
S. integrifolius (campestris)	o	[Betula alba]	l
Carlina vulgaris	l	[B. tomentosa]	l
Cnicus acaulis	o	Fagus sylvatica	vr
[C. arvensis]	o	Orchis pyramidalis	r
Centaurea nigra	a	O. maculata	vr
C. Scabiosa	a	Carex caryophyllea (= C. verna)	o
Picris hieracioides	r	C. flacca (glauca)	a
Crepis capillaris (virens)	o	Phleum pratense	o
Hieracium Pilosella	va	Agrostis tenuis (vulgaris)	l
Hypochæris radicata	r	Avena pratensis	ls
Leontodon hispidum	o	[A. fatua]	r
Sonchus arvensis	o	*Arrhenatherum elatius (avenaceum)	l
Tragopogon pratense	r	Koeleria gracilis (cristata)	a
Campanula glomerata	o	*Dactylis glomerata	l
C. rotundifolia	o	Briza media	va
*Primula veris	o	†Festuca elatior	l
[Anagallis arvensis]	r	Brachypodium pinnatum	l
Gentiana Amarella	r	[Picea vulgaris]	r
Echium vulgare	l	[Pinus sylvestris]	la
Linaria vulgaris	r	Juniperus communis	r
L. minor	r	Taxus baccata	vr
Euphrasia officinalis (agg.)	o	*Brachythecium purum	la
Thymus Serpyllum	a	*Hypnum molluscum	la
Calamintha Acinos	r	*Thuidium abietinum	la
Prunella vulgaris	o	*Fissidens taxifolius	la
*Plantago media	l	etc.	
*P. lanceolata	a		
[Fagopyrum sagittatum]	vr		

The species marked * occur mainly or exclusively in the ditch where the soil is deeper and not so dry. † under trees. The names enclosed in square brackets are those of species clearly alien to the vegetation of the dyke, whether native to the country or not.

The following species were present in one square metre of the association examined in detail. The order is that of decreasing frequency.

Festuca ovina: dominant, in considerable tufts, largely excluding other plants.
Poterium Sanguisorba: co-dominant with the last.
Hieracium Pilosella: very abundant, especially on patches of bare chalk, which it readily colonises.

Carex flacca: subdominant.
Helianthemum Chamæcistus: subdominant.
Briza media: in fair quantity.

Koeleria gracilis	Plantago lanceolata
Avena pratensis	Scabiosa Columbaria
Thymus Serpyllum	Pimpinella Saxifraga
Hippocrepis comosa	Centaurea nigra
Lotus corniculatus	Linum catharticum
Leontodon hispidum	Campanula rotundifolia
Asperula cynanchica	Euphrasia officinalis
Galium verum	Anthyllis Vulneraria
Cnicus acaulis	Carlina vulgaris
Plantago media	Pinus sylvestris (one seedling)

THE SUBFORMATION OF MARLS AND CALCAREOUS SANDSTONES

The vegetation to be described as belonging to this sub-formation occurs on calcareous soils derived from rocks containing a smaller proportion of lime than the limestones. It is therefore intermediate in character between limestone vegetation on the one hand and the vegetation of the non-calcareous rocks on the other. Owing to the presence of calcicole species by which this vegetation is characterised, it is treated as a sub-formation of the formation of calcareous soils, but naturally all transitions occur to the formations of non-calcareous soils.

Ash-Oakwood association.

General characters.

The woodland association of this sub-formation is characterised by the constant presence of the ash in considerable quantity and is thus related to the typical ashwood association of limestone soils. On the other hand oaks are always present, not only as occasional constituents, as in ashwood, but typically co-dominant with the ash. On marls and calcareous clays the oak is always *Q. Robur*, but on some, though not all calcareous sandstones, *Q. sessiliflora* occurs, either alone or mixed with *Q. Robur*.

The general ecological conditions are in many respects closely similar to those of the damp oakwood, but the floristic composition usually shows greater variety.

Shrubs.

The shrubs of the ash-oakwood are as varied as those of the ashwood and consist of most of the same species. The hazel (*Corylus Avellana*) is commonly the dominant shrub, and the flora includes such characteristic calcicole plants as *Viburnum Lantana*, *Euonymus europæus* and often *Clematis Vitalba*. Such species as *Cornus sanguinea*, *Ligustrum vulgare* and *Acer campestre* are generally more abundant in individuals than in the damp oakwood.

Ground vegetation.

Very much the same is true of the herbaceous ground flora. "The general list is mainly the same as that of the damp oakwood, but such plants as *Carex digitata*, *Paris quadrifolia*, *Colchicum autumnale*, *Iris fœtidissima*, *Helleborine media*, *H. purpurata*, *Viola sylvestris*, *Primula elatior* (an East Anglian species), *Lithospermum purpureo-cæruleum* (a south-western form), and *Campanula Trachelium*, which are almost or quite absent from woods of the oak type, are characteristic of the ash-oak association. Dog's mercury (*Mercurialis perennis*) is much more generally dominant in the ground vegetation, and [various] other species, such as *Hypericum hirsutum*...are more generally abundant[1]."

Ash-oak-hazel copse.

Some ash-oakwoods (*e.g.* many of those on the calcareous Upper Greensand of East Devon and of the Isle of Wight) have been left in an almost natural condition, and the close relationship of these to the ashwoods, into which they sometimes pass imperceptibly, is obvious enough. But the great majority of the ash-oakwoods of the southern half of England are treated as coppice with standards, the oaks being left as standard trees, while the ashes and

[1] Moss, Rankin and Tansley, 1910, p. 139.

practically all the other trees, as well as the various shrubs, are coppiced. Hazel is generally the dominant shrub in such coppice, and in many cases gaps are filled up with planted hazel. In this way the wood comes to resemble superficially the oak-hazel copse derived from damp oak-wood. It is only by noting the abundance of ash in the coppice and the variety and abundance of shrubs other than the hazel, as well as the presence of calcicole species in the ground vegetation, that the wood can be assigned to its proper type (Figs. 6 and 7).

Another factor tending to the assimilation of the oak-hazel copses derived respectively from the ash-oakwood and from the damp oakwood is the frequent poverty in lime of the surface soil of the former to a depth of some inches, probably due to leaching. This, with the accumulation of humus, leads to the establishment on the floor of the wood of shallow-rooting plants not found on soils rich in lime. The more deeply-rooting plants, therefore, such as the shrubs, are often a better index than the herbaceous species of the essential character of these woods.

Oak-hazel woods of the type described are well developed on the Triassic and Jurassic marls of Somerset and on the chalky boulder-clay of Cambridgeshire. They have been described by Moss[1] and by Adamson[2] respectively, to whose memoirs the reader is referred for lists of species. Adamson has also studied in some detail the ecological conditions determining the occurrence of different sub-associations of the ground vegetation.

The ash-oakwoods on calcareous sandstones and sandy limestones have not been described, but so far as they have been investigated they show similar intermediate

[1] Moss, 1907, pp. 51—56. These woods are described by Moss simply as "oak-hazel woods"; they certainly belong to the ash-oakwood association, an association which has since been recognised. See Moss, Rankin and Tansley, 1910, p. 138.

[2] Adamson, 1911.

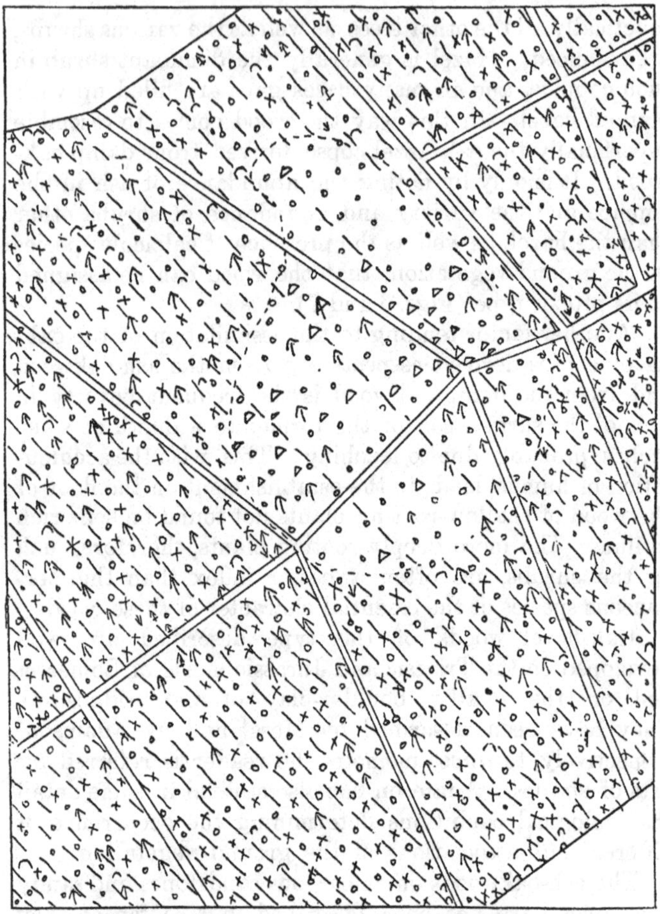

Scale ⅛ *mile*

EXPLANATION OF SYMBOLS

○ Quercus Robur	♁ Salix cinerea
× Fraxinus excelsior	↓ Viburnum Opulus
\\ Corylus Avellana	⇑ V. Lantana
⌒ Acer campestre	△ Betula *spp.*
↑ Populus tremula *var.* villosa	T = Pyrus torminalis

FIG. 6. Chart of the trees and shrubs of part of Gamlingay wood, Cambs., showing an island of the dry oakwood association (on sandy loam) enclosed within Ash-oak-hazel wood (on calcareous boulder-clay).

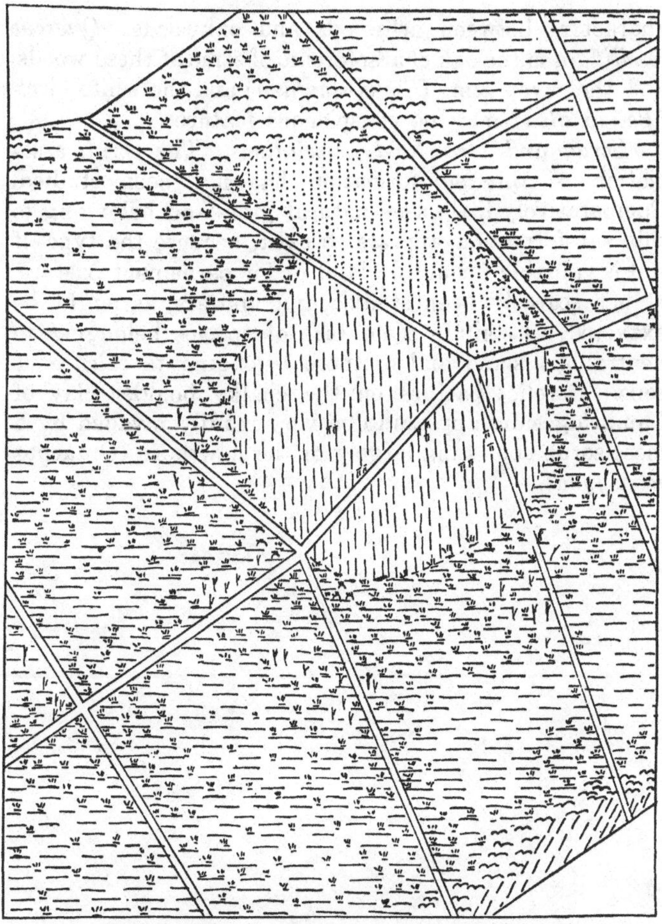

Scale $\frac{1}{8}$ mile

EXPLANATION OF SYMBOLS

Spiræa Ulmaria society

Spiræa-Deschampsia-cæspitosa society

Fragaria vesca society

Mercurialis perennis society

Pteris aquilina society

Holcus mollis society

Carex spp.

Juncus spp.

Polytrichum gracile

FIG. 7. Chart of the societies of the ground vegetation of part of Gamlingay wood, Cambs. (see Fig. 6).

characters between ashwoods and oakwoods. *Quercus sessiliflora* is the oak characteristic of some of these woods, and the wych elm (*Ulmus montana*) and the white birch (*Betula alba*) may be conspicuous features.

Scrub and Grassland associations. The scrub association of this sub-formation has not been studied. Comparatively little semi-natural grassland belonging to this sub-formation appears to exist, because the type of soil is valuable for arable land or for permanent pasture. It has scarcely been studied, but appears, as might be expected, to contain a mixture of species belonging to neutral grassland and of those belonging to calcareous soils. Derelict pasture on the chalky boulder clay of Cambridgeshire appears to be very readily invaded by a loose scrub consisting largely of the hawthorn (*Cratægus monogyna*).

CHAPTER VII

AQUATIC VEGETATION[1]

COMPREHENSIVE studies of aquatic vegetation, on the lines followed in this book, have not yet been carried out in this country; and consequently the material for a systematic treatment of the vegetation scarcely exists. For the sake of completeness the subject has not been omitted altogether, but the following pages must be regarded rather as suggestive of the lines along which future work may proceed, than as expressing a satisfactorily established scheme of classification[2].

Fresh and saltwater.

The first and most obvious division of aquatic vegetation is into the vegetation of saltwater and the vegetation of freshwater, for the presence of dissolved salt in any proportion near that of seawater is well known to change completely the aquatic flora. The vegetation of the sea is entirely omitted from this book, though the "maritime" land communities inhabiting the coasts and coming within the zone of direct influence of the sea are dealt with in a later chapter.

In considering the vegetation of freshwaters we meet with very wide differences in habitat, but nevertheless it

[1] The aquatic vegetation of the Norfolk Broads is dealt with in Chapter X, on account of its close relation to the vegetation of the fen formation.

[2] For some of the ideas contained in this chapter the Editor is indebted to Dr C. E. Moss.

seems clear that the differentiating or master factor which separates aquatic from all other habitats is the presence of the water itself, that is of free water constantly surrounding the whole or a very considerable portion of the plant-body. Consequently it is proposed to consider all freshwater vegetation as belonging to one formation.

The Freshwater Aquatic Formation

Differentiating factors.

The freshwater aquatic formation includes the plant life of all freshwater aquatic habitats, such as rivers, lakes, ponds, pools and ditches; and these cover so wide a variety of conditions of life, that it must be a matter of some difficulty to arrive at a satisfactory classification. It seems probable, however, that two factors, the aeration of the water, and the amount and nature of the dissolved mineral salts and other substances it contains, are of leading importance as differentiating factors, though our knowledge is insufficient to do more than suggest, very tentatively, the way in which these factors should be used in classification.

1. *The sub-formation of foul waters.*

In very stagnant pools containing an abundance of decaying organic matter and with deficient aeration, the vegetation is practically confined to the lower forms of plant life, and mainly to bacteria, with some Cyanophyceæ. The conditions render it impossible for plants requiring a supply of free oxygen for respiration to exist in such a habitat, and probably the substances formed by the activity of the anaerobic bacteria also inhibit the growth of the higher plants. The vegetation of these foul waters has not been studied ecologically, but it is clear that it belongs to a separate division of aquatic vegetation.

Of the waters which are not markedly foul the best primary division is probably based on the amount of nutritive mineral salts contained, though this division is founded at present on inference rather than on actual determinations.

Amount of dissolved salts.

The waters of most lowland lakes and rivers are more or less rich in dissolved mineral salts, which pass into solution as the water travels through various rocks and soils. On the other hand the lakes of mountain regions, like the soils of the surrounding country, are frequently poor in mineral salts.

The vegetation of the former waters is much richer in species than that of the latter, while the vegetation of mountain lakes possesses some species not found in the richer waters of the lowlands.

With regard to the associations into which the sub-formations of aquatic vegetation should be divided, those proposed by various authors are not, for reasons which need not be entered into here, altogether satisfactory, nor is our knowledge of the sub-formations, as developed in this country, sufficiently advanced to enable us to propose new association-names.

Zonation of associations.

In the case of a sufficiently large pond, lake, river or even canal, possessing a well-developed aquatic vegetation, it is clear, however, that the associations are zonally arranged, and in the case of a lake of any size we may distinguish a *plankton association* consisting of microscopic organisms floating freely in the middle of the lake[1] from the *associations of higher plants and algæ* occupying the margin, and sometimes, when the lake is not too deep, covering its entire floor. These associations we do not at present attempt to classify any further than is involved in separating the submerged forms and the forms with floating leaves from the *reed-swamp association* occupying the shore line. This last we include in the aquatic

[1] See pp. 196–203.

formation on grounds explained in Chapter X (p. 223). The rich algal vegetation of many of these habitats is of course very important. Part of this belongs to separate associations, the other part should no doubt be included in the various associations of higher plants with which the species of algæ are associated. The work of properly classifying these associations has not yet been undertaken on a comprehensive scale by phycologists.

2. *The sub-formation of waters relatively rich in mineral salts.*

(*a*) *Nearly stagnant waters.*

In the nearly stagnant waters of lowland ponds, ditches, the backwaters of rivers and some parts of the shallow lakes called the Norfolk Broads[1], a fairly rich vegetation of plants with submerged and floating leaves occurs, among which are the following species:—

Submerged or nearly submerged plants.

Elatine hexandra	o	U. intermedia	r
E. Hydropiper	vr	U. minor	o
Hippuris vulgaris	f	Ceratophyllum submersum	o
Myriophyllum verticillatum	o	C. demersum	o
M. alterniflorum	f	Elodea canadensis	f
M. spicatum	f	Stratiotes Aloides	o
Callitriche *spp.*	f	Potamogeton coloratus	o
Hottonia palustris	f	P. plantagineus	o
Utricularia vulgaris	o	etc.	

Plants with floating leaves.

Ranunculus peltatus, etc.	f	L. minor	a
Nymphæa lutea	f	L. gibba	o
Castalia alba	o	L. polyrhiza	o
Polygonum amphibium	f	Wolffia arrhiza	r
Hydrocharis Morsus-ranæ	o	Potamogeton natans	f
Sagittaria sagittifolia	f	Glyceria fluitans	f
Lemna trisulca	f		

[1] See Chapter X.

Reed-swamp association. In the lowlands a reed-swamp association, in which the common reed *Phragmites vulgaris* is often dominant, is characteristic of nearly stagnant as opposed to slowly flowing waters. Associated are:—

Lythrum Salicaria	f	Lysimachia vulgaris	f
Epilobium hirsutum	f	Symphytum officinale	a
Sium latifolium	o	Stachys palustris	o
Œnanthe aquatica (Phellandrium)	f	Rumex Hydrolapathum	f
		R. conglomeratus	o
Œ. Lachenalii	f	Juncus *spp.*	f
? Senecio palustris	vr	Typha angustifolia	ld
? S. paludosus	vr	Scirpus lacustris	ld
Sonchus palustris	vr	Carex riparia etc.	f

(*b*) *Slowly flowing waters.*

In the slowly flowing waters of many lowland streams plant communities occur which differ from those of stagnant waters enumerated above.

Thus in the river Cam at Cambridge the following species occur:—

Ranunculus circinatus
Œnanthe fluviatilis (and subm. form)
Elodea canadensis
Sparganium simplex *f.* longissima (Fries.)
Sagittaria sagittifolia *f.* vallisneriifolia (Coss. et Germ.)
Potamogeton lucens
Potamogeton lucens *f.* acuminatus (Fries.)
P. prælongus
P. perfoliatus
P. crispus
P. crispus *f.* serratus (Huds.)
P. densus
Scirpus lacustris (submerged form)

In shallower water at the edge.

Veronica Beccabunga (subm. form)
Myosotis scorpioides (subm. form)
Apium nodiflorum (subm. form)
Epilobium hirsutum (subm. form)
Callitriche stagnalis

Species with floating leaves are typically absent because of the current, but in the quiet water at the edge of the river, especially within the zone of protection of the reed-swamp association, the yellow water-lily

(*Nymphæa lutea*) and species of duckweed (*Lemna*) occur.

Reed-swamp association.

Glyceria aquatica d.

Thalictrum flavum	Sparganium simplex
Barbarea vulgaris	Iris Pseudacorus
Spiræa Ulmaria	Alisma Plantago-aquatica
Epilobium hirsutum	Carex vulpina
Lysimachia vulgaris	C. Pseudocyperus
Myosotis scorpioides (palustris)	C. elata (stricta)
Veronica Beccabunga	C. hirta
Typha latifolia	C. riparia
Sparganium erectum	Phalaris arundinacea

3. *Sub-formation of waters relatively poor in mineral salts.*

In the lakes and tarns of mountain districts, formed of hard siliceous rocks, and in which the ground waters are poor in dissolved salts, the aquatic vegetation is characteristic and much poorer in species, so far as the higher plants are concerned. That this vegetation in all probability owes its distinctive features rather to the poverty in mineral salts than to other factors such as altitude is shown by the fact that its characteristic species also occur in lowland pools in the midst of the barren soils of heath land and therefore also poor in dissolved salts[1].

The species most typical of this sub-formation are submerged forms; the quillwort (*Isoetes lacustris*), *Lobelia Dortmanna,* and the awlwort (*Subularia aquatica*), which are confined to such waters. The pillwort (*Pilularia globulifera*) and the shore-weed (*Littorella lacustris*) are also characteristic of this habitat, and in the extreme west of Ireland, *Eriocaulon septangulare*. Among forms with floating leaves the small yellow waterlilies *Nymphæa lutea* var. *intermedia* and *N. pumila* are apparently confined to this habitat, and also the floating bur-reed

[1] Graebner, *Die Heide Norddeutschlands*, 1901, p. 183.

(*Sparganium affine* [*natans*]), while among reed-swamp forms *Carex elata* is characteristic.

The type of lake surrounded by moorland or by hard siliceous rocks has been recognised by several Scottish workers[1] under the name of "highland" as opposed to the "lowland" loch, though, as already stated, it is probable that the character of the vegetation has very little to do with altitude. The water of these lochs is also characterised by having a very rich Desmid plankton[2] (see p. 197 *et seq.*), far richer than that of lowland lakes.

The following list is rearranged from Mr George West's lists appearing under the heading of "peaty moorland lochs" in his second paper on the flora of the Scottish lakes[3].

Submerged species.

†Subularia aquatica
Myriophyllum alterniflorum
Callitriche intermedia (hamuluta)
†Lobelia Dortmanna
Utricularia vulgaris
†U. intermedia
Littorella uniflora (lacustris)
Juncus bulbosus *var.* fluitans
Potamogeton lucens
P. angustifolius (Zizii)
P. prælongus
P. perfoliatus
P. crispus
P. pusillus
Eleocharis acicularis
Scirpus fluitans
†Pilularia globulifera
†Isoetes lacustris
Fontinalis antipyretica
F. squamosa
Blindia acuta
Eurhynchium rusciforme
Scapania undulata
Nardia emarginata
N. compressa etc.
Chara fragilis
var. delicatula
Nitella opaca

Species with floating leaves.

Castalia alba
Nymphæa lutea
†N. lutea *var.* intermedia
†N. pumila
Apium inundatum
Polygonum amphibium
Potamogeton natans
P. polygonifolius
P. alpinus (rufescens)

[1] Sturrock, R. Smith, W. G. Smith, 1905.

[2] W. and G. S. West, "The British Freshwater Phytoplankton, etc.," *Proc. Roy. Soc.* B, Vol. LXXXI. 1909.

[3] George West, "A further contribution to a comparative study of the

Reed-swamp association.

Menyanthes trifoliata	†Carex elata
†Sparganium affine (natans)	C. inflata (rostrata)
Eleocharis palustris	†C. filiformis
†E. multicaulis	Phragmites vulgaris
Scirpus lacustris	Glyceria fluitans
†Cladium Mariscus	Equisetum limosum

The † marks those species not recorded from "non-peaty lowland lochs."

4. *The sub-formation of quickly-flowing streams*[1]

By C. E. Moss

The streams of hill and mountain slopes are characterised by their quick current and their shallowness. A result of their quick current is that the water is well aerated; and the combined factors are related to the absence of a large number of lowland aquatic flowering plants. The streams of siliceous hills are deficient in humous acids, except in the case of the streams which drain upland peat moors; whilst those of limestone hills are not only deficient in humous acids, but are actually alkaline in reaction. Correlated with these facts, we find that the siliceous streams are poor in dissolved mineral salts, and that the calcareous streams are rich in such salts. For example, the waters of the siliceous soils near Halifax contain only 5·25 grains per gallon (0·075 gram per litre) of total solids in solution, whilst the calcareous waters of the mid-Pennines contain 14 to 16 grains per gallon (0·2 to 0·23 gram per litre) of dissolved matter. The former waters are very "soft," ranging from two to three "Clark's degrees[2]," whilst the latter are very "hard,"

Calcareous and non-calcareous streams.

dominant Phanerogamic and Higher Cryptogamic Flora of aquatic habit in Scottish Lakes." *Proc. Roy. Soc. Edin.* xxx. 1910, pp. 98, 99.

[1] This account is based on observations made in the southern Pennines. See Moss, 1911.

[2] A "Clark's degree" is one grain of carbonate of lime ($CaCO_3$), or its equivalent of other lime (calcium) compound, per gallon.

containing 10 to 12 grains per gallon of dissolved calcium salts.

These facts enable us to distinguish two associations of this sub-formation, and accordingly we may speak of the association of non-calcareous waters and that of calcareous waters. However, the marked common characteristics of all rapidly-flowing waters seem to prevent our giving these communities any higher rank than that of associations.

(*a*) *The association of non-calcareous waters.*

On the whole, flowering plants are not abundant in quickly-flowing streams on siliceous soils; but this is compensated by the great abundance and variety of liverworts and mosses. *Ranunculus Lenormandi* is locally abundant in such waters in the west and north of England; but the allied *R. hederaceus* is much rarer. *Glyceria fluitans, Callitriche stagnalis, Montia fontana*, and *Stellaria uliginosa* are all locally abundant. *Nitella opaca* and *Lemanea fluviatilis* are rather rare and local. Of Hepaticæ, *Scapania undulata* is abundant, *Aneura multifida* is rare, and *Jubula Hutchinseæ* local and very rare. Of mosses, *Sphagnum* is sometimes abundant, but, as is well known, it prefers more stagnant and acidic waters; but *S. crassidulum* is locally abundant in streams. Other common mosses are *Dicranella squarrosa, Rhacomitrium aciculare, Aulacomnium palustre, Philonotis fontanæ, Fontinalis antipyretica, F. squamosa, Brachythecium rivulare, Hylocomium flagellare, Hypnum fluitans, H. commutatum* and *H. stramineum.*

(*b*) *The association of calcareous waters.*

In the calcareous waters, flowering plants are still few as regards number of species; but these often grow in great masses, obscuring the bed of the stream. In particular, water-crowfoots, especially *Ranunculus fluitans*, are often abundant, as well as species of *Chara, e.g., C.*

vulgaris. As in the non-calcareous streams, mosses are abundant, especially *Cinclidotus fontinaloides*, and *Eurhynchium rusciforme*; and the following (among others) may also occur:—*Fissidens crassipes*, *Orthotrichum rivulare*, *Philonotis calcarea*, *Leskea polycarpa*, *Amblystegium fluviatile*. Liverworts are apparently less common than in the non-calcareous streams, but the following occur:—*Chiloscyphus polyanthus*, *Aneura multifida*, *A. pinguis*.

THE BRITISH FRESHWATER PHYTOPLANKTON

By G. S. WEST

General characters. The freshwater phytoplankton of the British Islands is chiefly developed in the four lake-areas which are situated in the west and north-west[1]. Malham Tarn, West Yorkshire, and Lough Neagh in north-east Ireland, are isolated lakes, and plankton of a much adulterated kind can be obtained in the Norfolk Broads and in the large pools of the English Midlands.

"Casual" and "true" constituents.

The plankton of the British lakes, although rarely of any great bulk, is often very rich in species. Many of them are *casual* constituents, consisting of shore-forms or bog-species which have been carried into the plankton by the rains. Such species are rarely able to withstand for very long the new conditions of environment, and soon die. Amongst them, however, are others which are either much more abundant in the plankton than in any other situations, or which exist only in the plankton. These are the *true* constituents of the plankton.

Diatoms.

The Diatoms (or Bacillarieæ) are abundant, but do not usually occur in great quantities. The greatest variety of species is met with in the Scottish and Irish lakes, whereas the plankton of the

[1] Scotland, the English Lake district, Wales, and western Ireland.

Welsh lakes is remarkable for the small number of Diatoms contained in it. Centric Diatoms are not very conspicuous, although in some lakes certain species of *Cyclotella* and *Rhizosolenia morsa* are abundant. The latter is a very delicate Diatom and is easily overlooked even when present in quantity. The genus *Melosira* occurs only in a few of the British lakes, and is represented by *M. varians* and *M. granulata*, the latter sometimes attaining a large maximum. The Pennate Diatoms are much more numerous than the Centric forms, the genera *Tabellaria* and *Asterionella* being conspicuously abundant. The star-dispositions of *Tabellaria fenestrata* (var. *asterionelloides*) are fairly common except in the Welsh lakes. Species of *Fragilaria* are uncommon, but the large species of *Surirella*, especially *S. robusta* var. *splendida*, are often quite a feature of the British plankton-collections.

Green Algæ.

The Green Algæ (or Chlorophyceæ) are a prominent feature of all the lake-areas. *Botryococcus Braunii*, *Sphærocystis Schroeteri*, species of *Oocystis*, and *Dictyosphærium pulchellum*, are fairly general, and sometimes abundant. Of the remaining members of the Protococcales which occur in the British plankton, most are found in the shallower lakes and pools. Sterile species of *Mougeotia* occur in most of the lakes, more especially in those of sub-alpine districts, and in these situations the filaments frequently become spirally coiled.

Desmids.

The most interesting feature of the plankton of the western British lake-areas, however, is the abundance of Desmids. In the majority of the lakes they are conspicuous, and in many they are the dominant features of the plankton. In the lowland districts of England, such plankton as exists contains very few Desmids, and only those which are characteristic of shallow pools and ponds.

The Blue-green Algæ (Myxophyceæ) are only of secondary importance in the majority of British lakes, although many of the shallower and more lowland lakes, and also those of the west of Ireland, contain in the summer and autumn an abundance of *Anabæna* and *Cœlosphærium*. It is the sudden appearance of large quantities of Blue-green Algæ in certain of the shallower pools and lakes which causes the phenomenon of "water-bloom" and which is responsible for the "breaking of the meres." *Oscillatoria tenuis* and *O. Agardhii* are frequent in the larger lakes, but never occur in great abundance, and *Gomphosphæria lacustris* and *Chroococcus limneticus* are not at all uncommon.

Blue-green Algæ.

Flagellates are well represented by various species of *Mallomonas* and *Dinobryon*, and by the Peridinieæ. Species of *Dinobryon* sometimes entirely dominate the spring plankton of the lakes, *D. cylindricum* and its var. *divergens* being much the most general forms. The Peridinieæ are represented by several species of *Peridinium* and *Ceratium*. *Peridinium cinctum* and *P. Willei* are the most general, the former in lowland areas and the latter in the lake-areas of Scotland, the English Lake District, and Wales. *P. Westii* is known only from the Scottish lakes. *Ceratium hirundinella* is general, but is curiously rare in the Welsh lake-area, and is entirely absent from many other British lakes. In the western areas of Scotland and Ireland one form of it exists which is known from no other part of the world.

Flagellata.

Comparison with continental areas. The features of the Central and Northern European lakes are to some extent combined in the plankton of the British lakes, with the addition of other features which tend to mark off the British lakes from either of those groups. They really form part of a north-western European lake-area, which would embrace the British Islands and part of Scandinavia.

Abundance of Desmids.

The plankton of the British lake-areas is distinguished primarily by the abundance of Desmids, a fact which has been shown to be coincident with the presence of a rich Desmid-flora in the drainage-basins of the lakes, although it must not be forgotten that the true plankton-species of Desmids differ markedly from those of the bogs of the drainage areas. Of a total of 162 species of phytoplankton observed in the Welsh lakes, over 62 per cent. belonged to the Desmidiaceæ, as compared with 11 per cent. of Diatoms and 7·4 per cent. of Blue-green Algæ.

The plankton of the pools of the low-lying districts in the midlands and east of England contains numerous Diatoms and relatively few Desmids.

The western lakes often contain few Diatoms, but the large species of *Surirella*, which appear to be absent from the plankton of other European lakes, are general and conspicuous. *Rhizosolenia morsa* is now known from all four British lake-areas, but so far has not been recorded from Central Europe. On the other hand, *Attheya Zachariasi*, which is not infrequent in Central and Northern Europe, has not yet been found in the British plankton.

The Blue-green Algæ are of no great importance except in the shallower lakes; and although many species of the Protococcales occur in the various lakes, they are never very numerous and rarely abundant.

It should be remarked that in the British lake-areas the winter temperatures are relatively high, nearly all the larger lakes remaining free from ice throughout the winter.

Distribution of plankton in British lakes, and its periodicity. The actual bulk of plankton in the British lakes is never very great as compared with some of the continental lakes, and only rarely does it give a decided colour to the water. In the four lake-areas

Desmids are conspicuous and often dominating constituents of the plankton, sometimes occurring in such quantity that the lake may be said to possess a Desmid-plankton. Such a Desmid-plankton is never of great bulk. In these lakes Diatoms are usually few in number, with the possible exception of *Tabellaria*, *Asterionella*, or *Melosira*. In the eastern and low-lying parts of England the plankton is much more contaminated by shore-species and bottom-species, and concurrently with the appearance of much more numerous and varied Diatoms the Desmids are represented by only a few of the commoner pond-species.

The Desmids, and, indeed, all the Green Algæ, attain a maximum during the autumnal fall of temperature, whereas many of the Diatoms have two maxima, one in the spring and the other in the autumn, the former being as a rule the greater of the two. It would appear, however, that to a great extent the plankton-species of Diatoms attain their maxima at different periods of the year. Similarly, some of the species of *Peridinium* are active only in the warm, summer temperatures, whereas others attain their greatest activity in the cold months of early spring.

Factors which determine richness or poverty of the plankton. The terms "rich" and "poor" when used in reference to plankton collections are only relative, and a collection which is rich in one group may be poor in another. For instance, the rich Desmid-plankton, such as that of some of the Welsh and Scottish lochs, is poor in Diatoms and Blue-green Algæ. It has been shown that the rich Desmid-areas correspond geographically with the outcrops of the older Palæozoic and Precambrian rocks, with their associated igneous masses, and it is probable that this abundance of Desmids is due to the purity of the water which drains from these comparatively hard strata; or, in other words, that the richness of the Desmid-flora is due to the poorness of the water in dissolved mineral salts. In contrast to this, the water which drains from the newer and softer strata contains a much greater quantity of dissolved mineral salts, and while greatly reducing the Desmid-flora, favours a more prolific growth of Diatoms and of many of the aquatic Blue-green Algæ.

Amount of dissolved salts.

It has also been shown that a greater bulk of plankton occurs in the waters of those lakes which are slightly contaminated owing to the presence of farms and villages on their shores. This fact very strongly supports the suggestion that the presence of available food-constituents in the form of dissolved salts is the principal determining factor in the quantitative development of plankton.

Some special instances of British plankton. The three following statements refer to widely separated areas:—

Norfolk Broads and East Anglian Pools. The phytoplankton of this area agrees with that of shallow low-lying lakes, with the addition of a large admixture of littoral species. A few pond Desmids occur rather sparingly, but the bulk of the plankton consists of various members of the Protococcales and Diatoms. Of the former group, species of *Pediastrum* and *Scenedesmus* are common, and the genera *Oocystis*, *Nephrocytium*, *Crucigenia* and *Closteriopsis* are all represented. The Diatoms are more numerous, and the genera *Fragilaria*, *Synedra*, *Asterionella*, *Diatoma*, *Melosira*, and *Cyclotella* are particularly abundant.

Blue-green Algæ are not very common.

Several species of *Peridinium* occur, of which *P. cinctum* is the most general.

In those broads in which the water is brackish (Hickling Broad, Horsey Mere, Heigham Sound, etc.) the phytoplankton consists only of one or two species of *Peridinium* and a few individual Diatoms. It is of interest to note that some of these Diatoms are marine or brackish water species, *e.g. Nitzschia longissima, N. reversa*, and several marine or brackish water species of *Navicula* (Heigham Sound).

Scottish Lochs. The phytoplankton consists largely of Chlorophyceæ, and is in most cases characterised by the abundance of Desmids, which are represented by a great variety of species. The most noteworthy of these are *Staurastrum Ophiura, St. Arctiscon, St. lunatum* var. *planctonicum, St. jaculiferum, St. anatinum*, and *Xanthidium subhastiferum*. The Protococcales are not very abundant, although *Sphærocystis Schroeteri* and *Botryococcus Braunii* are fairly general.

Diatoms are often conspicuous, but mostly owing to the occurrence in quantity of a few species. *Surirella robusta* var. *splendida, Asterionella gracillima*, and *Tabellaria fenestrata* var. *asterionelloides*, are the most frequent.

The Blue-green Algæ are only feebly represented in the deeper lochs, and are by no means abundant in the smaller, shallower lakes. *Dinobryon* occurs in quantity in the spring and summer. The western lochs, and those of the Outer Hebrides, are much the richest in species.

Desmids and Diatoms are abundant in the mixed plankton of Loch Tay, and there is also an abundance of *Cœlosphærium*. The following are fairly regular constituents of this plankton:—several species of *Closterium, Euastrum verrucosum* var. *reductum, Cosmarium depressum* var. *achondrum, C. abbreviatum* var. *planctonicum, C. Botrytis, Xanthidium antilopæum* and its var. *depauperatum, Arthrodesmus convergens, Staurastrum jaculiferum, St. cuspidatum* var. *maximum, St. lunatum* var. *planctonicum, St. pseudopelagicum, St. paradoxum* and its var. *longipes, St. Ophiura, St. furcigerum* and its forma *armigera, St. Arctiscon, Spondylosium planum, Hyalotheca mucosa, Eudorina elegans, Scenedesmus quadricauda, Ankistrodesmus falcatus, Sphærocystis Schroeteri, Tabellaria fenestrata, T. flocculosa, Synedra Ulna, Asterionella*, sp. of *Eunotia*, and various members of the Naviculaceæ.

Irish Loughs. Like the Scottish plankton, that of the Irish lakes is to a large extent Chlorophyceous, but the Diatoms, Peridinieæ and Myxophyceæ are more abundant. Especially is this the case with the Blue-green Algæ, species of *Anabæna*, *Oscillatoria*, *Gomphosphæria*, *Cœlosphærium*, and *Chroococcus* being much more conspicuous.

The Desmids are numerous and include many of the characteristic western types, but there is a curious absence of *Staurastrum Ophiura*. Diatoms form a relatively large part of the Irish plankton, and Centric Diatoms are more numerous than in the Scottish plankton. The narrower species of *Synedra* are also deserving of special mention.

Connemara lakes.

Desmids, if not more abundant in the Connemara district than in the Kerry lakes, are certainly more varied. The smaller lakes of Galway contain a rich Desmid-flora equal to that of the west of Scotland. There are also many Diatoms and Blue-green Algæ, and widely different plankton may be obtained from lakes in close proximity to each other.

Kerry lakes.

The Kerry lakes are moderately rich, and Lough Currane is the best of them. Lough Leane (the Lower Lake of Killarney) is the poorest in Desmids, although the Diatoms are numerous. On the whole, the lakes of the south-west of Ireland differ from the Connemara lakes in the greater abundance of *Cosmarium subarctoum*, *C. abbreviatum* var. *planctonicum*, *Xanthidium subhastiferum*, *Staurastrum jaculiferum*, *St. curvatum*, *St. cuspidatum* var. *maximum*, *St. lunatum* var. *planctonicum*, *St. pseudopelagicum*, *St. Arctiscon*, *Spondylosium planum*, *Botryococcus Braunii*, *Dictyosphærium pulchellum*, *Sphærocystis Schroeteri* and *Tabellaria fenestrata*. There is, of course, an absence of many species which occur in the Connemara lakes. A greater bulk of plankton occurs in the Kerry lakes, due to slight contamination, while the small bulk of the phytoplankton of many of the Connemara lakes is due largely to the complete absence of contamination.

CHAPTER VIII

THE MARSH FORMATION

On wet undrained soils of various kinds marsh is developed, supporting a characteristic vegetation sometimes dominated by trees, sometimes by herbaceous plants. The term *marsh* as used here is confined to the vegetation of soils on which peat is not formed generally and in quantity, and on which consequently the vegetation is determined by the saturated character of the soil rather than by the peculiar qualities of peat. The peat soils bear the moor and fen formations (Chap. IX) which have special characters, though the latter has much more in common with the marsh formation than it has with the moor formation.

In accordance with the principles laid down in the Introduction the various marsh communities are treated as a distinct plant formation because they occupy a special kind of habitat and show common features. They include communities dominated by trees, and others dominated by herbaceous plants, as in the case of most of the other great formations, but they have not as yet been studied with sufficient closeness in this country to admit of a very satisfactory treatment, and the following account will certainly require to be thoroughly revised at a later date. Possibly the associations will have to be subdivided according to different types of soil.

The marsh formation shows transitions to the formation of clays and loams through the Juncetum communis

(p. 86), and through the ash and alder societies of the damp oakwood association. It also shows transitions to the aquatic formation through the reed-swamp association and other fringing associations of ponds, lakes and rivers.

Owing to the extensive draining of the low-lying alluvial soils, the area occupied by the marsh formation in this country has been very greatly restricted, and only small fragments remain here and there from which to construct the characters of the formation.

Alder-willow association.

The woods of this association are frequently dominated by the alder (*Alnus rotundifolia*); often they show a mixture in various proportions of this tree with various willows (*Salix*) and the ash (*Fraxinus excelsior*). The pedunculate oak (*Quercus Robur*) also frequently occurs, but is not dominant. These woods are often coppiced, and alder and ash are frequently planted for coppice, as they both yield woods useful for a variety of purposes. The osier (*Salix viminalis*) is also commonly planted in the same way. Such plantations are frequent in some parts of the country on alluvial riverside soil whose surface is only a few inches above the ground water level. The ground vegetation consists of many typical marsh plants. When the soil is partially drained the vegetation often passes over into the damp oakwood type.

Of rare or local trees apparently belonging to this association we may mention the hoary and black poplars (*P. canescens* and *P. nigra*) and the small-leaved elm (*Ulmus sativa* Miller).

Trees and shrubs.

Alnus rotundifolia	d	S. triandra	r
Salix fragilis	a	S. purpurea	r
S. caprea	a	S. alba	la
S. cinerea	a	and various Salix hybrids	
S. viminalis	f	Populus canescens	l

P. nigra	l	Betula tomentosa	o
Ulmus sativa Miller	l	Acer campestre	f
Fraxinus excelsior	a	Cratægus monogyna	f
Quercus Robur	f	Prunus spinosa	o
Viburnum Opulus	f	Sambucus nigra	o

Ground vegetation.

Ranunculus repens	a	Lysimachia vulgaris	f
R. Ficaria	a	L. Nummularia	f
Caltha palustris	a	Symphytum officinale	a
Radicula Nasturtium-aquaticum (=Nasturtium officinale)	a	Myosotis cæspitosa	a
		M. scorpioides (palustris)	a
		Calystegia sepium	a
R. sylvestris	o	Solanum Dulcamara	a
R. palustris	o	Scrophularia aquatica	a
R. amphibia	o	S. nodosa	a
Barbarea vulgaris (agg.)	o	Veronica Anagallis	o
Cardamine amara	o	V. Beccabunga	f
C. pratensis	a	V. Chamædrys	f
C. flexuosa	f	Mentha *spp.*	a
Lychnis dioica	la	Scutellaria galericulata	f
L. Flos-cuculi	la	Stachys palustris	o
Stellaria aquatica	o	S. sylvatica	f
Hypericum tetrapterum	f	Lamium album	f
Spiræa Ulmaria	ld	Rumex Hydrolapathum	f
Chrysosplenium oppositifolium	lf	Humulus Lupulus	o
		Urtica dioica	a
C. alternifolium	o	Helleborine palustris	f
Epilobium hirsutum	la	Orchis latifolia	f
E. parviflorum	f	O. maculata	la
E. palustre	o	Iris Pseudacorus	a
Sium erectum	f	Arum maculatum	o
Anthriscus sylvestris	f	Scirpus sylvaticus	f
Œnanthe crocata	a	Carex paniculata	la
Angelica sylvestris	la	C. elongata	f
Galium Cruciata	la	C. pendula	la
G. palustre	a	C. Pseudo-cyperus	la
G. Aparine	a	C. vesicaria	f
Valeriana sambucifolia	a	Phalaris arundinacea	f
Dipsacus sylvestris	f	Calamogrostis epigeios	o
Cnicus palustris	f	C. lanceolata	r
Sonchus palustris	r	Deschampsia cæspitosa	f

Holcus lanatus	f	Lastrea spinulosa	o
Catabrosa aquatica	f	Osmunda regalis	o
Poa trivialis	f	Equisetum maximum	f
Glyceria fluitans	f	E. sylvaticum	o
G. aquatica	f	E. hyemale	r

Herbaceous marsh associations.

When the trees and shrubs are removed an herbaceous marsh association, consisting of most of the ground vegetation of the alder-willow association, is left. The increase of light increases the vigour of the herbaceous vegetation, which generally closes up, and resembles the fen association in many respects (see p. 230). Since this marsh association has not been specially studied, no list will be given.

If alluvial land is drained, good semi-natural permanent grassland may be produced, which in this country is generally used as pasture, and is very characteristic of the low-lying alluvial flats bordering the slow streams. It has the vegetation of neutral pasture (see p. 85) with a preponderance of damp-loving species. Alluvial meadows cut for hay generally have a richer flora than alluvial pasture, presumably because the constant pasturing of the latter exterminates a certain number of species. The lighter alluvial soils are frequently used for market gardening in the neighbourhood of the great cities situated on the flood plains of the larger rivers.

CHAPTER IX

THE VEGETATION OF PEAT AND PEATY SOILS—MOOR, FEN AND HEATH

PEAT is one of the most distinctive of soils and the plant-communities which it bears are extremely well-marked. For this reason the tracts of peatland in north-west Europe possess very old distinctive names. Language rarely separates the common name of a well-characterised type of ground from that of the vegetation which covers it,—the common people make, and rightly make, one entity of the two together—and the English and German word *Moor*, like the French *tourbière*, applies alike to the peat soil and to the plant covering which it bears. The German word "Moor" however applies to all soils of deep, nearly pure, peat, and to those only, while the corresponding English word in ordinary language as it is used to-day is both more restricted and wider in its meaning. We speak of moors and moorland not only when we mean heathy or boggy land with a deep covering of peat, but also when there is only a shallow layer of surface peat much mixed with mineral substances derived from the underlying rock. The "moors" of the Cleveland district of the North Riding of Yorkshire and many of the Scottish "moors" are an instance of the latter kind. Such moors are intermediate between the moors on deep peat and the *heaths* on sandy soil which are characteristic of the drier climate of south-

The term "moor."

Moor and heath.

eastern England, and of similar sandy soil in Belgium, Holland, Denmark and north-west Germany (see p. 98). On such heaths, as we have already seen, there is a formation of dry peat (Trockentorf) on the surface of the soil, but deep peat is never formed except locally, as in hollows above an impermeable substratum, or where the course of a river has been checked; and in these places they form true moor peat with characteristic moor plants (wet heath sub-association, p. 106). The line between moors and heaths is necessarily rather an arbitrary one, and exactly where we draw it is largely a matter of convenience. Weber[1] has suggested a limit to the use of the term moor, namely, the presence of a layer of peat at least 8 inches (20 cm.) thick in the dry state, and the absence of visible mineral constituents (corresponding with about 40 °/。 of ash). From the ecological standpoint, however, it seems better to decide by the general character of the vegetation, and accordingly some of the Scottish "moors" which have a depth of peaty soil extending to 12 inches (30 cm.) in extreme cases, have been considered as heaths, in cases where the underlying soil is sandy or gravelly (p. 114). The proper correlation of the vegetation with the physical conditions of the soil is only now being undertaken in this country, and at present it is impossible to furnish satisfactory definitions based on quantitative results.

On the other hand, not all tracts of deep, relatively pure, peat are "moors" in the English language of to-day.

Moor and fen. The great tracts of peat laid down in the upper parts of old estuaries, and round fresh-water lakes, fed by water relatively rich in lime and other salts, bear plant associations quite distinct from those

[1] C. A. Weber. Die wichtigsten Humus- und Torfarten und ihre Beteiligung an dem Aufbau norddeutscher Moore. Festschrift zur Feier des 25-jahrigen Bestehens des Vereins zur Förderung der Moor-Kultur im Deutschen Reich. Berlin, 1908, p. 91.

which characterise the moor soils, fed by waters poor in lime and in mineral salts generally, often by *aerial* as opposed to *telluric* water. Such areas are not called moors in current English, as they are in current German (though there is evidence from old place-names that some of them did once bear this name), and it would be stretching the language too far to extend the English word to these very distinct soils and their characteristic plant-communities. Such peat-soils, relatively rich in lime, are most extensively developed, so far as the British Isles are concerned, in East Anglia, where the word *fen* or " black fen " is applied to them.

These two types of soil and their corresponding plant-communities are very distinct, and it seems correct to adopt the two terms *moor* and *fen* as English technical terms in ecological plant-geography, since this can be done without unduly straining their meaning in ordinary language.

" Hochmoor " and " Flachmoor."

Moor and fen correspond in a very general way with the German " Hochmoor " and " Flachmoor," as used by most continental writers, but the correspondence is not exact. The term " Hochmoor " is derived from the fact that certain types of moor rise from the edges towards the middle, *i.e.* they show a convex upper surface in section. This is due to the peculiar habit of growth of the bog moss (*Sphagnum*) which forms the basis of such moors. The term " Flachmoor " on the other hand is applied to moors which have a flat, or even a slightly concave surface. The distinction based on the chemical characters of the soil, namely the presence of a ground water rich in mineral salts, especially lime, and giving an alkaline reaction, as opposed to water poor in salts, with practical absence of lime and an acid reaction, does not coincide exactly with the distinction based on the shape of the surface.

It is the former distinction which determines the plant-communities, and though it is often used in the definitions of the two types of "moor" distinguished by continental writers[1], there will always be a tendency to include some of the types of moor which show no convexity of surface, but which are poor in salts and whose water gives an acid reaction, in the "Flachmoor" series. Weber's term "*Niedermoor*[2]," however, corresponds more nearly with our English fen.

Moor, then, possesses a relatively wet peat soil of considerable depth, fed by water, often aerial water[3], poor in mineral salts, so that the acids characteristic of the humous substances produced in the process of peat formation are not neutralised by bases, and the soil water is consequently *acid in reaction.* The plant community is independent of the underlying rock, which is very different in different cases.

Fen possesses a peat or peaty soil fed by water, always telluric water, relatively rich in mineral salts, and the ground water is *alkaline in reaction*[4].

These differences[5] have a profound effect on the vegetation borne by the soils in question, and clearly indicate a distinction between the *fen-formation* and the *moor-formation.* Under appropriate conditions, however, the fen-formation passes over into the moor-formation, by the gradual growth of plants above the reach of the alkaline ground water and the consequent accumulation of peat poorer in mineral salts, so that moor plants begin to

[1] Cf. for instance the definitions given by Früh and Schröter, *Die Moore der Schweiz*, Bern, 1904, pp. 12–13; and by Warming, *The Œcology of Plants*, Oxford, 1909, p. 197. Warming, however, states that "moor soil is probably always acid" (*loc. cit.* pp. 195–6). This is certainly not the case in fen.

[2] Weber, *loc. cit.* p. 95.

[3] Moor in this sense is typically developed in regions of high rainfall.

[4] This is the case in our typical East Anglian fens; whether plant-communities on peat with neutral ground water may fairly be included in the fen formation we cannot yet say.

[5] See note on p. 35.

supplant fen plants. The succession observed under such conditions has often been described by continental writers. The most recent description is that of Weber relating to the North German "moors[1]." Weber distinguishes the "eutrophic peat" (fen peat) of the earlier stages of such a succession from the "oligotrophic peat" (moor peat) of the later stages. He also distinguishes intermediate stages of the formation of "mesotrophic peat" as the characteristic vegetation of the "Uebergangsmoore"—transition moors—of various authors. These intermediate stages of the succession can also be recognised in the British Isles, but their presence should not be allowed to obscure the importance of the fundamental distinction between moor and fen[2]. Accurate quantitative estimation of the soil characters of the two formations and their transition stages are as yet lacking.

The Fen Formation

Distribution of the fen formation.

The fen formation is developed most extensively in East Anglia. The name "Fens" or "Fenland" is the old historic name, still in full use, applied to the tract of almost perfectly level country some 1300 square miles in extent, for the most part but a few feet above sea level, round the Wash, which is the name given to the common estuary of the rivers Witham, Welland, Nene and Great Ouse. This great tract of country is formed towards the sea of marine and estuarine silt[3]. Towards its southern

[1] Weber, *op. cit.*

[2] It is quite possible that the moor and fen formations as distinguished here are extreme types, and that a further knowledge of intermediate cases will necessitate a certain readjustment of the classification. See notes on pp. 251 and 261.

[3] This seaward part of the Fenland with a silt soil is sometimes known as "the Marshland," as opposed to the Fenland proper, with peat soil, further inland.

or landward extremity, however, it is composed of deep peat (fenland proper). Almost the whole of this region is now under cultivation, though a few small patches of peat fen remain in a natural or semi-natural condition[1].

The common estuaries of the Bure, Yare and Waveney in east Norfolk form a similar though much smaller area; a much larger portion of the East Norfolk fenland is, however, in a natural condition, and forms by far the best and most extensive area of existing natural fenland in the country. This is described in the succeeding chapter.

No doubt other old British estuaries were originally partly occupied by fen; indeed, there is ample evidence from the plant remains of which the peat occurring in some of these estuaries is composed, that this was the case[2], but the present surface vegetation in all cases that have been investigated, where this is still in an approximately natural condition, now belongs to the moor formation[3].

[1] *E.g.* Wicken Fen, eleven miles from Cambridge, and the valley fens in the valleys of rivers debouching into the Fenland, such as Chippenham Fen, and Woodwalton and Holme Fens in Huntingdonshire. See R. H. Yapp, "Wicken Fen" [Sketches of Vegetation at home and abroad], *New Phytologist*, vol. VII. 1908, p. 61.

[2] See pp. 253, 258. There is no direct evidence that the ground water was alkaline.

[3] See p. 249. Cf. also the landward portions of the Somerset levels occupying the sites of the old estuaries of the Parrett and Brue as described by Moss (1907, p. 29).

CHAPTER X

THE RIVER-VALLEYS OF EAST NORFOLK: THEIR AQUATIC AND FEN FORMATIONS

By MARIETTA PALLIS

Character of the district.

THE district (map, Fig. 8) in which the broads and fenland of East Norfolk are situated covers an area of about 360 square miles (93,312 hectares), and is roughly triangular in shape, bounded on the east by the North Sea and on the north-west and south-west by lines drawn from near Norwich to the sea. "The area consists of a gently undulating plain, nowhere so much as 200 feet (60 m.) above sea-level. It is intersected by the valleys of the Bure, the Yare, the Waveney and their tributaries[1]," which form one inter-connected river system. The Ant and the Thurne are important tributaries of the Bure.

The broads, which are a feature peculiar to the river valleys of East Norfolk and East Suffolk, are shallow lakes connected with the rivers but not lying in their direct courses. In the waters and also on the land in the river valleys the indigenous vegetation still flourishes, in spite of the fact that the uplands of the district form one of the oldest cultivated regions in England. The special features of these river valleys have produced a type of

[1] H. B. Woodward, "The Geology of the country around Norwich," pt I. *Memoirs of the Geological Survey of England and Wales.*

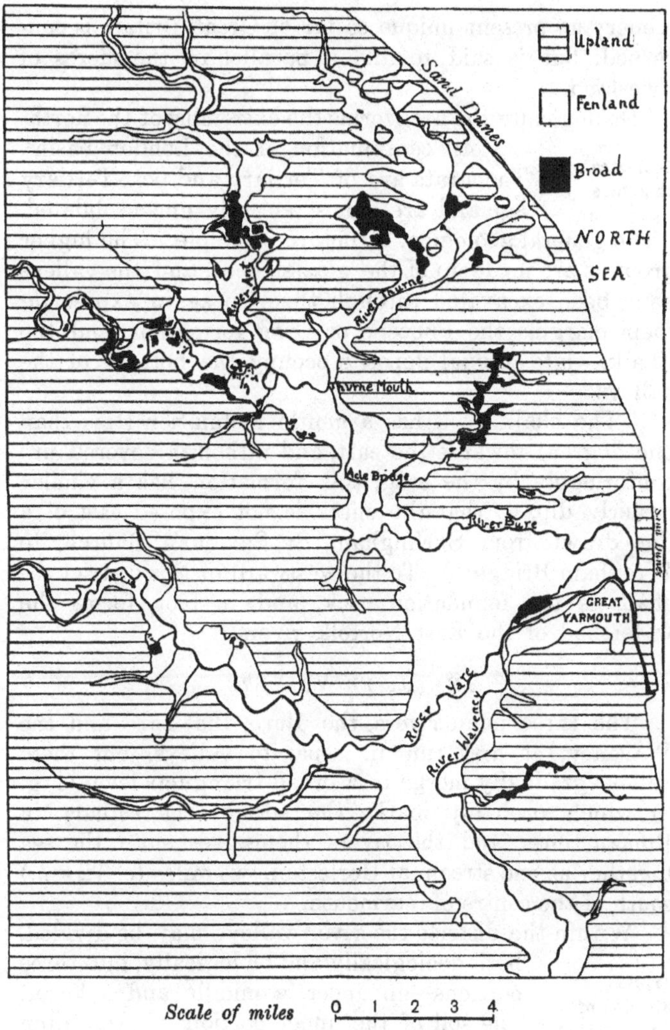

FIG. 8. Sketch-map of the River Valleys and Broads of East Norfolk.

scenery at present unique so far as Great Britain is concerned, but is said to find a parallel in the meres of Friesland.

Geological structure.

Geologically the area forms the extremity of the north-eastern continuation of the London basin[1]. The strata are of Tertiary and post-Tertiary age and are almost entirely unconsolidated. "The general geological structure is simple. The higher grounds are made up of the glacial beds, and the valleys have been excavated through them so as to expose on their margins the Pliocene or Crag formation, and the Chalk, while alluvial deposits occupy the bottoms of the valleys.

"The chalk itself has a gentle inclination (less than one degree) towards the east, and although covered unconformably by the crag, this formation has a similar easterly dip, so that the chalk is not exposed east of a line drawn from Surlingham, St Saviour's Church, to Wroxham Bridge[2]." To the westward of this district the chalk is the foundation rock, and, in fact, forms the watershed of the East Norfolk rivers.

The River Valleys

The three main rivers, the Bure, the Yare and the Waveney, at first run in separate valleys, but these valleys gradually merge in a broad triangular level (Fig. 8) which probably marks the site of the estuary in Roman times, and the rivers themselves enter the sea together as one stream at Gorleston, two miles (*c.* 3·2 km.) south of the centre of Yarmouth.

Triple division of river-valleys.

Within the district the river valleys may be divided, both geologically and biologically, into three portions—an upper, a middle and a lower. The soil of the upper portions is peat, often more than 18 feet (*c.* 5·4 m.) deep, and the

[1] See p. 56.

[2] H. B. Woodward, *loc. cit.* pp. 4 and 5.

vegetation is of the peculiar type known as *fen*. *All the broads are situated in these upper portions.*

In the middle portions the soil is a bluish unctuous ooze, and the vegetation a poor type of grassland. The rivers still run in separate valleys.

In the lower portions the valleys have merged, the

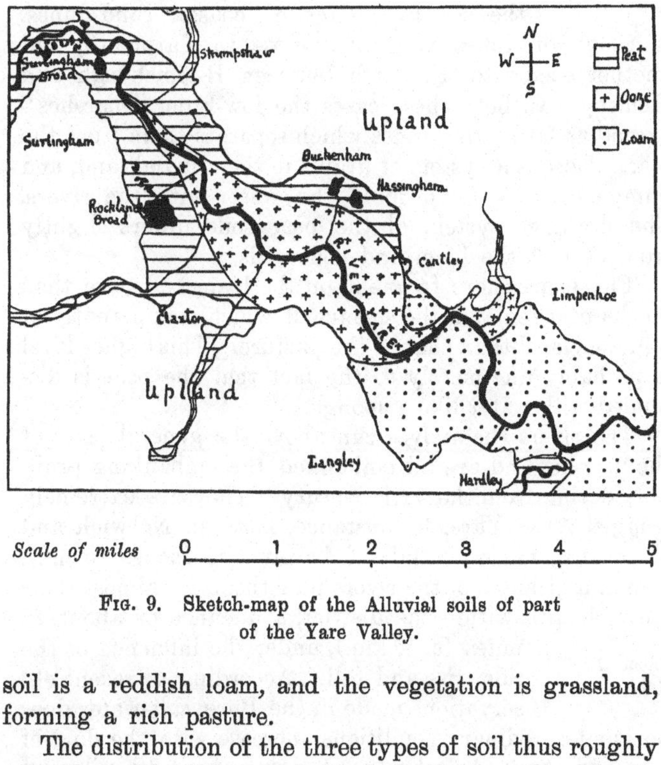

FIG. 9. Sketch-map of the Alluvial soils of part of the Yare Valley.

soil is a reddish loam, and the vegetation is grassland, forming a rich pasture.

The distribution of the three types of soil thus roughly coincides with definite physical features and with definite types of vegetation. Fig. 9 shows the distribution of the three types of soil in the valley of the Yare. There is naturally an overlap both of the soils and of the vegetation, and there are also other varieties of soil not shown in the sketch-map; roughly speaking, however, the three-fold division prevails.

Relative levels of land and water.

The bottoms of the river valleys are below the level of high tide and parts of the pasture land are no higher than low-tide level. The valleys are, however, protected from the invasion of the sea by chains of coastal sand-dunes. A range of dunes is present close to Yarmouth, and another occurs to the north between Happisburgh and Hemsby. In both these cases the low-lying "marshes" extend as far as the dunes which separate them from the sea. There is a system of drains in the alluvial land, and pumping mills which force the water into the rivers. The drainage system of the peat soils differs slightly from that of the loam and ooze soils.

Floods are more frequent on the loam and ooze than on the peat fenland, the surface of which is apparently at a higher level than that of the pasture. This higher level is probably due merely to the fact that the peat is distended with water like a sponge.

Effect of tides.

The rivers themselves run above the general level of the marshes and are all embanked, the embanking probably dating from the 12th century. They are extremely sluggish; the Yare, for instance, between Norwich and Yarmouth, having a fall of barely two inches (5 cm.). This sluggishness of the rivers puts them, for almost their entire length within the district, a distance of about 25 miles (*c.* 40 km.), under the influence of the tidal rise and fall. According to recent observations made in the Bure, salt-water does not, under ordinary conditions, increase with the flow of the tide above Acle bridge, a point about 12 miles by river (*c.* 19 km.) from Yarmouth[1].

[1] Eustace and Robert Gurney, *The Sutton Broad Freshwater Laboratory*, Norwich, 1905.

The Yare is more strongly affected by the tide than any of the other rivers. It is subjected to a greater rise and fall, and has also, for an East Norfolk river, an unusually strong tidal current. As is well known locally, northerly or north-westerly winds coinciding with spring tides give rise to unusually high flood tides which hold back the river water and cause it to distribute itself over parts of the valleys. Under exceptional conditions, also, salt-water may pass for considerable distances up the rivers.

These and other physical features indicate clearly that the river system of East Norfolk is still in a late estuarine phase of evolution.

Origin of the Broads

Professor Gregory has advanced an explanation[1] of the origin of the peculiar physical features of the river system and the broads belonging to it. These peculiar features he attributes to the course of the silting process that has prevailed in these valleys. He sketches the probable course taken by this process and points out how upon it the origin of the broads may depend. Professor Gregory's theory is briefly as follows:—

Blocking and silting of estuaries.

The silting of the river valleys has proceeded in the first instance from the mouth of the estuaries backwards, along the courses of the rivers. This backward silting appears to have been due to three causes: (1) to the southward flow of the tide along the eastern shores of Great Britain, (2) to the existence on the Norfolk coast of strata yielding a large amount of detritus to the sea, and (3) to the fact that the power of the east coast rivers is insufficient to sweep any considerable obstruction from their mouths.

The interaction of these three factors has caused a

[1] J. W. Gregory, "The Physical Features of the Norfolk Broads," *Natural Science*, 1892.

southerly drift of sand down the coast and the laying down of a spit or bar extending from the north side across the mouth of the estuary, so that this has been shifted southward and considerably narrowed. The mouth of the Yare has thus been shifted about four miles (*c.* 6·5 km.) to the south, from near Caister where it used to enter the sea, to Gorleston, where it now emerges (Fig. 8). The most striking example, however, is that of the River Alde in Suffolk, whose mouth has been driven 10 miles (*c.* 16 km.) south of its original point of emergence.

The effect of such a bar at the mouth of an estuary is to check the river current and thus to decrease its carrying power, and this would lead to the deposition on the inner side of the bar of some of the suspended material brought down by the river. The result of the continuation of such a process would be the formation and backward creep of alluvial land behind the bar. In the course of time branches of the estuary corresponding with the entrances of tributary streams would be encountered by the accumulating alluvium, and these branches would also in course of time be narrowed by the continuation of the same process, so that the outlets of the tributaries into the estuary would be narrowed or altogether closed, and their waters, unable to escape, would form broads in the tributary valleys. Fritton Decoy, which lies in a lateral valley of the Waveney and is wholly disconnected from the river, and the three connected broads of Filby, Rollesby and Ormesby, which discharge into the Bure by the Muck Fleet dyke, are examples of broads formed in this way (Fig. 8).

Blocking of tributaries.

This explanation does not, however, apply to broads which lie in the main river valleys, separated from the river itself by a narrow band of alluvium. Wroxham Broad, Hoveton Great Broad, Hoveton Little Broad, Salhouse Broad and Decoy Broad, all in the Bure valley, are examples of the latter type.

Phot. F. F. Blackman

a. View across the Bure Valley from the Upland: the river Bure is separated from Salhouse Broad (nearest the observer) by a narrow strip of alluvium, and from Hoveton Great Broad (in the distance) by a broader strip bearing carr.

Phot. F. F. Blackman

b. Reedswamp of *Typha angustifolia* with *Myriophyllum spicatum* in front. Heigham Sound.

Aquatic and Fen formations.

According to Professor Gregory all these broads once formed a single sheet of water, the extent of which is indicated by the area occupied by alluvium in Fig. 10. Through this water the river Bure of course flowed; indeed, at an earlier period it was probably part of the Bure estuary. At

Delta formation.

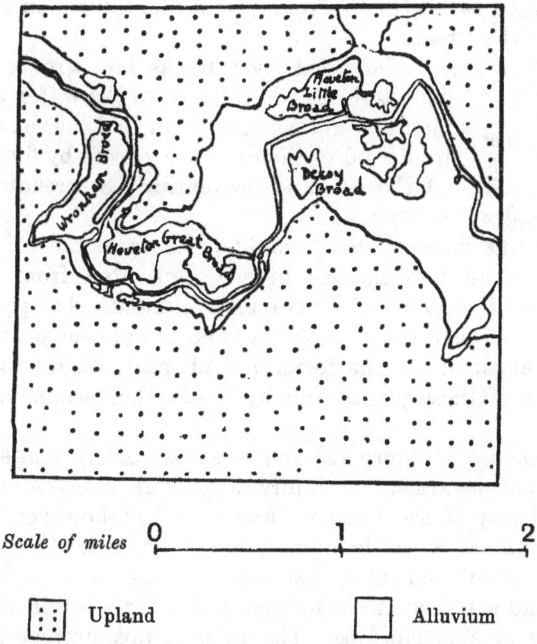

FIG. 10. **Sketch-map of the Wroxham-Hoveton group of broads in the Bure valley, showing position of the broads on each side of the course of the river.**

the point where the river struck the broad a delta would gradually be formed, owing to the checking by the sheet of water of the river current, which must at that period have been very much stronger than it is at present. In times of flood the delta would be spread further and further outwards, the river continuing to flow through its centre. The growth of the delta would naturally be greatest towards the middle, along the line of maximum current, so that the river channel would in time be separated from sheets of dead water on either side, that is to say a natural embankment would be formed on either side of the stream.

Lateral separation of broads.

When these submerged river banks had grown to a definite height in relation to the water level, plants would colonise them, and the rhizomes would aid in their further growth by filtering off the silt, and preventing its spreading to the dead water beyond.

In this manner there would arise from the original large broad two smaller broads, separated from each other by the river and its two embankments. The present configuration of the river valleys and broads has probably been attained by the formation of peat, owing to the growth of these plants, and by the further accumulation of silt[1].

Professor Gregory's theory is an extremely suggestive one, and is almost certainly a partial explanation of the history of the broads. Evidence has, however, been accumulated, since the publication of Professor Gregory's paper, which indicates that other causes have probably also had a considerable influence in giving origin to the broads of East Norfolk. The topic is not, however, ripe for further discussion.

[1] It is known from the evidence of old ordnance survey maps that many of the broads have decreased greatly in size during recent years. Some have been nearly or completely obliterated, *e.g.* Surlingham Broad and Chapman's Broad (Fig. 12) [Editor].

THE AQUATIC FORMATION

Characters of the waters.

The aquatic formation occurs in the river-valleys of East Norfolk in the rivers, broads and dykes—wherever, in fact, free water is present all the year round. The waters of certain parts of the area are brackish, but throughout the area they give an alkaline reaction[1]. It is probable that this alkalinity of the waters is the factor which differentiates the vegetation of the East Norfolk river-valleys, that is to say, alkalinity is probably the factor which separates the aquatic associations of East Norfolk from those of districts with ground waters which are neutral or acid in reaction and relatively poor in soluble salts[2].

Associations.

In the following account the general terms *submerged leaf*, *floating leaf*, *open reed swamp* and *closed reed swamp* associations[3] are used for the four communities of aquatic vegetation which have been recognised, but fuller comparative knowledge will almost certainly show that these "associations" are really groups of associations. Indications that these units ought to be subdivided have already been found, but the determining factors are not yet clear. The names employed are therefore to be regarded as provisional.

Status of reed-swamp.

The inclusion of the open and closed reed-swamps in the aquatic formation has been determined upon on the strength of two facts—one relating to the environment and the other to the conditions of life of the species of the reed-swamp associations. First, the reed-swamp associations are in-

[1] Due no doubt to the fact that they drain strata possessing large quantities of alkaline salts [Editor].

[2] In other words the East Norfolk aquatic associations belong to the sub-formation of waters relatively rich in mineral salts, provisionally distinguished on p. 190 [Editor].

[3] See Warming, *The Œcology of Plants*, 1909, p. 165 etc.

cluded in the aquatic formation because the general environment is essentially the same as that of the submerged leaf and floating leaf associations, for all four associations included in the aquatic formation can occupy definite portions of the same habitat. The actual distribution of the four associations is partly, at least, determined by the "invasional" factor. Close to the land, this apparently operates in favour of those members of the reed-swamp which germinate on land, and which afterwards, as an association, increase by vegetative propagation. Secondly, the conditions of life of certain members of the reed-swamp make it difficult to separate them from aquatic vegetation. *Scirpus lacustris* is the most aquatic of the reed-swamp dominants, occupying the zone farthest from the land. The flowering shoot of this plant alone extends above the surface of the water, and lasts only during part of the year. It is probable that this species can germinate from submerged seed at a considerable distance from the land as indicated by the scattered clumps of *Scirpus lacustris* shown in Fig. 11.

Variation of conditions.

The aquatic vegetation of the district varies considerably in the different river-valleys, both qualitatively and quantitatively, the variation corresponding with more or less obvious differences in the waters. Thus the Thurne alone has brackish waters, and the Yare is more tidal than the other rivers, for it has, compared with them, a strong tidal current. The waters of the Bure and the Ant are fresh, and subject to a slight tidal rise and fall, which causes very little current. The broads of the Bure are comparatively deep; they average six or seven feet (*c.* 2 m.) and may be as much as fifteen feet (*c.* 4·5 m.) in depth. Those of the Ant are shallower, not exceeding, as a rule, 4½ feet (*c.* 1·5 m.) in depth.

The difference of the fen vegetation of the different valleys is less marked than that of the vegetation of the

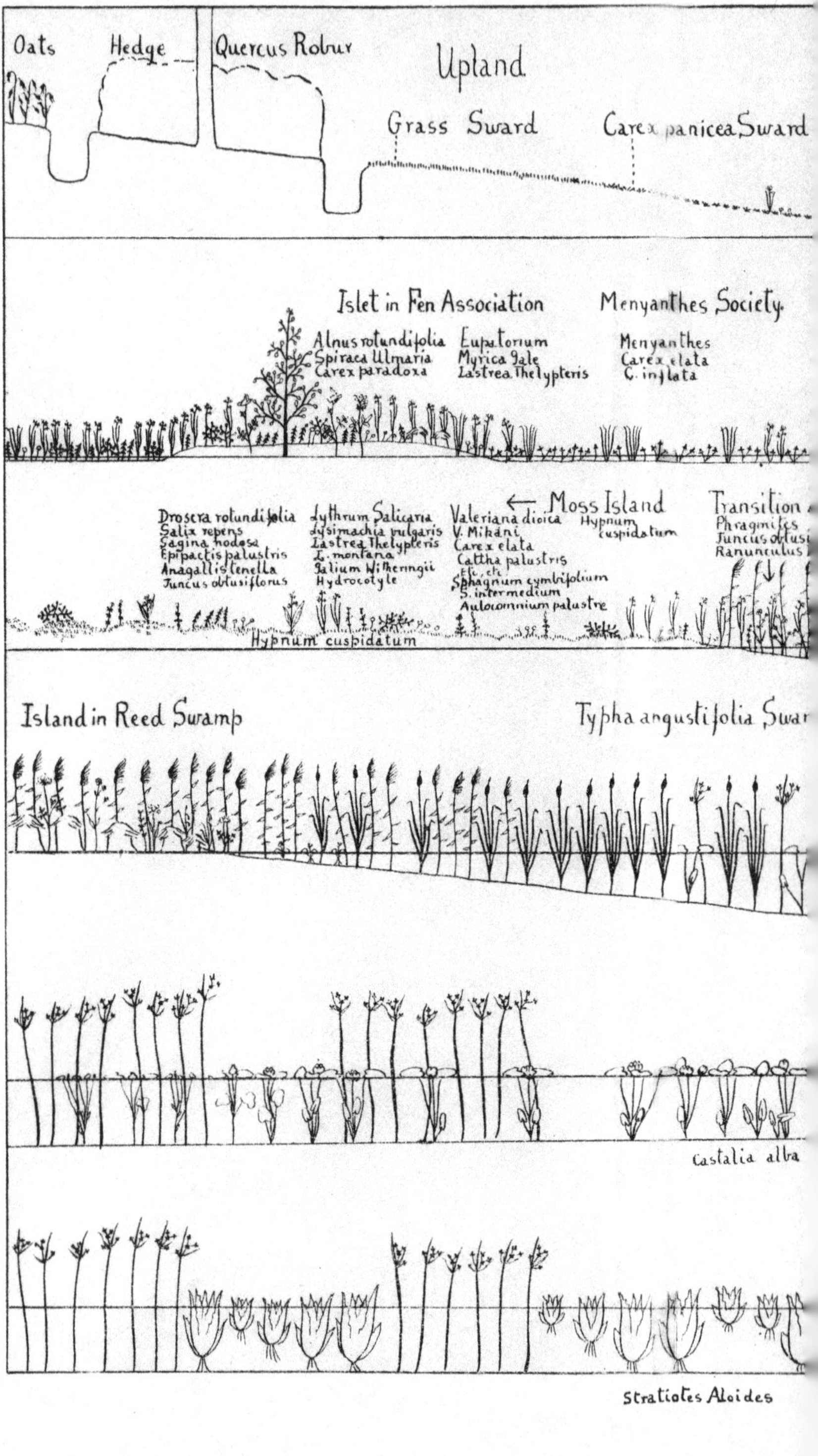

Oats
Hedge
Quercus Robur
Upland
Grass Sward
Carex panicea Sward
Islet in Fen Association
Menyanthes Society.
Alnus rotundifolia
Spiraea Ulmaria
Carex paradoxa
Eupatorium
Myrica Gale
Lastrea Thelypteris
Menyanthes
Carex elata
C. inflata
← Moss Island
Transition
Drosera rotundifolia
Salix repens
Sagina nodosa
Epipactis palustris
Anagallis tenella
Juncus obtusiflorus
Lythrum Salicaria
Lysimachia vulgaris
Lastrea Thelypteris
L. montana
Galium Witheringii
Hydrocotyle
Valeriana dioica
V. Mikani
Carex elata
Caltha palustris
Etc, etc
Sphagnum cymbifolium
S. intermedium
Aulocomnium palustre
Hypnum cuspidatum
Phragmites
Juncus obtusi
Ranunculus
Hypnum cuspidatum
Island in Reed Swamp
Typha angustifolia Swar
Castalia alba
Stratiotes Aloides

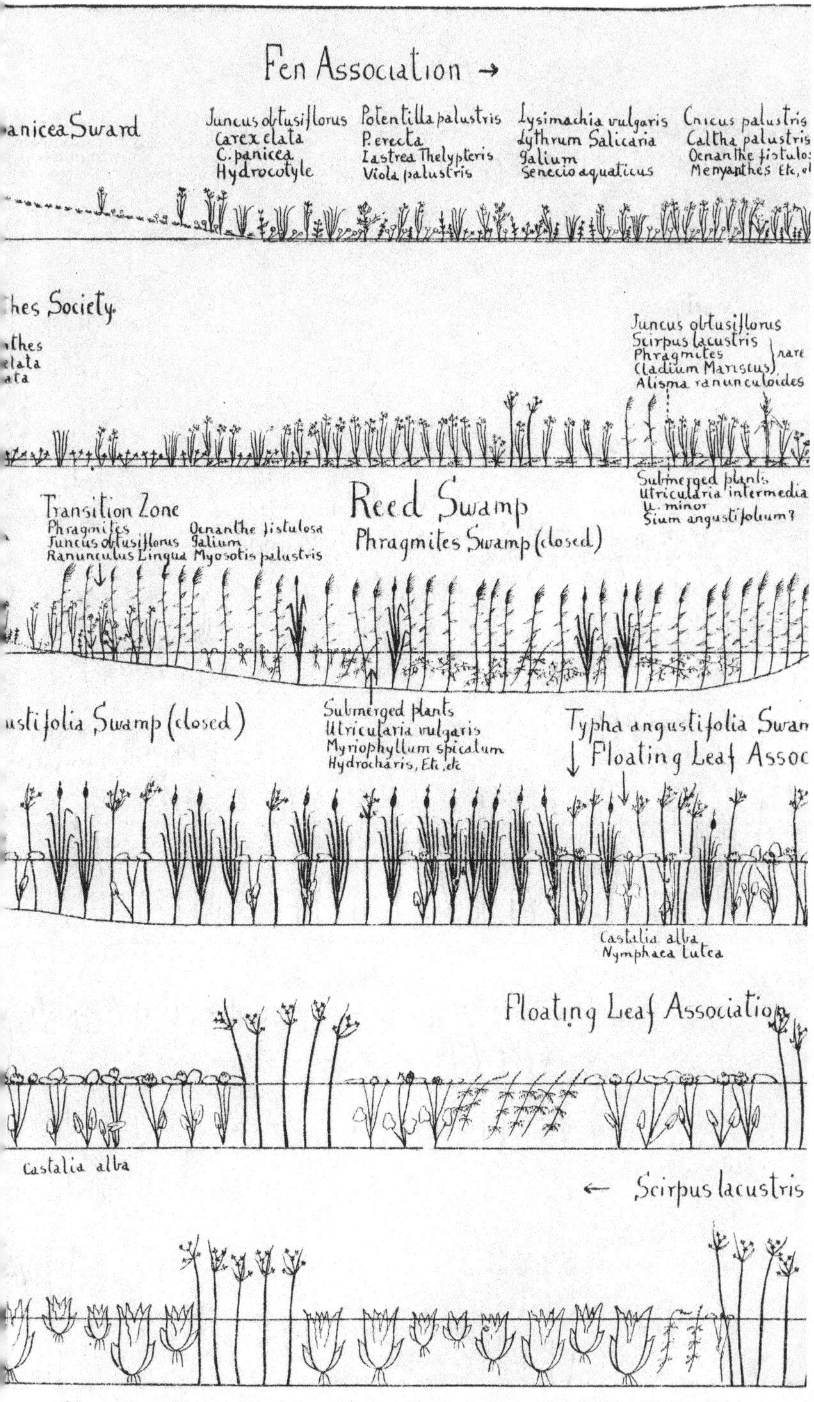

Fig. 11. Transect taken from Barton Broad to the Upland. Horizontal scale about 1 : 200; vertical scale about 1 : 150.

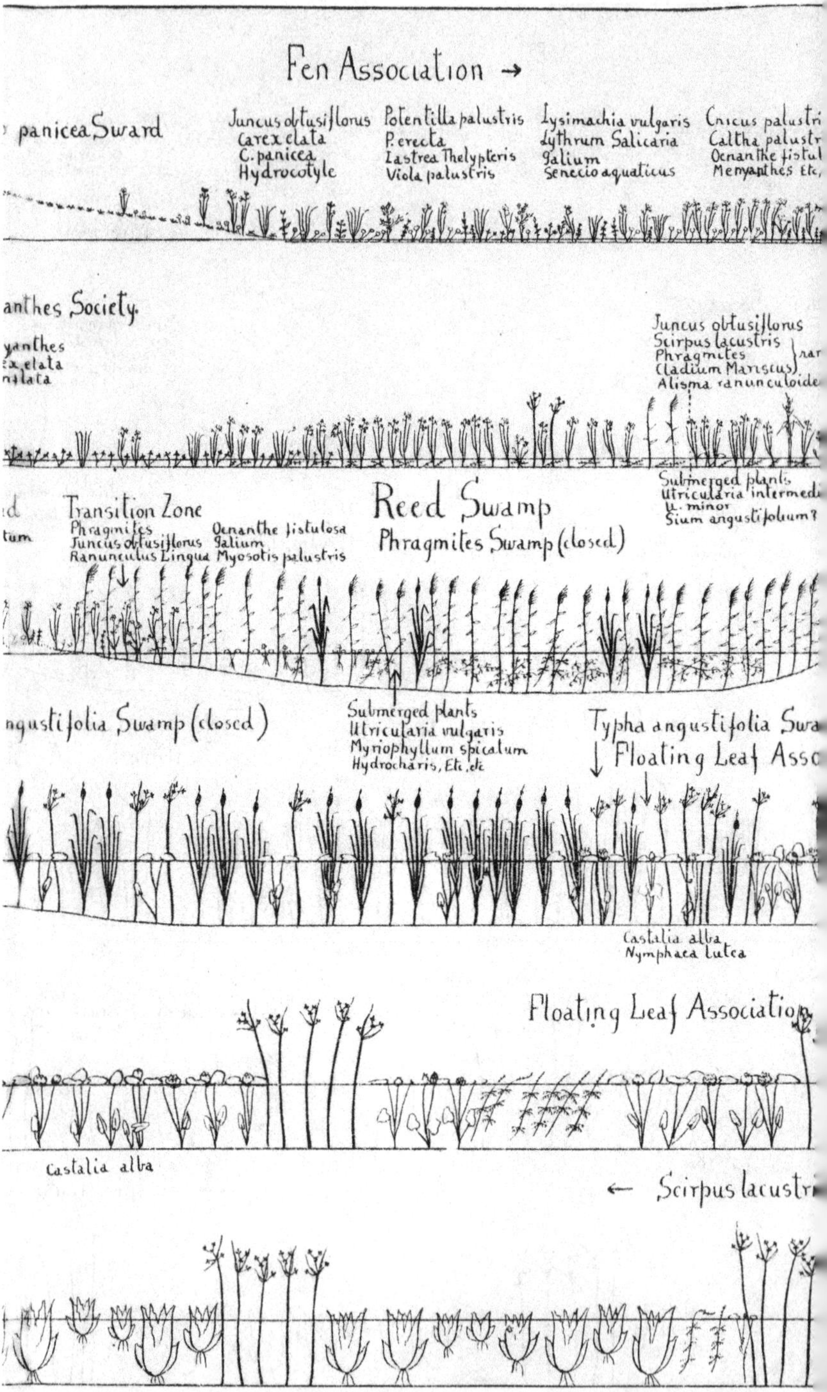

Fig. 11. **Transect taken from Barton Broad to the Upland. Horizontal scale about 1 : 200; vertical scale about 1 : 150.**

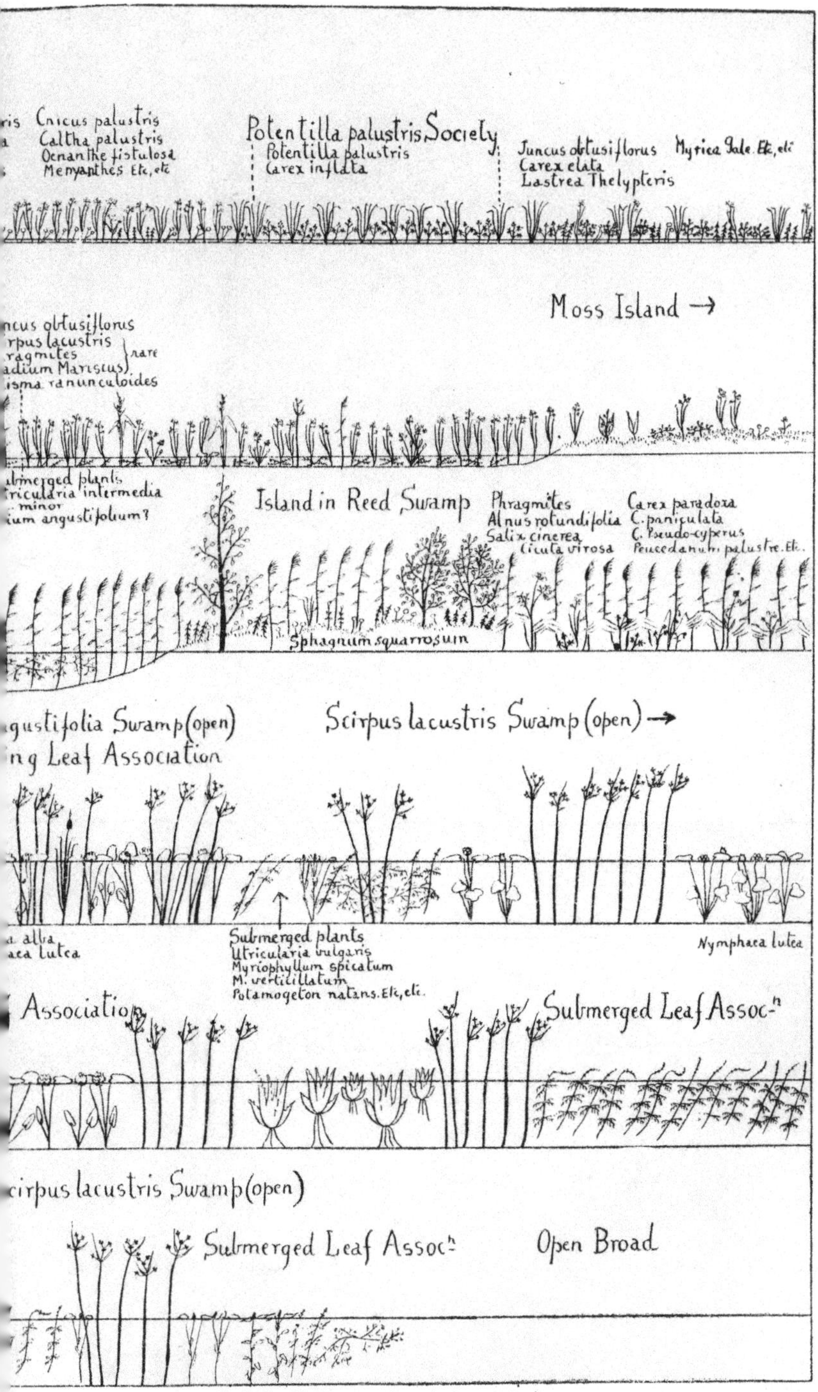

Cnicus palustris
Caltha palustris
Oenanthe fistulosa
Menyanthes Etc, etc
Potentilla palustris Society
Potentilla palustris
Carex inflata
Juncus obtusiflorus
Carex elata
Lastrea Thelypteris
Myrica Gale. Etc, etc
Moss Island →
ncus obtusiflorus
rpus lacustris
ragmites
adium Mariscus
isma ranunculoides
rare
ubmerged plants
tricularia intermedia
minor
ium angustifolium?
Island in Reed Swamp
Phragmites
Alnus rotundifolia
Salix cinerea
Cicuta virosa
Carex paradoxa
C. paniculata
C. Pseudo-cyperus
Peucedanum palustre. Etc.
Sphagnum squarrosum
gustifolia Swamp (open)
ng Leaf Association
Scirpus lacustris Swamp (open) →
a alba
aea lutea
Submerged plants
Utricularia vulgaris
Myriophyllum spicatum
M. verticillatum
Potamogeton natans. Etc, etc.
Nymphaea lutea
Association
Submerged Leaf Assocn
cirpus lacustris Swamp (open)
Submerged Leaf Assocn
Open Broad
Stratiotes Aloides
Potamogeton natans
P. praelongus
P. lucens
Myriophyllum spicatum
M. verticillatum
Utricularia vulgaris. Etc, etc.

broads. The table on p. 241, summarising the aquatic and fen associations of the four valleys, indicates these differences.

The facts suggest that the differences in vegetation are due primarily to the physical and chemical characters of the ground waters, since they become progressively less evident as we pass from the water to the land (fen). The differences in the waters, and consequently in the vegetation, probably have an historical basis, since it is highly probable that the Yare, which stands furthest apart from the other rivers, has had a different history from the Bure and its tributaries.

The conditions affecting the aquatic associations, so far as they can be deduced from direct inspection, are three:—

Circulation, aeration and silting.

First there is the circulation of the water, to which are incidental differences in aeration, in the thickness of silt covering the bottom and in the presence of the gaseous products of putrefaction which the silt (largely organic) conditions. The broads may be classed under three heads so far as this first group of factors is concerned: (1) those of the Yare in which the circulation of water is comparatively great, (2) those of the Ant and Thurne in which there is moderate circulation, and (3) those of the Bure, in many of which, owing to the closing of the dykes connecting them with the rivers, there is very little circulation indeed. The smaller broads of the Bure may in extreme cases be silted up to within about six inches (15 cm.) of the surface of the water.

The submerged-leaf association is most strongly affected by this group of factors. In the last-mentioned case it is practically absent. The floating leaf-association is also strongly affected, a few plants of *Castalia alba* and *Nymphæa lutea* alone occurring in these shallow broads which are almost silted up. The reed-swamp

Aquatic Vegetation of the Broads of the different River Valleys.

Association		Thurne	Ant	Bure	Yare
Submerged-leaf		*Dominants* Chara aspera C. hispida C. polyacantha Cladophora ægagropila Caucheria dichotoma Zygnema stellinum Spirogyra *spp.* Potamogeton interruptus *Associated plants, abundant* Ranunculus circinatus Myriophyllum spicatum (Plate XXI b) Elodea canadensis *Less abundant* Hippuris vulgaris Stratiotes Aloides Lychnothamnus stelliger *Rare* Naias marina	*Dominant* Stratiotes Aloides *Associated plants, abundant* Ranunculus circinatus Myriophyllum verticillatum M. spicatum Ceratophyllum aquaticum (agg.) *Less abundant* Sparganium simplex *forma* zosterifolium Chara hispida Lychnothamnus stelliger	*Association very poorly developed:* same species present as in Yare and Ant broads, but in smaller numbers	*Dominants* Potamogeton interruptus Zannichellia palustris (agg.) *Associated plants, abundant* Myriophyllum verticillatum Ceratophyllum aquaticum (agg.) Enteromorpha intestinalis *Less abundant* Stratiotes aloides *Rare* Various Characeæ
Free floating-leaf					*Dominants* Lemna trisulca L. minor L. gibba L. polyrhiza *Associated plants* Nymphæa lutea Hydrocharis Morsus-ranæ Potamogeton lucens *forma* acuminatus P. perfoliatus *Rare* Castalia alba [Utricularia vulgaris *absent*]

Open reed-swamp and Rooted floating-leaf	α-type	*Dominants* Typha angustifolia Phragmites vulgaris *Associated plants* A few floating leaf and submerged leaf aquatics of the same species as above and as occur in the open reed-swamp of the Ant broads	*Dominants* Typha angustifolia Phragmites vulgaris *Associated plants* as below	*Dominants* Typha angustifolia Phragmites vulgaris *Associated plants:* small numbers of the same species as occur in the Ant and Thurne broads	*Dominants,* *poorly developed* Typha angustifolia Phragmites vulgaris *Associated plants: as before*
	β-type		*Dominants* Scirpus lacustris (Pl. XXIIa) Nymphæa lutea Castalia alba *Associated plants, abundant* Lemna trisulca L. minor L. gibba L. polyrhiza Potamogeton natans P. lucens *f.* acuminatus P. perfoliatus P. crispus Riccia natans		
	γ-type			*Tussock swamp* (Pl. XXIIIb) Carex paniculata C. acutiformis	*Tussock swamp* Carex paniculata C. acutiformis
Closed reed-swamp	α-type	*Dominants* Typha angustifolia Phragmites vulgaris *Associated plants, very few* Utricularia vulgaris Myriophyllum spicatum M. verticollatum, etc.	*Dominants* Typha angustifolia Phragmites vulgaris *Associated plants* As in Thurne broads *On the free water side* Ranunculus Lingua Epilobium hirsutum Cicuta virosa Sium latifolium Myosotis palustris Solanum Dulcamara Mentha *spp.* Rumex Hydrolapathum Iris Pseudacorus Sparganium ramorum Acorus calamus Typha latifolia	*Dominants* Typha angustifolia Phragmites vulgaris *Associated plants* As in Thurne and Ant broads *On the free water side* As in the Ant broads	β-type *Dominants* Phalaris arundinacea Glyceria aquatica *Associated plants: absent except* *On free water side: as in Ant and Thurne broads: also a stooled form of* Phragmites vulgaris

association, however, flourishes, though not to the same extent as in the Thurne and Ant.

Depth of water and light. The second group of factors depends on the depth of the water, which affects the aquatic formation. The submerged and floating aquatics occur (in the Norfolk Broads) in water not exceeding six or seven feet (*c.* 2 m.) in depth, the reed-swamp associations in water up to about four feet (*c.* 1·2 m.) in depth.

The light factor, depending on the depth of the water and the amount of suspended material it contains, probably limits the occurrence of the submerged aquatics. The waters of the broads are very dark, owing to the quantity of suspended material; even on a sunny day the bottom of a broad six or seven feet deep is hardly visible. The broads of the Bure are either too deep or too stagnant for the submerged leaf and rooted floating-leaf associations; hence probably the very poor development of the two associations in these broads.

The depth of the water also directly limits the occurrence of the floating-leaf and reed-swamp associations, owing to the limited potentiality of growth in length of the shoots and petioles of these plants.

Shelter. The third condition is that of shelter. This affects the floating-leaf association, and depends on the existence of certain plants—the reed-swamp dominants, when growing in open association. In the large broads there is no separate rooted floating-leaf association, but a mixed association of open reed-swamp and floating-leaf plants, the latter extending only a foot or two beyond the edge of the former. This is strikingly developed in the broads of the Ant, where *Scirpus lacustris*, which forms the widest open reed-swamp in the district, is the dominant (Plate XXII a). Where, in the larger broads, the open reed-swamp forms a narrow belt only (*Typha angustifolia* and *Phragmites vulgaris*), the plants of the

Phot. M. Pallis

a. Open reed swamp association: *Scirpus lacustris* and *Castalia alba*. The dark line in the background is fen association. Sutton Broad.

Phot. M. Pallis

b. Detail of fen association: *Phragmites vulgaris*, *Juncus subnodulosus*, *Lastrea Thelypteris*, *Helleborine longifolia* (*Epipactis palustris*), *Lysimachia vulgaris*, etc.

Aquatic and Fen formations.

floating-leaf association are scarce, for they hardly extend beyond the shelter of the reeds (Table pp. 226–7).

The Fen Formation

The fen formation is characterised by an assemblage of over a hundred species of woody and herbaceous vascular plants, by a peat soil, and by alkaline ground waters whose level varies from a few inches below the surface of the soil in summer to a few inches above in winter and early spring.

Fen and carr associations.

Two associations have been identified with the fen formation, and these have been called the *fen association* and the *carr association*[1]. The vegetation of the fen association consists chiefly of herbaceous plants. It is dominated by grasses and sedges and averages about five feet (1·5 m.) in height. The carr association is a wet type of wood developed on peat.

Two varieties of fen and two varieties of carr have been recognised. The two varieties of fen are topographically separated within the district[2]; one occurs in the valleys of the Bure and of its tributaries the Ant and the Thurne (Bure valley fen association), whereas the other is confined to the valley of the Yare. The two varieties of carr, on the other hand, are not topographical but genetic varieties[3], which exist side by side in the same valley.

[1] The word "carr" (or "car") is applied locally to the fenwoods. In other parts of the country, *e.g.* in Lincolnshire and Yorkshire, it is used for a marshy or fenny tract of country. Both meanings are of very ancient use. The former appears to be connected with the Icelandic *kjarr*, the latter with Swedish *kærr* (Murray's *New Eng. Dict.*). [Editor.]

[2] They are apparently correlated with different edaphic conditions, and may, perhaps, be considered as sub-associations. [Editor.]

[3] *i.e.* stages in succession; they would be considered as different associations according to the terminology of this book. [Editor.]

Fen association.

The fen association is the initial association of the fen formation, and has been developed directly from closed reed-swamp. It occupies the river valleys between the water and the upland, coinciding with the formation of land. The fen association is a mixed association, and, considered as a whole, is dominated by a variety of species; locally single species become dominant. Various species of Cyperaceæ and of Gramineæ are the most abundant constituents of the vegetation (see list on pp. 233–4); on some fens, however, trees and shrubs are so numerous that the fen appears from a distance to be almost a wood. Such trees and shrubs as *Alnus rotundifolia, Betula tomentosa, Salix cinerea, Myrica Gale*, etc., occur scattered in this way on some fens. (See list of young fen carr on p. 240.)

General characters.

The fen association is constantly invading the reed-swamp along the zone where land and water meet. The gradual rise in level of the soil below the surface of the water by the continuous accumulation of organic debris and the consequent shallowing of the water provides conditions suitable for the establishment of the plants of the fen association, which are thus enabled to invade the reed-swamp and to displace the reed-swamp plants. The consequent change in the boundary of land and water is so rapid that a mapped boundary cannot long remain correct. Between the closed reed-swamp and the fully developed fen association there is a strip of transitional vegetation which is neither reed-swamp nor fen, but represents a loosely disposed invading vanguard of fen plants among the plants of the reed-swamp.

Succession from reed-swamp.

Though the process of the replacement of water by land involved in the succession from the plant communities of open water to those of the fen is perfectly gradual and continuous, the zone in which fen replaces reed-swamp is to be regarded as the zone of greatest change, because it is in this zone that land replaces water. It is for this reason that we separate the aquatic formation from the fen formation—the formation in which water is the differentiating factor of the habitat from that in which the shoots of the plants are sub-aerial, and peat with alkaline ground water is the differentiating factor of the habitat. Thus the invasion and displacement of the reed-swamp association by the fen association is to be regarded as the invasion and displacement of the aquatic formation by the fen formation.

The surface of the fen is uneven, because the growth of the peat is unequal, owing to the differences in habit and mode of growth of the different plants of the fen association. This uneven growth of the fen peat reacts on the plants in various ways, for instance through the initiation of local differences in the relation of soil surface and water level; thus a locally varying habitat is provided, and the fen vegetation has no common dominant but is made up of numerous small societies.

Varieties of fen.

The existence in the district of two varieties of the fen association, the Bure valley fen and the Yare valley fen, has already been alluded to.

The differences in the vegetation can be seen both in the societies of social plants dominating the association and also in the non-social species. The Bure valley fen shows in summer a dense growth of grasses and sedges (such as *Phragmites vulgaris, Molinia cærulea, Cladium Mariscus,* etc.) comparatively unrelieved by dicotyledonous plants. The Yare valley fen, at the same season, is a rich flowery water-meadow, bright with forget-me-not (*Myosotis scorpioides*), ragged robin (*Lychnis Flos-cuculi*), valerian (*Valeriana officinalis*), flag (*Iris Pseudacorus*), yellow meadow rue (*Thalictrum flavum*), etc., and the inflorescences of the local dominants, *Glyceria aquatica*, *Phalaris arundinacea* and *Poa trivialis*.

The contrast between the two varieties is perhaps even more striking in winter. The Yare fen is then conspicuously chequered, owing to the patches of colour formed by the sub-aerial shoots of the social plants. *Glyceria aquatica* dies down but remains green, *Phalaris arundinacea* turns light yellow, *Juncus subnodulosus* (*obtusiflorus*) turns a reddish brown (Plate XXIII a).

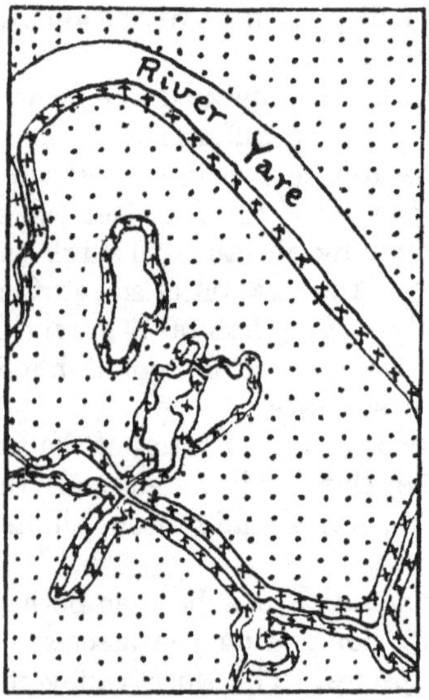

FIG. 12. Part of the remains of Surlingham Broad, most of which has been obliterated by the growth of peat. The river, dykes, and the pools of water representing all that remains of the broad are bordered by a narrow belt of reed-swamp.

Scale 1 : 10560.

Fen.

Reed-swamp (*Glyceria aquatica* and *Phalaris arundinacea*).

The Bure fen in winter, on the other hand, is much more uniform in colour. Straw-coloured expanses are due to *Phragmites vulgaris*, reddish tracts to *Juncus subnodulosus* (*obtusiflorus*) and greenish ones to *Cladium Mariscus*.

A rather striking difference shown by the two varieties of fen is that in the Yare fen association the important

Phot. M. Pallis

a. Yare valley fen association in winter: society of *Phalaris arundinacea*; the dark streaks in the distance represent societies of *Juncus subnodulosus*; in the foreground clumps of *J. glaucus* (not belonging to fen association) on edge of fen. *Glyceria aquatica* (light in colour) on nearer edge of dyke and in foreground.

Phot. M. Pallis

b. Tussock swamp with *Carex paniculata*: Swamp-carr with *Alnus rotundifolia* behind: on the right *Salix cinerea*. Cockshoot Broad (Bure valley).

Aquatic and Fen formations

a. Part of one wheel of the Cart, showing method of fixing the Thymus (?) Cheese, and Fish, etc., as offerings, and the method of coupling the Thymuses. The Man driving sits on the Cross-beam, on the place of [????] to which [????] of offering is attached.

b. Funeral Couch, with Cloth Canopy above, being drawn on a Sledge; Canopy painted on the Wall in [??] corner. Section of Wood Bier [????]

Aqqal and Ttu [??]mations

social species, *Glyceria aquatica* and *Phalaris arundinacea*, are the dominant plants of the association—the closed reed-swamp preceding the fen in succession (Fig. 12). The Bure fen, on the other hand, has only one species, *Phragmites vulgaris*, dominant in the fen and also in the closed reed-swamp. The other dominants of the Bure fen are confined to the fen association.

The difference in composition of the two varieties of fen association is shown in the following lists:—

BURE VALLEY FEN (Plate XXII b).

Dominants.

Phragmites vulgaris	
Cladium Mariscus	
Juncus subnodulosus (= J. obtusiflorus)	
Large societies.	
Molinia cærulea	
Small societies.	
Carex filiformis	
[Sphagnum with accompanying plants]	
Eriophorum angustifolium	
Accompanying plants.	
Rhamnus Frangula	f
Myrica Gale	f
Ligustrum vulgare	o
Liparis Loeselii	r
Drosera intermedia	r
Pyrola rotundifolia	r

YARE VALLEY FEN.

Large societies.	
Glyceria aquatica	
Phalaris arundinacea	
Small societies.	
Poa trivialis	
Accompanying plants.	
Lychnis Flos-cuculi	va
Spiræa Ulmaria	va
Galium palustre	va
Valeriana officinalis	va
Myosotis scorpioides	va
Ligustrum vulgare	f
Thalictrum flavum (agg.)	f
Eriophorum angustifolium	r
Cladium Mariscus	r
Rhamnus Frangula	o
Myrica Gale	o

The following is a representative list of species of the fen association common to all the river-valleys:—

Caltha palustris	f
Viola palustris	o
Lychnis Flos-cuculi	a
Hypericum quadrangulum	f
H. elodes	o
Lathyrus palustris	f
Spiræa Ulmaria	a
Angelica sylvestris	a
Peucedanum palustre	a
Œnanthe fistulosa	a
Œ. Lachenalii	a
Potentilla erecta	a
P. palustris	f
Lythrum Salicaria	a
Hydrocotyle vulgaris	va
Galium palustre	a

G. uliginosum	a	Molinia cærulea	ld
Valeriana dioica	a	Phragmites vulgaris	ld
V. officinalis	a	Calamagrostis lanceolata	ld
Lysimachia vulgaris	a	Carex disticha	f
Myosotis palustris	a	C. paradoxa	f
Menyanthes trifoliata	f	C. elata	f
Utricularia minor	o	C. panicea	f
U. intermedia	o	C. fulva	f
Epipactis palustris	f	C. flava (agg.)	f
Orchis incarnata	a	C. lasiocarpa	f
Juncus subnodulosus (= J. obtusiflorus)	ld	Ophioglossum vulgatum	f
Cladium Mariscus	ld	Lastrea Thelypteris	f

The mosses growing at or below the surface of the ground water are species of Hypnum, *H. giganteum* and *H. scorpioides*.

Sphagneta and moor associations.

Sphagneta composed of *Sphagnum cymbifolium, S. intermedium* and *S. squarrosum* are fairly abundant on the fens of the Bure, Ant and Thurne, though apparently absent from those of the Yare. They occupy levels in the association which are high relatively to the surface of the ground water, for they occur either on the edge of the fen close to the upland, or round stools of *Phragmites*, or round the bases of shrubs.

Close to the village of Potter Heigham there is a moor association on what has apparently been fen, close to the upland, dominated by Sphagnaceæ and by *Calluna vulgaris*.

The following are the more conspicuous plants of the association :—

Calluna vulgaris	Eriophorum angustifolium
Erica tetralix	Sphagnum *spp.*
Potentilla erecta	Polytrichum commune
Drosera rotundifolia	Aulacomnium palustre

The position of this and other moor associations close to the upland, suggests that they may be regarded as the wet continuations of heaths which probably occupied the upland soil before the days of cultivation. Even now a heath association with *Calluna vulgaris*, *Ulex europæus* and *Pteris aquilina* is not uncommon on the upland slopes bordering the valleys.

The position of the Sphagneta on the stools of *Phragmites* and round the bases of bushes accords well with the recent work of Paul[1] who has shown that the Sphagnaceæ secrete acids which are essential to these plants. The alkaline ground waters of the fen would tend to neutralise these acids, and hence we have a possible explanation of the position of the Sphagna above the general level of the ground waters.

The establishment of Sphagneta and the presence of moor associations on the fen suggest that the fen association may pass into the moor formation[2]. In the writer's opinion, however, this does not appear probable in the case of the fens of this region. The cushions of *Sphagnum* found on the fen are often bleached in summer, probably owing to insufficient rainfall, and the further growth of *Sphagnum* would in all likelihood be checked as soon as the plants reached a definite height above the general level of the fen.

The natural evolution of the fen formation at present results, not in the production of moor, but in the production of woodland (carr), and perhaps also, in certain places on the boundary of the fen, in the production of grassland. The Sphagneta are rather to be regarded as a sporadic development of moor in certain favourable spots.

[1] H. Paul, "Die Kalkfeindlichkeit der Sphagna und ihre Ursache," etc., *Mitt. d. k. bayrischen Moorkulturanstalt*, 1908, p. 78.

[2] Similar associations are well known to pass regularly into moor on the continent, and there is evidence from the peat records that they have done so in the past in this country. See p. 246 *et seq*. [Editor.]

Carr association.

Trees and shrubs, both scattered, and also in patches and belts, occur in the fen association of the different river-valleys. These patches and belts represent a transitional phase in the passage of the fen association into fen carr (Plate XXIV). They furnish a good example of true progressive scrub, as opposed to the retrogressive scrub so often seen in other formations.

In the fully formed carr some of the woody plants of the initial stage become dominant, while others become scarce, and others again disappear altogether. For instance, the fen association plants *Myrica Gale* and *Salix repens* persist in the loose scrub stage, but disappear when the carr is fully formed.

Two types of carr, the *fen carr* and the *swamp carr*, have been distinguished.

Fen carr and swamp carr.

The fen carr occurs in the fen and probably represents the ultimate stage of development of the fen formation of East Norfolk. That it succeeds and displaces the fen association is not merely inferred from the relations existing between the herbaceous and woody plants at the present day. Proof is furnished by some of the ordnance maps for which the survey was made in 1880, and which show fen where there are now carrs of considerable size.

Swamp carr occurs on the edge of open water, often fringing a broad. Typically it is a swamp wood, partially floating over the water. Swamp carr may develop independently of fen carr, or it may be formed by the extension over swamp of the trees and shrubs from a neighbouring fen carr. Similarly it may be the continuation over water of an upland wood (Fig. 13 a); in that case it often has a shallow peat soil mixed with sand and gravel.

Phot. M. Pallis

Young fen-carr developing on fen (Bure valley type). *Fraxinus excelsior*, *Betula tomentosa*, *Salix cinerea*, *Myrica Gale*. The fen association consists largely of *Phragmites vulgaris* and *Juncus subnodulosus* with many other species intermixed.

Fen formation

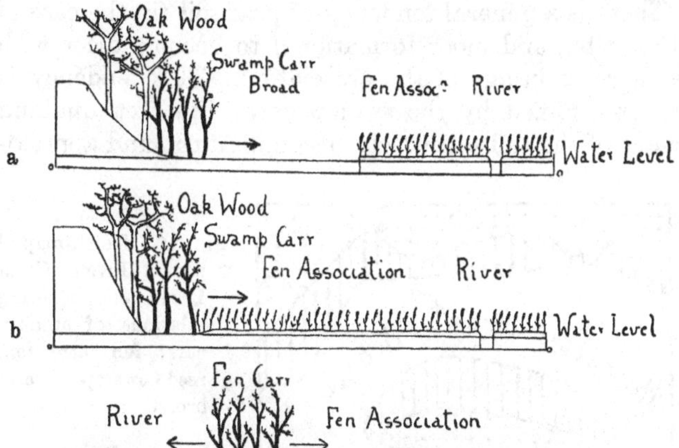

Fig. 13. Diagrammatic sections illustrating probable origin of carrs. Horizontal scale 1 : 21120, vertical scale 1 : 600. The arrows indicate the direction of the extension of the carrs.

Origin certain
- a. Swamp carr bordering on upland oakwood. It is apparently increasing at the expense of the Broad.
- b. Swamp carr abutting on fen; further extension as fen carr.
- c. Fen carr surrounded by fen. Originated as fen carr.

Origin uncertain
- d. Fen carr on edge of broad. It may have replaced a swamp carr. At present it is entending as fen carr only.
- e. Fen and swamp carr extending on the fen side only as fen carr.
- f. Fen and swamp carr extending as such on the two sides.

There is a general tendency of peat soil (in the case of both the fen and moor formations) to become drier with the increase in age of the association. This tendency is well exemplified by the swamp carr; the wet quaking surface of the soil ultimately becoming firm, and approxi-

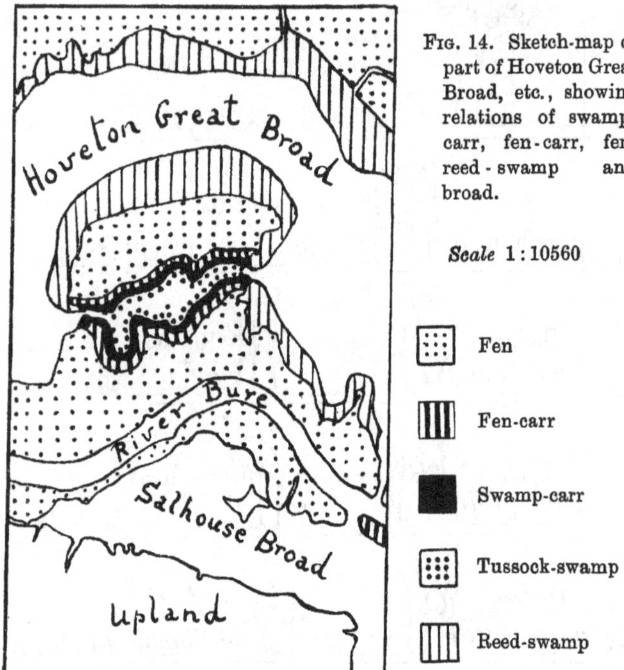

FIG. 14. Sketch-map of part of Hoveton Great Broad, etc., showing relations of swamp-carr, fen-carr, fen, reed-swamp and broad.

Scale 1:10560

mating in dryness to that of the fen carr. With this

Convergence of their vegetation.

increase in dryness the vegetation of the two types of carr tends to converge, so that the origin of a particular carr in a late stage of development cannot always be determined by inspection. Position is sometimes a sufficient indication of the origin of a carr, as, for instance, of those represented in Fig. 13 a, b, c. On the other hand the position of the carrs shown in Fig. 13 d, e, f, leaves their origin uncertain.

Evidence of origin.

The depth and nature of the underlying peat furnishes important evidence on this point. A layer of amorphous peat generally covers the bottoms of the broads, so that if a carr with

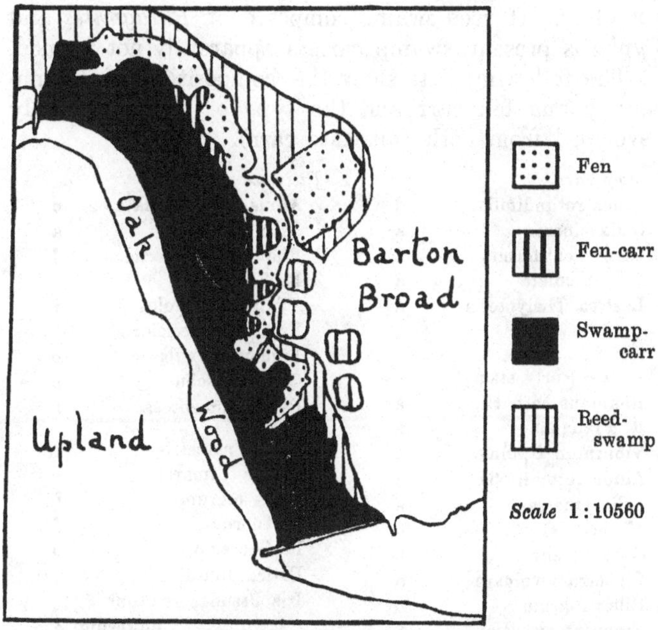

FIG. 15. Sketch-map of Heron's Carr, Barton Broad. Swamp-carr and fen-carr growing side by side.

the characteristics of fen carr is growing on amorphous peat, this fact would point to the fen carr having originated as a swamp carr, whereas if the carr were growing on fen peat the conclusion would be that the fen association had preceded it and that therefore the carr was genetically fen carr.

Association of carr with tussock swamp.

Swamp carr often occurs associated with a narrow reed-swamp consisting chiefly of tussocks of *Carex paniculata* and *C. riparia* (Fig. 14 and Plate XXIII b). Whether the swamp carr invades the tussock swamp or whether the two develop contemporaneously is, however, not yet clear. If reed-swamp composed of *Phragmites* and *Typha* is present, swamp carr is apparently not formed.

The following lists show the composition of swamp carr, young fen carr and the type of carr ultimately developed from both (ultimate carr).

Swamp carr.

Alnus rotundifolia	d
Salix cinerea	a
Carex acutiformis	a
C. paniculata	a
Lastrea Thelypteris	a

Fen carr (early stage).

Rhamnus catharticus	a
R. Frangula	a
Viburnum Opulus	a
Alnus rotundifolia	a
Salix cinerea	a
Myrica Gale	la
Salix repens	la
Ligustrum vulgare	o
Ribes nigrum	o
Fraxinus excelsior	f
Betula tomentosa	a
Quercus Robur	o

The undergrowth is formed by many herbaceous fen plants.

Ultimate carr.

Alnus rotundifolia	d
Salix cinerea	a
Rhamnus catharticus	f
R. Frangula	f
Viburnum Opulus	f
Fraxinus excelsior	f
Ligustrum vulgare	o
Quercus Robur	o
Betula tomentosa	f
——	
Caltha palustris	f
Spiræa Ulmaria	f
Ribes nigrum	f
R. rubrum	f
R. Grossularia	o
Urtica dioica	a
Iris Pseudacorus (not flowering)	a
Carex paniculata	a
C. acutiformis	
Lastrea Thelypteris	a
Osmunda regalis	r
Mnium hornum	a

SUMMARY OF THE ASSOCIATIONAL DIFFERENCES IN THE FOUR RIVER-VALLEYS

	Association	Bure	Ant	Thurne	Yare
Fen formation	Fen carr	×	×	Many trees, almost becoming fen carr	×
	Swamp carr	×	×		×
	Fen	*a*	*a*	*a*	*β*
Aquatic formation	Closed reed-swamp	*a*	*a*	*a*	*β*
	Open reed-swamp and Floating leaf	*Carex acutiformis* and *C. paniculata*	×	×	same as Bure
		Phragmites vulgaris and *Typha angustifolia* with Rooted floating-leaf association (poorly developed)			×
			Scirpus lacustris and Rooted floating-leaf association (well developed)		
					Free floating association
	Submerged leaf	Poorly developed	Chiefly *Stratiotes*	Submerged-leaf association of brackish water	Well developed (resembles Bure)

× = present

Other Associations

Grass fen.

Although the normal final association of the fen formation is carr, there are indications that on the boundaries of the fen area the fen association may pass into grassland. Three types of grass fen have been distinguished:—

(α) Fen pasture with *Poa trivialis*. This association has the appearance of fen, but grasses (other than the characteristic fen grasses), and especially *Poa trivialis*, are abundant. The association is developed on peat about six feet (*c.* 1·8 m.) in depth, overlying ooze. The water level is just below the surface of the peat. The type occurs in the Yare valley, and except for the abundance of *Poa trivialis* and the scarcity of trees is practically Yare fen.

(β) This type occurs on the boundary of the fen association in the Bure valley. *Phragmites vulgaris* is the dominant plant, pasture grasses occur but are much scarcer than in (α); trees and shrubs are equally scarce. The soil is here also shallow peat overlying ooze.

These two types of grass fen are each related in a similar way to the typical fens of their respective valleys.

(γ) This type of grass fen occurs on humified peat, which is sometimes shallow and sometimes of great depth. It is, comparatively speaking, dry. Grasses such as *Agrostis vulgaris*, *Anthoxanthum odoratum*, *Holcus lanatus*, *Sieglingia decumbens* and *Briza media* are abundant, and scattered among the grasses are sharply defined societies of *Juncus subnodulosus* (*obtusiflorus*). Some of the less characteristic plants found in the fen association also occur. The two most typical plants of this association are *Sieglingia decumbens* and *Linum catharticum*; in the Yare valley *Thalictrum flavum* (agg.) is present. The soil of this association is drained and provides poor pasture. It is rather characteristic of the Waveney valley and also occurs near Wicken Sedge Fen in Cambridgeshire.

Submaritime fen pasture association.

This association occurs in the neighbourhood of the sea between Horsey Mere and Hickling Broad and the line of coastal sand dunes. It is developed on peat, which may be as much as 20 feet (*c.* 6 m.) in depth. In the peat are buried trunks of *Betula* and of *Salix cinerea* (or ? *S. caprea*). The remains of *Phragmites* and seeds of *Menyanthes trifoliata* are also abundant.

The ground water is alkaline and brackish. The vegetation consists chiefly of grasses, especially *Agrostis vulgaris,* and of rushes, *Juncus effusus* and *J. maritimus.* Scattered plants of the fen association and local societies of bog plants, *e.g. Sphagnum* and *Eriophorum angustifolium,* also occur. The soil of the association is very infertile and patches which are bare of vegetation occur. The soil is drained, and the association is to some extent pastured.

The existence of small isolated areas of fen, having *Juncus effusus* and *J. maritimus* as commonly associated plants, together with the frequent occurrence of stunted *Phragmites* and the remains of fen plants in the peat, suggest that this association is possibly derived genetically from the typical fen association.

The following is a list of representative plants of the submaritime fen pasture association:—

Dominants.
- Juncus effusus
- J. maritimus
- Agrostis tenuis (= A. vulgaris)

Small societies.
- Juncus Gerardi
- Eriophorum angustifolium
- Phragmites vulgaris (stunted)

Abundant.
- Holcus lanatus
- Sieglingia decumbens

Scattered.
- Viola palustris
- Spergularia salina
- Peucedanum palustre
- Potentilla erecta
- P. reptans
- Plantago maritima
- P. Coronopus
- Triglochin palustre
- T. maritimum

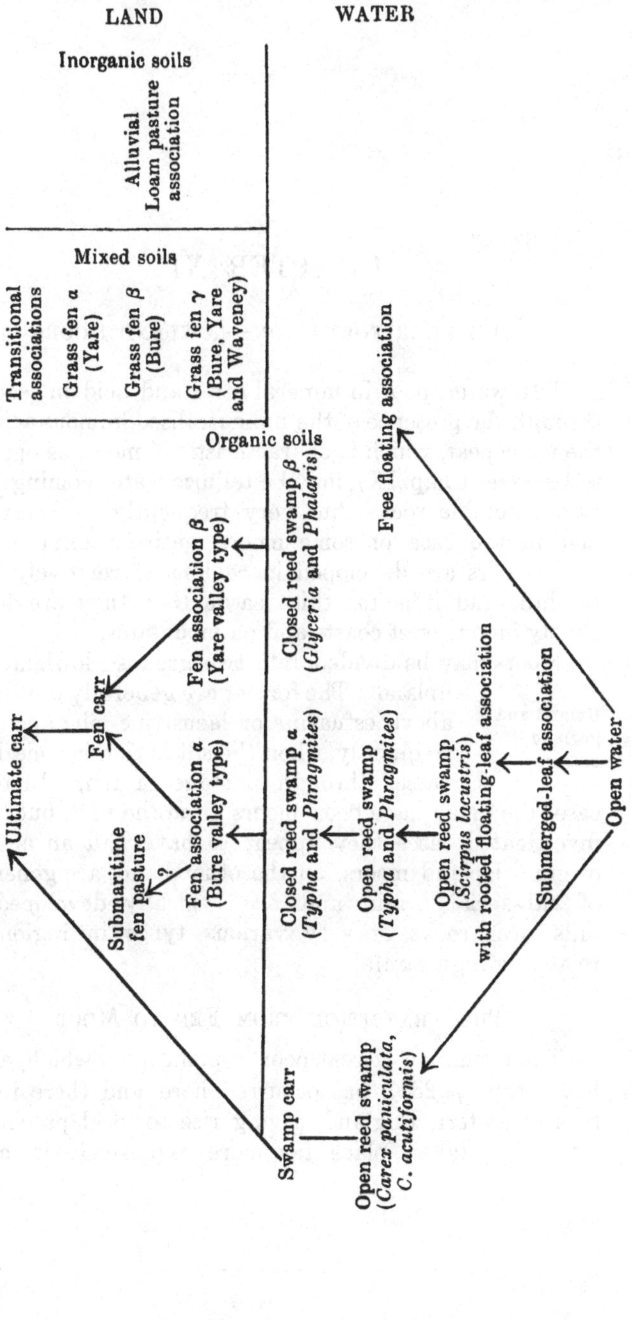

DIAGRAM SHOWING THE PROBABLE GENETIC RELATIONSHIPS OF THE AQUATIC AND FEN ASSOCIATIONS OF THE RIVER VALLEYS OF EAST NORFOLK.

CHAPTER XI

THE MOOR FORMATION—LOWLAND MOORS

THE water, poor in mineral salts, and acid in reaction through the presence of the unneutralised humous acids of the moor peat, which is characteristic of moors as opposed to fens (see Chap. IX), may be telluric water coming from very insoluble rocks, but very frequently it is mainly, and in the case of some moors, entirely aerial water. Such moors are developed in regions of relatively high rainfall, and it is for this reason that they are found chiefly on our west coasts and on mountains.

Upland and lowland moors.

Moors may be divided into two groups—lowland and upland. The former are generally developed above estuarine or lacustrine silt, and very frequently, when the silt is rich in lime, have passed through a stage of fen. In other cases they may have been moors from the first, but in all investigated cases they appear to have had an aquatic origin. Upland moors, on the other hand, are generally of sub-aerial origin and are typically developed on hills over rocks of very various types in regions of relatively high rainfall.

THE TRANSITION FROM FEN TO MOOR

The formation of peat poor in plant-food, which, as we have seen (p. 234), has occurred here and there in the fens of eastern England, giving rise to local patches of moor, has taken place far more extensively in many localities. Weber[1] describes the typical succession of stages leading from fen to moor as recorded in the peat of the North German moors, and quite similar successions have been recognised in this country. A case that has been closely studied is the development of moors on deep peat near the coast of Morecambe Bay in North Lancashire and the adjoining parts of Westmorland. The lower layers of this peat show clear evidence that the area was formerly occupied by marsh vegetation on silt, and that this was succeeded by fen vegetation, which in its turn gave place to the moor type at present existing.

The Lowland Moors ("Mosses") of Lonsdale (North Lancashire), and their Development from Fens

By W. Munn Rankin

Estuarine and lacustrine moors.

Two cases of moor succession are represented in the peatlands that occur within a short distance of the coasts of Morecambe Bay: the first that of the "estuarine" moors developed on the level silt, chiefly maritime, which filled up the greater part of the long fiords or "drowned valleys" reaching from the coast many miles inland; the second that of the "lacustrine" moors which encroached upon the waters of shallow lakes (locally "tarns") held up among the inequalities of the ground moraine of an ancient ice-sheet which had brought fragments of slate, limestone, and igneous rocks from the north.

The first has resulted in moors (locally called "mosses") many square miles in extent; the second in scattered, comparatively small, areas of peat. The exploitation of many years has left various fragments of moorland, and in some places there are wide, almost untouched moorland wastes of the estuarine group; while

1 Weber, *loc. cit.*

all except a few patches of the lacustrine moors have been obliterated.

The "mosses" of the Gilpin and Winster valleys to the north-east of Grange-over-Sands, of which one, Foulshaw Moss, is still many hundreds of acres in extent, will be taken as examples of the estuarine group. The few acres of peat diggings below Yealand Storrs in the Kent valley will illustrate the formation of the peat which choked up the western part of a lake that formerly covered several hundred acres between Yealand Storrs and Holme.

The Estuarine Moors

Development of the moors. Nearly, if not quite, the whole surface of these are now occupied by associations belonging to the moor formation. In places, however, where the edge of the peat impinging on the hard land has been left untouched, a zonation representing an earlier phase of development of the moor as a whole can still be recognised. Sections through the peat of the fully developed moor afford unmistakable evidence that the earliest vegetation on the estuarine mud was that of a marsh which passed later into a fen. The fen in turn gave place to associations belonging to the moor formation, which at length terminated in a *Sphagnum*-moor. The succession following this stage was at least partly brought about by the activity of man in removing peat and digging drains, thus leading to a marked desiccation of the surface of the moor, and to its colonisation by associations characteristic of much drier conditions. A subordinate succession also is shown in the colonisation of the bare peat surfaces formed by the waste fragments of peat thrown down in the process of digging.

The present vegetation of the moors. Foulshaw Moss, the largest and most representative example, now occupies about two square miles (*c.* 5·2 sq. km.), though formerly, with other "mosses," it reached from the sea for seven miles inland. Its surface slopes gently up from the edge, where the moss abuts on the hill slope, but the greater part of it is practically level. It bears for the most part an association of cottongrass (*Eriophorum vaginatum*) mixed with *Scirpus cæspitosus*. On the tussocks formed by these plants grow shrubby ling (*Calluna vulgaris*), cross-leaved heath (*Erica Tetralix*), and *Andromeda Polifolia*; while mosses characteristic of wet moors, chiefly *Sphagnum*, spread between them. The occasional pools are filled with a matrix of *Sphagnum*, traversed by the rhizomes of *Eriophorum angustifolium* and *Rhynchospora alba*, and the lanky stems of the cranberry (*Oxycoccus quadripetala*); here also the bog asphodel (*Narthecium ossifragum*), and the sundews (*Drosera rotundifolia*, *D. intermedia* and—most rarely—*D. anglica*) find a sufficiently firm substratum for their growth.

Gradual drying of the moors.

There is evidence, however, that Foulshaw Moss is much drier than it was, say, seventy years ago, when it had a greater expanse of bog-moss and was quite impassable, except perhaps by skating. Where the peat has been cut the sections show that beneath the few inches of recent peat there is everywhere many feet of loose *Sphagnum*-peat. The occasional patches of existing bog-moss, with its associated plants, are probably the relics of that earlier condition. The gradual desiccation, greatly aided if not determined by trenching and removal of peat, has resulted in the appearance of the Eriophoretum. On the tussocks of the cottongrass and so elevated a little above the water level inhabited by the bog plants, grow the species of heath, which become more abundant towards the drier places, such as the edges of deep drains.

As the peat is cut back and drained, its mass contracts at the edges, and becomes fissured by deep cracks. On such drying peat the cottongrass wages an unequal struggle and gives way to the ling. At the stage of drying marked by the mixed association of *Eriophorum vaginatum* and *Calluna,* such as obtains widely on Foulshaw Moss, the conditions appear to be suitable for the invasion of birch and Scots pine, whose seeds are blown from neighbouring woods and plantations. As the moor passes still further under the dominance of the ling, the trees increase and form a scrub. Later on small thickets concentrate about the drier spots, and by the time the cottongrass has disappeared, are sufficiently advanced to be called woods. Here, in addition to the pioneer species, *Betula tomentosa* and *Pinus sylvestris,* we find the mountain ash or rowan (*Pyrus Aucuparia*), the yew (*Taxus baccata*), the alder (*Alnus rotundifolia*) and the sallows (*Salix caprea* and *S. cinerea*). Occasionally a stage is reached in which the ground vegetation of the wood is dominated by *Molinia cærulea* mixed with ling and bilberry (*Vaccinium Myrtillus*) and other species of heathy woods.

Invasion by trees.

Thus we have the following succession represented on these moors:—

Betuletum tomentosæ (Birchwood)
↑ Birch thicket
| Birch scrub
Callunetum vulgaris (Heather moor)
↑ Eriophoro-Callunetum
| **Eriophoretum vaginati** (Cottongrass moor)
Sphagnetum cymbifolii (Sphagnum moor)

Foulshaw Moss is the nesting ground of fairly numerous lesser black-backed gulls (*Larus fuscus*), and the dung of the birds together with the scraps of shell-fish brought in from the bay for food have completely changed the character of the vegetation over considerable areas. Heather, cottongrass and bog moss have given place in these spots to a mixture of *Holcus lanatus,* rushes (*Juncus conglomeratus*), willow herb (*Epilobium angustifolium*), and other plants more or less alien to moor habitats.

Marginal associations. Where the edge of the moor abuts on the adjacent slate hills it carries associations of a different type.

The extreme edge is occupied by an alder wood, in which the dominant tree (*Alnus rotundifolia*) is accompanied by sallow (*Salix caprea* and *S. cinerea*), alder buckthorn (*Rhamnus Frangula*), mountain ash (*Pyrus Aucuparia*), yew (*Taxus baccata*), oak (*Quercus sessiliflora*), juniper (*Juniperus communis*), and birch (*Betula tomentosa*). The floor of this wood is very swampy, with many pools, in which grow bog-bean (*Menyanthes trifoliata*), marsh cinquefoil (*Potentilla palustris*), the great stooled sedge (*Carex paniculata*) and other species of *Carex*, *Phragmites vulgaris*, and ferns (*Lastrea aristata*, *L. spinulosa*, *Blechnum Spicant*). *Gentiana Pneumonanthe*, *Hottonia palustris* and *Osmunda regalis* also formerly occurred in this association, but they are now extinct.

Among the rotting timber of fallen trees are tussocks of *Molinia cærulea*; while further away from the extreme edge of the marsh this grass becomes dominant and the firmer substratum presented by its tussocks serves as a *nidus* for plants of ling (*Calluna vulgaris*) and tormentil (*Potentilla erecta*). *Sphagnum* appears between the tufts of *Molinia*, and with the bog moss, species of *Drosera* and *Narthecium ossifragum*. Here the Molinietum is seen in transition to the typical moor. Further towards the centre of the moor *Molinia* gives place to *Eriophorum vaginatum* and the region of the Eriophoretum already described is reached. *Myrica Gale* accompanies the *Molinia* and extends beyond it into the Eriophoretum.

Here then we have the following zonation:—

Eriophoretum vaginati
Molinietum cæruleæ[1]
Marsh formation with alder wood (Alnetum rotundifoliæ).

[1] The conditions on the edge of the moor are clearly different, and give rise to a different vegetation, from those in the middle of the

Stratigraphical Succession of the Peat.

A typical section shows a thick layer of peat 12 to 15 feet (*c.* 4·5 m.) in depth, underlying the shallow superficial thready peat formed by the existing Eriophoretum, and resting on a basis of sandy clay. The lower quarter of this mass is composed of the remains of plants characteristic of fen, the upper three-quarters of bog moss and cottongrass, *i.e.* of plants characteristic of moor peat. A layer of wood peat, generally consisting of birch timber, occasionally of pine, invariably separates the fen peat from the moor peat. The moor peat, above the wood-layer, can be further separated into an upper more recent stratum, composed of bog mosses of the group of *Sphagnum cymbifolium*, and a lower stratum of the much rotted, red, slimy fragments of other species of *Sphagnum*. The upper layer of bog moss is always pared off when the peat is cut, and thrown into the ditch in front of the face of the peat section to make a basis of soil for cultivation. Elsewhere in the district it has been exploited as "peat-moss-litter."

The stratification may be summarised as follows:—

Moor peat	5	Surface Eriophorum peat	3 inches	·075 m.
	4b	Upper Sphagnum peat...	3 feet	·9 m.
	4a	Lower Sphagnum peat...	3 to 7 feet	·9 to 2·13 m.
Fen peat	3	Wood peat	9 inches	·22 m.
	2	Reed and sedge peat ...	3 to 4 feet	·9 to 1·2 m.
	1	Basement bed of sandy clay.		

moor. The difference is probably due to the drainage water from the neighbouring hills by which the edge of the moor is fed. The Alnetum with its marsh plants and the Molinietum succeeding it recall the zonation of the valley moors of the New Forest (p. 261). Probably, as in that case, the Molinietum is intermediate between the fen formation and the moor formation as defined in this book. Such associations and their conditions of life have not received adequate study in this country. The problem of the edaphic conditions of Molinieta is particularly interesting, because they occur both in fen and also in acid peaty soils. [Editor.]

The different strata will now be considered in rather more detail, beginning at the base.

1. *The basement clay.* This is a deposit of estuarine silt of unknown depth, and is largely composed of the waste of banks of glacial debris originating from limestone, slate and various sandstones, which formed and still form much of the shores of parts of the neighbouring Morecambe Bay. The particles are sufficiently fine to form a stiff plastic mass when wet, easily loosening into dust on drying. The top layers are much mixed with plant remains, and it is evident from the occurrence of vertical roots in the clay, and from the character of the plant remains occurring immediately above, that at one time this silt formed the soil of a marsh.

2. *The reed-swamp and fen peat.* Timber, chiefly of oak, and sometimes of great length, is occasionally found resting on the bottom clay. Near the edge of the peat lands the lowest peat is full of wood, not only oak, but also yew and hazel. The marsh and fen peat of this lowest layer forms a spongy compost containing fragments of stems, leaves, and rhizomes, embedded in a more or less amorphous paste, from which seeds may be separated. Fragments which lie horizontally are penetrated by vertical roots, indicating that the peat accumulated *in situ*. The remains include *Phragmites vulgaris*, *Cladium Mariscus*, *Scirpus fluitans*, *S. pauciflorus*, *Carex* spp., *Molinia cærulea*, and *Menyanthes trifoliata*, indicating the existence of a reed-swamp (with local patches perhaps of open water) and of fen succeeding it.

3. *The wood peat.* This very constant layer is generally composed of well-preserved birch bark and rotten birch wood. Near the head of the Gilpin estuary the birch is accompanied by the Scots pine (*Pinus sylvestris*), whose bark and timber are both very well preserved. These trees must have colonised the peat of the mixed

fen association, probably when the fen peat had become drier, and perhaps poorer in mineral content, by growth of the surface above the water-table. The pine trunks show 55 annual rings. The further growth of the pine was stopped by an engulfing bog of *Sphagnum* and

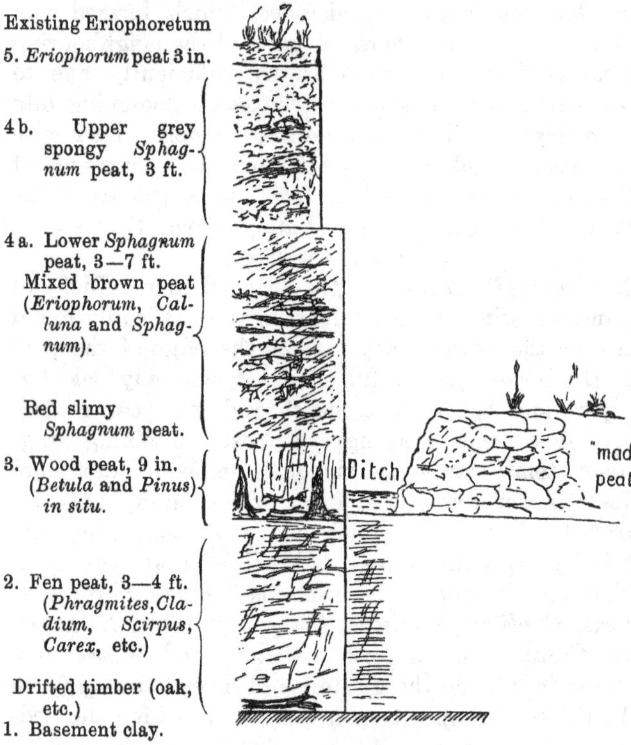

FIG. 16. Generalised section of the peat of the Lonsdale Estuarine Moors.

Eriophorum. Where the stumps are still erect they are two to three feet high and tapered sharply to a point. This probably represents the depth of bog which sufficed to arrest the growth of the pine wood, causing rotting of the tree trunks at the surface of the bog, and the toppling over of the shafts of the trees. Some stumps show signs of charring, but none shows the mark of an axe.

4 *a*. *The lower moor peat* is composed chiefly of varying proportions of *Sphagnum*, *Eriophorum* and *Calluna*. *Calluna* is more abundant near the wood-layer than at higher levels. This drier layer, in which the ling is abundant, is succeeded in places by several feet of a red slimy peat composed almost entirely of much decayed bog moss in which the stem cores (mechanical tissue of the medulla) are the only well-preserved parts. Mixed with this are seeds of *Rhynchospora alba* and stems of *Oxycoccus quadripetala*. This moss peat dries hard and black and burns well. Above, and often replacing it, is a mixed peat characterised by the tough sheaths of cottongrass, and the tough gnarled twigs of heather, with pockets of spongy bog-moss. This brown peat also burns well, though when much mixed with moss it easily crumbles.

4 *b*. *The upper moor peat* is often sharply defined from the lower. In some places an actual "unconformity" is seen, and this, taken together with the occurrence of an ancient "corduroy" road[1] at the level of separation of the upper and lower peats, makes it highly probable that at this time there was a check in the growth of the moor, possibly a drier stage, followed by the resumed growth of the wet moor corresponding with the upper peat. Unfortunately there is no evidence as to the date of this check, which perhaps made the construction of the "corduroy" road possible. This upper peat is composed of species of *Sphagnum* of the *cymbifolium*-group and is very well preserved.

5. The upper *Sphagnum* peat passes up into the thin layer of *Eriophorum* peat corresponding with the existing moor vegetation.

[1] A road made of logs of wood placed transversely.

The Lacustrine Moors

The peat that still remains in the basin of the ancient "Burton Lake," which has long completely disappeared, also shows the primary succession from fen to moor. The surface vegetation has, however, been entirely destroyed, as well as the surface layers of the newer peat. A few sections are nearly, though none is wholly, complete.

Stratigraphical Succession of the Peat.

The following stratification may be distinguished:—

		Constitution of peat.	Corresponding plant-community.
7	b	Upper *Sphagnum* peat	Moor
	a	Lower *Sphagnum* and *Eriophorum* peat	
6		Wood peat...	Birch wood
5	b	Upper sedge and reed peat	Fen and reed-swamp
	a	Lower sedge and reed peat	
4		*Hypnum* peat	Aquatic formation (Open water of lake)
3		Amorphous peat	
2		Shell marl...	
1		Lake mud...	

1. *Lake mud.* This is a fine grey silt, probably washed out of the ground moraine of the ancient ice-sheet, and is largely composed of fragments of limestone and slate. The upper layers show horizontal intercalations of organic matter and are apparently stained with humus, but are nowhere penetrated by roots.

2. *Shell marl.* This occupies a considerable portion of the lake basin and has an average depth of 18 inches (·5 metre). It is not now found immediately below undisturbed peat. The marl contains 95 per cent. of calcium carbonate and is almost wholly composed of the shells and fragments of shells of the common pond snail (*Limnæa pereger*); the general appearance of the shells suggests rather severe conditions of life. The lake waters, judging from those of the two streams which now drain the basin,

one of which has a total hardness of 29·7, the other of 20·7 parts per 100,000, must have been highly calcareous.

3. *Amorphous peat.* This layer, which may reach a thickness of six inches, is yellow, compact, and easily

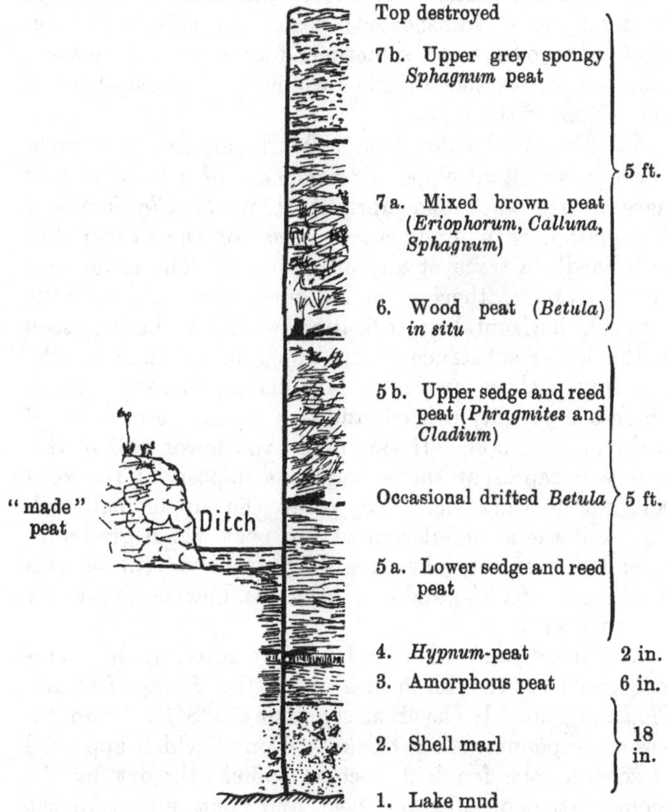

FIG. 17. Generalised section of the peat of the Lonsdale Lacustrine Moors.

separable into thin flakes. The matrix of fine particles of peat contains rhizomes and haulms of *Cladium* and *Phragmites*; while seeds of *Castalia alba*, of Potamogetons and of the smaller Carices may be picked out. This peat burns with a smoky flame and leaves a considerable proportion of ash, varying from 7 °/₀ to 13 °/₀ of the dried mass.

4. *Hypnum peat.* A few inches of compact *Hypnum* peat enclosing seeds of *Cladium* sometimes occur above the amorphous (transported) peat. At this level fragments of wood are also sometimes found, possibly derived from the trunks and branches which were rolled about on the bottom of the lake.

5. *Reed and sedge peat.* This layer, five feet thick where it is well developed, is composed of a loose spongy mass of rhizomes and upright stems of *Cladium* and *Phragmites*, with innumerable seeds of the former, but with hardly a trace of any other plant. The lower half shows least weathering on its outer face, and exhibits a rough horizontal stratification which is hardly seen in the looser substance of the upper half. Between the two layers there are, in a few places, pieces of birch which are possibly drifted, and may mark the water level of the ancient lake. If this be so, the lower half of this peat will represent the subaqueous deposit of the reed belt surrounding the lake, while the upper half will represent the accumulation of fen peat which gradually extended over the lake area. The ash content of this peat is between 3 °/₀ and 5 °/₀, which is, however, very low for fen peat.

6. *Wood peat.* A thin layer composed of the stems and roots of the birch lies above the *Phragmites* and *Cladium* peat. It has an ash content of 2·8 °/₀. It doubtless corresponds with a thicket of birch which appeared as soon as the fen had become sufficiently dry by the accumulation of subaerial peat, and continued until the growth of *Sphagnum* put an end to its existence.

7. *Sphagnum and Eriophorum peat.* This layer resembles the peat which lies above the layer of wood in the peat of the estaurine moors, though the lowest moss

layer is wanting, and the stratum of grey broad-leaved *Sphagnum* of the *cymbifolium*-group is probably much deeper than in the case of the estuarine moors. It is, however, difficult to decide this question, because the upper layers have been removed over the whole area, not more than five feet of this upper peat remaining anywhere on the moor. The lower (*Eriophorum*) peat gives 1·7 °/。 of white ash, while the upper grey moss peat gives only 1·16 °/。 of ash. These low percentages very clearly mark the difference between the vegetation dependent on the highly calcareous lake water and the moor vegetation independent of the bottom water, and receiving its supplies from the air.

The moor peat occurs in places only a few inches above limestone rock, on what were the old limits of the moor. The rotten and stained surface-layers of the limestone show the corrosive action of the peat acids on the finer limestone matrix, though the calcite of the fossils (crinoids and *Productus*) is untouched[1].

THE VALLEY MOORS OF THE NEW FOREST

BY W. MUNN RANKIN

Physical geology of the area

The moors of the New Forest owe much of their distinctive character to the special type of drainage system that has developed in that region. The Forest is a low plateau built up of the several successive sands and clays

[1] The peat successions described above have much in common with those given by Weber (*op. cit.*) for the North German moor peat, in which a succession of amorphous and silty peat based on post-glacial soil is followed by Phragmites peat, wood peat and various types of moor peat in order. They also correspond, in a general way, with the succession of existing vegetation observed in the aquatic and fen formations of the Norfolk Broads region described in Chapter X; in that region, however, the development of moor succeeding fen is very local. [Editor.]

of the Eocene period and spread over with a more or less regular covering of plateau gravel of a later period. This plateau is tilted towards the coast on the southern edge. Small streams have eroded the soft rocks into a complex system of valleys, which in their upper reaches are narrow and relatively deep. The nature of the rocks is such that the base level of erosion of the valleys has been closely approached and the flow of the drainage water is consequently slow. This applies more particularly to the upper and middle stretches; for the valley of the Avon, into which the western streams flow, has been more deeply cut and widened, owing to other reasons into which we need not enter, and as a consequence there is a marked increase in the gradients of the lower courses of its tributary streams.

Valley moors and spring moors.

In what may be regarded as the "hanging valleys[1]," within this area, the streams are either lined by peat bogs or wholly lost in them. These peat bogs or wet moors are especially well developed near Burley in the west and Beaulieu Road in the east. They extend broadly for a mile and more following the sinuosities of the valleys, and contrast strikingly with the dry *Calluna*-heath that spreads over the valley slopes and broad ridges of the dissected plateau. In addition to these valley moors there are other, much smaller moors, formed about the springs that issue on the sides of the valleys along the junctions of the different strata. The valley moors are composite; the spring moors are comparatively simple and correspond to a few only of the associations of the larger moors. The two classes of moor may therefore be dealt with together.

A typical valley moor shows several well-defined plant

[1] "Hanging valleys" are valleys whose drainage system has become disconnected, so to speak, from that of the larger valleys into which their streams ultimately flow.

communities, which are either the evident relics of earlier stages of development or progressive associations belonging to later stages of the same succession. Thus within one moor may be seen (1) a line of alder thicket, either entire, or, as is more commonly the case, broken into detached clumps, and corresponding with the wet wood which occupied the alluvium of the stream before its flow was hindered; (2) the swamps on either side of the thicket; and (3) the quaking peat bog of sodden moss which stretches away from the centre of the valley and swells upon the rising land of the slopes on either side. Such a moor consists of several associations, and shows not only the manner of succession within the limits of one formation, but also the transition from one formation to another. The following associations are represented in a valley moor:—

1. Alder thicket (*Alnetum rotundifoliæ* of the Marsh formation).
2. Reed-swamp associations.
3. *Molinia*-association[1].
4. *Sphagnum*-Moor and derivatives (Moor formation).

1. *The alder thicket.* This is generally seen in a state of degeneration, owing to the encroachment of the wet moor. In the earlier stages of degeneration the wood enclosed within the moors is of the usual alder type, though the ground-vegetation is largely peat-loving, with cotton grasses, *Eriophorum angustifolium* and the extremely rare *E. gracile.* Degeneration is most marked on the edges against the swamps.

[1] This association, which recalls one of the marginal associations of the Lonsdale moors (p. 251), cannot be regarded as belonging to the fen formation described in Chapter X. It probably represents a type of "Flachmoor" (see p. 210) intermediate in character between the fen formation and the moor formation as defined in this book. This vegetation, and especially the chemistry of its soil, has not been sufficiently studied in this country to permit of its further characterisation. [Editor.]

2. *Reed swamp*. Pools mark the course of the stream in its sluggish progress through the moor, and wide shallow "plashes" are scattered over the flood plain. In the deeper waters reed-grasses (*Phragmites vulgaris* and *Phalaris arundinacea*), together with reed-mace or bull-rush (*Typha angustifolia*), and bur-reed (*Sparganium erectum*) form high growths, while in the shallower waters the sedges (*Carex vulpina*, *C. rostrata*, *C. Goodenowii*, *C. paniculata*, etc., and *Scirpus palustris*) and the marsh horsetail (*Equisetum limosum*) constitute an active land-winning vegetation. In both, as well as in the less encumbered waters of the pools and plashes, there are many accessory plants, of vigorous growth and often with showy flowers: the bog bean (*Menyanthes trifoliata*), the marsh St John's wort (*Hypericum elodes*), which often completely hide wide surfaces beneath their foliage and bloom, the marsh cinquefoil (*Potentilla palustris*), the marsh lousewort (*Pedicularis palustris*) and other plants of a similar ecological type. The smaller bladderwort (*Utricularia minor*), with its delicate yellow flowers borne just above the water, is also a characteristic associate, and hardly less so is the pillwort (*Pilularia globulifera*), which covers wide patches of mud with its grass-like shoots. Rarely the uncommon bladderwort (*Utricularia intermedia*) occurs, and the sedges *Carex filiformis* and *C. limosa*. The absence of *Hottonia* and *Hydrocharis* from these swamps is no less noteworthy.

On the limits of the area of inundation immediately against the drier soil of the *Calluna*-heath, sweet gale (*Myrica Gale*) is the most striking feature.

3. *Molinia association*—Molinietum cæruleæ. Tussocks of moor grass, *Molinia cærulea*, are often associated with the lower plants of the reed-belt, among which they appear as pioneers from the drier zone behind, where *Molinia* is the dominant plant. The Molinietum is much drier than the reed-swamp, the water table of which is

either quite close to the surface or actually above it. Even when there is open water among the tussocks of *Molinia*, these are raised some little distance above its surface. *Schœnus nigricans*, which is especially common on the western edge of the Forest, plays a similar part. Generally the tussocks of the grass are close together and form a more or less closed association over which one can walk dry-shod. The Molinietum is restricted to a narrow belt, which forms a transition from the reed-swamp on the one side to the "moss-moor" on the other. The tussocks of *Molinia* serve as a convenient *nidus* for such plants of the heath formations as ling (*Calluna vulgaris*), heaths (*Erica Tetralix*), and tormentil (*Potentilla erecta*), which seem to be independent of telluric water. Pockets of bog-moss also accumulate between and eventually overwhelm the tussocks, and the Molinietum passes into the *Sphagnum* moor.

4. *The Sphagnum moor*—Sphagnetum. The *Sphagnum*-moor and its derivatives form the greater bulk of the New Forest moors, especially of the spring bogs.

The vegetation of the *Sphagnum* moors is dominated by *Sphagnum*, with which are abundantly associated *Eriophorum angustifolium*, *Scirpus multicaulis* and *Juncus articulatus*. The matrix of bog-moss is made firmer by the rhizomes and roots of these and similar plants, and in the web so formed, smaller and less gregarious species, such as *Pinguicula lusitanica* and *Malaxis paludosa*, take root. The sundews, *Drosera rotundifolia* and *D. intermedia*, and the bog asphodel (*Narthecium ossifragum*) often adorn wide patches. The nodding white plumes of the cotton grass make the spring moors in June particularly noticeable at a distance. When swollen with rain the edges of the moor overlap the heath and there is a distinct narrow edging zone inhabited by societies of *Lycopodium inundatum*, *Rhynchospora alba*, *R. fusca* (often very abundant), and the sundew, *Drosera intermedia*.

Where peat has been dug are deep pits filled with water and a progressive association of *Sphagnum*, tending to reconstruct the moor, with cottongrass rhizomes interwoven and plants of *Drosera intermedia* and *Utricularia minor* scattered throughout. There is a great contrast between this semi-aquatic vegetation and that of similar stations near the stream course described under the heading of reed-swamp.

On the drier moors *Calluna* is an abundant associate; *Eriophorum vaginatum* is, however, absent. Birch (*Betula tomentosa*) occasionally extends over such firmer bogs as an open scrub.

While the valley moors are often deep and treacherous, and are avoided by the cattle and forest ponies, the spring bogs are quite shallow.

The Lowland Moors of Connemara (West Ireland)

The lowland moors of Connemara, with their deep peat soil, cover many square miles of nearly flat country not many feet above sea level, and interspersed with innumerable lakelets. The margins of a few of these lakelets show narrow reed-swamps of *Phragmites vulgaris* and *Cladium Mariscus*, and these plants are abundant in small tracts of moor which probably represent the sites of lakelets now obliterated by vegetation. The association of these two species recalls the East Anglian fens, but the vegetation of the district resembles that of the fens in no other respect.

1. *Rhynchospora association* (Rhynchosporetum albæ). The wetter parts of the moor, which fully deserve the title of bog, are dominated by *Rhynchospora alba*. The following are the principal species:—

Drosera rotundifolia	f	Narthecium ossifragum	f
D. anglica	f	Eriophorum vaginatum	f
D. longifolia	f	E. angustifolium	f
Erica Tetralix	f	Rhynchospora fusca	r
Calluna vulgaris	l	Schœnus nigricans	sd
Menyanthes trifoliata	f	Cladium Mariscus	l
Myrica Gale	f	Phragmites vulgaris	l
Rhacomitrium lanuginosum	l	Sphagnum spp.	l
		Leucobryum glaucum	l

2. *Molinia association*—Molinietum cæruleæ depauperatæ. This is a slightly drier type of moor, dominated by the tussocks of *Molinia cærulea* var. *depauperata*, and is much more extensive than the Rhynchosporetum. It includes more *Calluna* and *Erica Tetralix*. The rare *E. Mackayi*, closely allied to *E. Tetralix* and replacing it over a considerable area, occurs only in this station in the British Isles. Like *E. mediterranea*, which also occurs (very locally) in the same association, it belongs to the Iberian (Lusitanian) flora.

Lowland moors occur in various other parts of the British Isles, but except those of the Somerset Levels[1], which, where untouched, are largely occupied by Callunetum, they have not been surveyed botanically.

[1] See Moss, 1906, p. 29.

CHAPTER XII

THE UPLAND MOORS OF THE PENNINE CHAIN

By F. J. LEWIS (Northern Pennines), and
C. E. MOSS (Southern Pennines).

THE Pennine Chain (sometimes called "the backbone of England"), extending from Northumberland to Derbyshire, forms the greater part of the high ground of northern England, and bears very extensive tracts of moorland. These moors, which have been carefully surveyed by British plant-geographers during the last ten years, may be taken as typical examples of English upland moorlands. The following account of moorland associations is based on the Pennine examples; but Scottish and Irish moorland associations, so far as these are known, are referred to in their proper places.

Depth of peat and extent of moorland.

The peat or peaty soil bearing the associations of the moor-formation varies from a few inches to many feet or metres (*e.g.* 30 feet = 9 metres) in depth, and overlies shales, sandstones, and in some cases limestones. The moor-formation occupies the hill slopes and lower summits from a height of 1000 feet (*c.* 305 metres) or even less, to 2200 feet (650 metres) or slightly higher. In the Central Highlands of Scotland it may extend up to 3000 feet (*c.* 910 metres). It is interspersed at various altitudes with grassland on the steeper slopes with fresher soils.

The following well-defined associations have been described by various authors. They are dealt with in the order of decreasing water content.

1. **Bog-moss association** (*Sphagnum* moor, moss-moor) (*Sphagnetum recurvi*). This association, in which various species of bog-moss are dominant, forming the groundwork of the moor, is, so far as has been observed, of quite limited extent in the British Isles. It occurs, however, in the northern Pennines (North Lancashire) and also in the far north of Scotland. It was first systematically described by Lewis (1904 *a*), who gives the following vascular plants as characteristic:—

Sphagnum *spp.*	d	Empetrum nigrum (generally scarce)
Rubus Chamæmorus	f	
Erica Tetralix	f	Calluna vulgaris (generally very scarce)
Oxycoccus quadripetala	f	
Eriophorum angustifolium (generally present in small quantities)		

Wheldon and Wilson[1] state that the dominant species of *Sphagnum* is *S. recurvum*, though other species occur. They add *Viola palustris, Juncus effusus, Carex stellulata, C. curta*, and *Rhynchospora alba* as "marginal or very subordinate" constituents of the association.

Sphagnum peat is rare in the upland Pennine moors, and this fact, together with the very local occurrence of the association at the present day, is interesting, seeing that the *Sphagnum* moor is generally considered the starting-point and type association of the moor ("Hochmoor") series.

2. **Cottongrass associations** (cottongrass moors): *Eriophoretum vaginati* and *Eriophoretum angustifolii*. These associations are apparently closely related to the much rarer *Sphagnum* moors. The former occupies very wide tracts throughout the Pennine Chain and is the

[1] *Flora of Lancashire*, 1907, p. 102.

characteristic wet moor of this region. The Eriophoretum angustifolii is local and restricted in the Southern Pennines, but occupies more extensive areas in the North.

The Eriophoreta occur on gently sloping plateaux at elevations varying from about 1200 feet (*c.* 370 metres)[1] to 2200 feet (675 metres)[2]. The cottongrass moors are locally known as "mosses," which is by far the commonest place-name of the Pennine summits. The peat on which they are developed is deep, rarely less than five feet (*c.* 150 cm.), frequently ten to fifteen feet (*c.* 300 to 450 cm.), and occasionally reaches a depth of 30 feet (*c.* 900 cm.). It is usually saturated and frequently supersaturated with water, though the superficial layer sometimes becomes very dry. The peat yields a small quantity only of ash.

These moors are dreary and monotonous in character. *Eriophorum vaginatum* is not only the dominant but frequently the only vascular plant over wide areas. Ferns and horsetails (Equisetaceæ) are absent, club-mosses (Lycopodiaceæ) extremely rare; and none of the species of non-vascular plants is of general occurrence (Plate XXV b).

It is a curious fact that these extensive cottongrass moors should have escaped definite record by British botanists till the method of vegetation survey was applied to the Pennine region, when they were recognised by Smith and Moss[3]. The peat on which the cottongrass moors are developed consists almost wholly of the leaf-bases and leaf-sheaths of *Eriophorum vaginatum*. In the Southern Pennines a layer of black structureless peat from six inches to a foot in depth (*c.* 15 to 30 cm.) underlies the *Eriophorum* peat, passing down, if the peat is developed over shale, into shale with numerous roots, and

Peat of cottongrass moors.

[1] Average lower limit on Southern Pennines.

[2] Average upper limit on Northern Pennines.

[3] Smith and Moss, 1903, p. 6.

Phot. W. B. Crump

a. Heather moor of Pennines (August).

Phot. W. B. Crump

b. Cotton grass moor. *Eriophorum vaginatum* in front. *Nardus stricta* on the drier edge.

Moor formation

finally into pure shale; if the peat is developed on sandstone, a layer of red ironstone (a kind of "moor-pan") underlies the structureless peat. In the Southern Pennines the remains of trees, chiefly birches, occur locally under the *Eriophorum* peat—usually in somewhat sheltered situations. There is no evidence in this region to warrant the conclusion that the whole of the extensive elevated tracts now occupied by cottongrass moors were formerly covered by continuous forests; though it appears highly probable that woodland existed on the less exposed portions of these tracts. Besides the birch (probably *Betula tomentosa*), remains of the oak (probably *Quercus sessiliflora*), of the alder (*Alnus rotundifolia*), of willows (*Salix cinerea* or *S. caprea*), of the hazel (*Corylus Avellana*), and of the Scots pine (*Pinus sylvestris*), have occasionally been met with under the peat of the Southern Pennines.

Plant remains in the peat.

The hill peat of the Northern Pennines, where more than 6—8 feet (*c.* 2 m.) in depth, generally contains a well-marked stratum of buried forest[1]. This can be well seen in Teesdale, Tynedale, Weardale and on Stainmoor.

The actual sequence of the stratification of the peat in these districts is as follows:—

4. Eriophorum, Sphagnum, Scirpus cæspitosus and Calluna.
3. Betula alba (agg.), Alnus rotundifolia, Viburnum Opulus, Elatine hexandra, Lychnis dioica, Ajuga reptans, Ranunculus repens.
2. Salix arbuscula, Empetrum nigrum, Viola palustris, Potentilla palustris.
1. Salix reticulata, Arctostaphylos alpina, embedded in sand and silt underlying the peat.

Strata 1 and 2 represent the primitive vegetation covering these hills shortly after glaciation, and contain several plants which are now entirely absent from the Pennine Range.

The peat of the forest bed in Stratum 3 consists of

[1] F. J. Lewis, "Peat Moss Deposits in the Cross Fell and Caithness Districts," *Brit. Assoc. Reports*, Sect. K, 1907.

large stools and trunks of birch and alder. This bed is found in exposed places at altitudes of 2450 feet (*c.* 745 m.) on Cross Fell, far above the present limit of woodland in the district. The presence of the seeds of *Elatine hexandra* in the forest bed in several sections cut at 2000—2400 feet (*c.* 610—730 m.) both in Teesdale and Tynedale is of interest, as the plant is now chiefly confined to the western regions of Britain and is not typical of peat bogs at high elevations. Of similar interest is the occurrence of *Viburnum Opulus* in this stratum from a section at about 2330 feet (*c.* 710 m.). Lagerheim has also recognised the pollen of elm, hazel, and pine from a specimen of peat taken from this forest bed at an altitude of 700 m. The presence of these plants implies entirely different conditions from those indicated by the basal arctic and sub-arctic vegetation.

In certain well-defined areas a thick bed of the rhizomes and leaves of *Phragmites vulgaris* occurs between the forest bed and the basal arctic vegetation, marking the sites of ancient lakes which were at length silted up with plant remains and overgrown by the forest represented in the forest bed.

These successive phases of the vegetation in this district form a marked contrast with the existing monotonous *Eriophorum*, *Vaccinium* and *Calluna* moors.

Floristic composition.

The following short list includes all the flowering plants that have been met with in the association of *Eriophorum vaginatum* in the Peak district of Derbyshire (Southern Pennines):—

Dominant species.
Eriophorum vaginatum.

Locally sub-dominant species.

Rubus Chamæmorus	Eriophorum angustifolium
Vaccinium Myrtillus	Molinia cærulea (rare)
Calluna vulgaris	Erica Tetralix
Empetrum nigrum	

Locally abundant species.

Scirpus cæspitosus	Carex curta

Local and rare species.

Narthecium ossifragum	Oxycoccus quadripetala
Andromeda Polifolia	Pinguicula vulgaris

In the Northern Pennines both the Eriophoreta occur, and there is frequent mixture of the two. *Erica Tetralix*, *Empetrum nigrum* and *Rubus Chamæmorus* are sparingly present; they nowhere become a marked feature of the vegetation.

Throughout the Pennine chain, there exist extensive areas of an association transitional between the Eriophoretum and the Callunetum. It is more characteristic of the drier moors east of the Pennine watershed than of the wetter moors to the west. It occupies, for example, the whole of the sub-alpine region of Teesdale and Weardale. The peat is drier than that of the pure Eriophoretum vaginati and very much drier than that of the Eriophoretum angustifolii. It is often much furrowed by denudation channels.

Distribution of cotton-grass moor.

While the cottongrass moors are so characteristic of the Pennine region, their occurrence appears to be quite limited in other moorland areas of the British Isles, so far as these have been examined. Comparatively small areas have been described by Robert and W. G. Smith in the Central and Eastern Highlands of Scotland, and the association occurs in abundance nowhere in the North-west Highlands or the Hebrides. Limited areas of Eriophoretum angustifolii are described by Pethybridge and Praeger in the Wicklow mountains of east Ireland. In these districts, its place is taken mainly by moors on which *Scirpus cæspitosus* is dominant.

3. **Association of Scirpus cæspitosus** (*Scirpetum cæspitosi*). This occurs, and has been described by

Pethybridge and Praeger[1], on the Wicklow mountains to the south of Dublin. It occupies the flatter and less well-drained slopes of these mountains between about 1250 feet and 2000 feet (*c.* 380—600 metres) of altitude, while the association of *Eriophorum angustifolium* occurs in limited adjacent patches near its upper limit.

This Scirpetum is developed on deep peat, which is still being accumulated, and is saturated with water at most times of the year. The dominant species is often mixed with a considerable amount of stunted *Calluna*, thus showing a transition to the heather moor. The association, like the Eriophoreta, is floristically poor and has the following composition:—

Dominant.
Scirpus cæspitosus

General.
Calluna vulgaris
Eriophorum angustifolium
Erica Tetralix
Narthecium ossifragum
Sphagnum *spp.*

Occasional.
Drosera rotundifolia
Andromeda Polifolia
Erica cinerea
Empetrum nigrum
Eriophorum angustifolium
Rhacomitrium lanuginosum
Cladonia rangiferina

The remains of Scots pine and birch have been found in the peat at about 1250 feet (*c.* 380 m.) and 1700 feet (*c.* 520 m.) respectively.

This association, as has been noted by Lewis[2], occupies very extensive areas in the north-west Highlands of Scotland, in the Hebrides and in Shetland, but its floristic composition in those regions has scarcely been studied.

4. **Bilberry Moor associations** (*Vaccinieta Myrtilli*): *Vaccinium Myrtillus* (Bilberry, Whortleberry, Blaeberry in Lowland Scots) occurs more or less abundantly on practically every part of the Pennine peat moors, except in the wettest places. It is often dominant or sub-dominant on the lower slopes of heather-moorland, where

[1] Pethybridge and Praeger, 1905, p. 162.

[2] "The British Vegetation Committee's Excursion to the West of Ireland," *New Phytologist*, vol. VII. 1908, p. 257.

the peat is very shallow and much mixed with silica. Such tracts are often characterised also by an abundance of *Erica cinerea*, and are here regarded as a society of the heather moor. Besides these local communities, three varieties of Vaccinietum Myrtilli are recognised.

(i) The typical association of *Vaccinium Myrtillus* on a peat soil which is much drier than that of the Eriophoreta, or of the Scirpetum cæspitosi, is characteristic of the ridges and summits of the southern Pennines. It was first recognised by Robert Smith[1] in the Central Highlands of Scotland, where it is extensively developed and reaches an altitude of 3000 feet (*c.* 900 m.)[2]; and it also occurs on the Wicklow mountains in Ireland. The characteristic "Vaccinium-ridges" and "Vaccinium-summits[3]" are high-lying wind-swept areas usually with rather shallow peat, through which bare sandstone rock frequently protrudes.

(ii) A modification of this type occurs on steep sandstone slopes and screes. The bilberry is dominant, but the greater shelter of such a slope allows of the development of a richer flora. *Erica cinerea* is often abundant; and locally the bearberry (*Arctostaphylos Uva-ursi*) occurs in great masses, while the local abundance of the bracken fern (*Pteris aquilina*) is indicative of greater shelter. The cowberry (*Vaccinium Vitis-Idæa*) is dominant in places.

In the northern Pennines, the large areas of *Vaccinium Myrtillus* occupying similar situations, generally above 600 m. (1950 feet), are mixed with much *Calluna* as a co-dominant, and form an association transitional to the typical heather moor or Callunetum; large areas are also covered with this mixed association in the Central Highlands of Scotland.

[1] R. Smith (1900). [2] See p. 301.

[3] W. G. Smith and C. E. Moss (1903, p. 8); W. G. Smith and W. M. Rankin (1903, p. 6).

Retrogressive bilberry moor.

(iii) The habitat of the third type is different. On the great plateau of the Peak in Derbyshire, the peat is largely covered with an association of this plant. This association represents a stage in the degradation of a

Fig. 18. Plant associations of the Peak of Derbyshire.

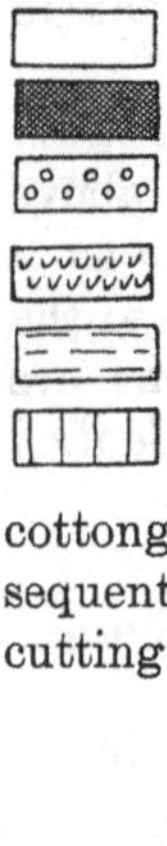

1. Land under cultivation.
2. Siliceous grassland.
3. Heather moor (Callunetum vulgaris).
4. Bilberry moor (Vaccinietum Myrtilli (i) and (ii), see p. 273).
5. Cottongrass moor (Eriophoretum vaginati).
6. Retrogressive Eriophoretum with *Vaccinium Myrtillus* (iii), *Empetrum nigrum*, *Rubus Chamæmorus* and bare peat.

cottongrass moor owing to progressive dessication consequent on the increased drainage of the plateau by the cutting back of the channels of the streams which leave its edges. The peat under the roots of the bilberry is entirely composed of the remains of *Eriophorum vaginatum*, clearly showing that the plant association was an Eriophoretum vaginati very recently. The stream channels as shown on the revised Ordnance maps published between 1870 and 1880 are nearly three-quarters of a mile (1·2 km.) longer than they were when the Peak was originally surveyed in 1830; and they are now about a quarter of a mile (·3 km.) longer than they were shown to be on the revised maps of 1879. The channels thus recently formed in the peat are fringed by banks of bare peat, while the surface of the peat above is clothed by the Vaccinietum. *Rubus Chamæmorus* is a very typical and abundant plant of this Vaccinietum, which also occurs in the northern Pennines. The smaller mounds left between the channels are often covered with the crowberry (*Empetrum nigrum*) and locally with *Lycopodium Selago*.

The following plants have been noted in this association of the Peak plateau:—

Dominant.
- Vaccinium Myrtillus

Abundant.
- Empetrum nigrum
- Calluna vulgaris (rare to abundant)
- Festuca ovina
- Eriophorum angustifolium
- Juncus squarrosus
- Rumex Acetosella
- Rubus Chamæmorus

Rare.
- Eriophorum vaginatum
- Lycopodium *spp.*

Occasional.
- Nardus stricta
- Aira flexuosa

Associations of *Vaccinium Myrtillus* are almost entirely absent from the north-west Highlands of Scotland, from the Hebrides, and from the Shetland Islands.

5. **Heather Moor association** (*Callunetum vulgaris*) (Plate XXV a). This is one of the best known and most extensive moorland associations in northern England. The heather "moors" of the Central and Eastern Highlands of Scotland belong rather to the heath than to the moor formation (Chapter IV). The heather moors are

the typical "grouse moors" and are generally strictly preserved for shooting this bird, the heather being "fired" every few years to promote young growth.

The Callunetum is developed on shallower drier peat and at lower altitudes than the types of moorland association already described. In the southern Pennines the association occasionally descends so low as 750 feet (228 m.), but as a rule it does not occur below 1000 feet (305 m.) On the lower hills it often stops at about 1250 feet (382 m.), though on the higher it ascends to about 1500 feet (456 m.). In the northern Pennines, however, it ascends to 2000 feet (610 m.), corresponding with the greater height of the hills. At higher levels the mixed associations of heather and cottongrass, or heather and bilberry, already alluded to, frequently occur.

Altitudinal range.

On the eastern slopes of the Pennines, the heather moors are very extensive; on the western slopes the zone of heather moor is much narrower and locally non-existent. This difference between the eastern and western zone of the heather moors of the Pennines is a reflection of the fact that the western slopes of the Pennines are, on the whole, steeper than the eastern slopes. The difference would appear to have little or nothing to do with the local variation in rainfall; and it certainly has nothing to do with aspect. Ostenfeld[1] states that the heather moors of the Faröes are always met with on slopes with a southern exposure; this is not the case with regard to the association on the Pennines.

Distribution.

The peat of the lower heather moors is invariably shallow, frequently not much more than a foot (30 cm.) in depth, and mixed with a large quantity of mineral matter, chiefly silica. The higher heather moors are sometimes found to occur on peat about four feet (*c.* 120 cm.)

Peat of heather moor.

[1] C. H. Ostenfeld, *The Botany of the Faröes*, 1908, p. 887.

in thickness, and such peat contains much less sand. It is a general rule that the more insoluble mineral matter is contained in peat, the lower is the water-content. Six samples of peat chosen from what were regarded as typical situations of the heather moors gave an average water-content of 53 per cent.

"Firing" and regeneration.

There is not a great deal of human interference with nature on the heather moors, although they are systematically fired by the keepers every few years. The length of time which elapses between the periods of firing varies locally, and determines the height to which the heather grows. On moors which are fired about every four years, the heather never grows much more than ankle high; while on the moors which are fired only every eight or ten years, the heather is frequently more than knee-deep. For one or two years after firing, the heather moor presents a desolate appearance, for the heather does not at once strongly reassert itself. The first plant to become conspicuous after firing is the bilberry (*Vaccinium Myrtillus*). This plant frequently occurs in a semi-etiolated condition under the dominant heather (*cf.* pp. 104, 112), where its habit simulates that of the rare *Listera cordata*. In such circumstances, the bilberry rarely flowers or fruits. *Deschampsia flexuosa* and *Nardus stricta* are also frequently conspicuous during the first summer after firing. In this district, however, seedlings of heather usually establish themselves immediately and in abundance; and so the complete and speedy rejuvenation of the heather moor is in most cases assured. The object of the firing is to retain the dominance of the heather, because heather moors are the most favourable for grouse.

Plant remains in peat.

It is a common thing to find remains of birch trees (probably *Betula tomentosa*) under the peat of the heather moors; and it seems highly probable that the present extensive areas of the Pennine moors now dominated by *Calluna vulgaris* were, at some former period, characterised by forests probably consisting largely of birch trees.

The following is a list of the plants that have been met with on typical heather moorland of the Peak district:—

Dominant—Calluna vulgaris.

Polygala serpyllacea	o	Scirpus cæspitosus	o
Ulex Gallii	la	Carex flacca (= C. glauca)	o
Genista anglica	r	Agrostis *spp.*	la
Lathyrus montanus	o	Deschampsia flexuosa	a
Potentilla sylvestris	o	Molinia cærulea	o
Pyrus Aucuparia (dwarfed)	r	Festuca ovina	o
Galium saxatile	a	Nardus stricta	o
Hieracium Pilosella	o	Pteris aquilina	la
Campanula rotundifolia	o	Blechnum Spicant	o
Vaccinium Myrtillus	ls	Lastrea aristata (= L. dilatata)	r
V. Vitis-Idæa	ls	Dicranum scoparium	o
Erica cinerea	a	Campylopus flexuosus	o
Melampyrum pratense *var.* montanum	o	Webera nutans	o
Rumex Acetosella	o	Leucobryum glaucum	o
Betula tomentosa (dwarfed)	r	Hypnum *spp.*	l
Salix repens	vr	Plagiothecium undulatum	l
Empetrum nigrum	ls	Lepidozia reptans	a
Juncus squarrosus	a	Cladonia *spp.*	o
Luzula erecta	o	Lecanora *spp.*	l
forma congesta	o		

Of rare or local plants, which have been noted on heather moors of other parts of the Pennines, may be mentioned *Lycopodium* spp., *Listera cordata*, *Pyrola media*, and *Trientalis europæa*.

Rush society (Juncetum communis). Here and there on the heather moors, small patches of wet peaty soil occur. Some of these mark the source of streams; and such tracts are almost invariably characterised by an association of the larger rushes (*Juncus effusus* or *J. conglomeratus*); and, in the issuing water, it is usual to find such plants as *Stellaria uliginosa*, *Ranunculus Lenormandi*, *Montia fontana*, *Potamogeton polygonifolius*, and *Galium palustre* var. *Witheringii*.

Society of stagnant hollows. Other wet peaty places on the heather moors are hollows where the water is stagnant, and such places are frequently rich in species. The following plants have been noted in stagnant hollows among the heather moors of Derbyshire :—

Locally sub-dominant species.

Calluna vulgaris
Erica Tetralix
Eriophorum angustifolium
Sphagnum *spp.*
Polytrichum commune
Hypnum *spp.*

Locally abundant species.

Oxycoccus quadripetala
Narthecium ossifragum
Juncus acutiflorus
J. lamprocarpus
J. supinus
Carex curta
C. flava (agg.)
C. panicea
C. stellulata
C. Goodenowii

Occasional species.

Ranunculus Flammula
Viola palustris
Cnicus palustris
Taraxacum paludosum
Pedicularis sylvatica
Eriophorum vaginatum

Rare or very rare species.

Drosera rotundifolia
Andromeda Polifolia
Pinguicula vulgaris
Orchis ericetorum

Zonation of the Moor associations.

The five upland moor associations and their transitional forms described in the preceding pages form a series showing a decreasing soil-water content. They do not, however, for the most part, represent a succession in time of associations following one another on gradually desiccating peat, as do the series of associations of the series of lowland moors previously described. They are rather to be regarded as co-ordinate associations adapted to different conditions of moisture and of altitude. Thus they show a fairly definite zonation round the summits of the mountain masses of the various regions.

The general zonation in the southern Pennines is as follows:—

	Altitude in feet.	Altitude in metres.
Vaccinietum Myrtilli	1700—2088 ...	c. 520—635
Eriophoretum vaginati ...	1200—2000 ...	c. 365—610
Calluno-Eriophoretum	1250—1800 ...	c. 380—540
Callunetum vulgaris	750—1550 ...	c. 230—470

In the northern Pennines the zonation appears to be less well marked. Thus we have:—

Calluno-Vaccinietum	2000—2250 ...	c. 610—685
Callunetum vulgaris	1000—2000 ...	c. 305—610

or

Sphagnetum	1250—1500 ...	c. 380—460
or Eriophoretum	(2000 or more) ...	c. 610—
Callunetum vulgaris	1000—1500 ...	c. 305—460

or

Calluno-Vaccinietum	2000—2250 ...	c. 610—685
Eriophoro-Vaccinio-Callunetum	2000—2250 ...	c. 610—685
Eriophoro-Callunetum	1650—2000 ...	c. 505—685

These variations are to some extent due to the height of the neighbouring summits as well as to varying local conditions of slope and moisture.

In the Wicklow mountains (Ireland):—

	Altitude in feet.
Scirpetum or Eriophoretum	1250—2250
Callunetum	1000—2000

The Callunetum occupies the lower, steeper and drier slopes, the Vaccinietum the better drained summits and steep slopes at higher altitudes, while the Eriophoretum is met with on the badly drained wet plateaux which are usually intermediate in position.

Degeneration (Retrogression) of Moorland.

The earlier stages of the degeneration of a cottongrass moor, in which the wetter Eriophoretum vaginati is replaced by the drier Vaccinietum Myrtilli owing to gradual

Phot. A. Wilson

a. Summit plateau of Fairsnape Fell, Lancashire. Peat-hags with *Eriophorum* left between network of water channels.

Phot. F. H. Graveley

b. *Rubus Chamæmorus*, ♂, *Vaccinium Myrtillus*, *Eriophorum vaginatum* on edge of cotton-grass moor. Head of Ashop Clough, Derbyshire.

Retrogressive Moor.

desiccation of the peat by improved drainage, have already been described. This process may be continued till the moor formation is completely destroyed.

In times of drought, the beds of the stream channels contain little or no water, and the peat banks dry and crumble, the dust being removed by the wind. After heavy rain-storms the channels are the paths of rapid torrents of brown peaty water. These carry away large quantities of peat; and they are continually widening and deepening their channels, as well as cutting them back further into the plateau. In many cases, the streams have cut right through the peat to the bare rock or into shale beneath. Numerous tributary streams are also formed in course of time; and eventually the network of channels at the head of a stream coalesces with a similar system belonging to a stream flowing down on the other side of the watershed. The peat moor, which was formerly the gathering-ground of both rivers, is thus divided up into detached masses of peat (known as "peat-hags"), and the final disappearance of these is only a matter of time (Plate XXVI a).

The first result on the vegetation is, as we have seen, the disappearance of the cotton-grass and the occupation of the peat surface by the bilberry (*Vaccinium Myrtillus*), the crowberry (*Empetrum nigrum*), and the cloudberry (*Rubus Chamæmorus*) (Plate XXVI b). As the process of denudation continues, this association gradually succumbs to changing conditions until the peat-hags become almost or quite destitute of plants. The peat, being no longer held together, is whirled about and washed away by every rainstorm or by the waters of melting snow.

In the end, the retrogressive changes result in the complete disappearance of the peat; and a new set of species begins to invade the now peatless surface. Such a denuded area occurs on several of the Pennine summits, *e.g.* on Bleaklow Hill a few miles north of the Peak. The plants of the surrounding peat moors can take no part in this invasion; the successful invaders are those species, such as *Nardus stricta* and *Deschampsia flexuosa*, which usually follow the moorland streams almost to their sources.

It is clear that the plant-formation of the moors has, in such cases, been destroyed. It is highly probable that every mountain moor is ultimately destined to be destroyed in some such way as has here been outlined[1]. The change of habitat may be compared with the destruction of sand dunes by the wind when a fresh supply of sand is no longer forthcoming.

Grass Moor association

By W. G. Smith

General characters. This is a widely distributed type of moorland association which covers wide areas on the glacial "tills" or boulder clay deposits of the hills of the Scottish Southern Uplands and of the Western Highlands. It is intermediate in character and composition between the moor of *Scirpus cæspitosus* (p. 271) and siliceous grassland (p. 131).

It is not so uniform in character as most other forms of moor, and the vegetation is mainly composed of a variety of grasses, rushes and sedges. *Molinia cærulea* forma *depauperata* and *Nardus stricta* are the most characteristic plants, and these are frequently dominant over considerable areas. *Deschampsia flexuosa, D. cæspitosa, Agrostis tenuis, A. alba, Anthoxanthum odoratum*, and *Sieglingia decumbens*; *Juncus squarrosus* and *Luzula multiflora*; *Scirpus cæspitosus, Eriophorum angustifolium, Carex*

[1] This process of peat destruction is taking place at present on very many of the hill-moors of these islands. The process depends to some extent on altitude and very markedly in some districts on the average level of the cloud-belt. It is rather less common in the extreme west. [F. J. L.]

vulgaris, *C. echinata*, and other Carices, are the most abundant species.

Soil. The soil is peaty, acid, and generally wet during most of the year. Typically, the surface peaty layer consists of a sod about 6 to 9 inches thick, made up of shoot bases and rhizomes matted with mosses and decaying herbage. This rests on an impervious subsoil, from which it is easily separable. Grass moor is sometimes formed on deep dark peaty humus, but it is generally absent from deep pure peat, the soil containing a higher percentage of mineral constituents than that of the other moorland associations, while it is more peaty than that of the siliceous grassland association. Thus in a moorland region the grass moor association frequently occurs on earthy deposits of rainwash, while in siliceous grassland it occurs where acid humus has collected as a result of impeded drainage and the consequent stagnation of the water.

Water-content. In water-content, also, the soil of grass moor is intermediate between that of the wetter types of moorland and that of siliceous grassland. Thus a series of drains cut through Eriophoro-Callunetum were running with water at a time when the drains cut through the grass moor were only moist and sticky on the bottom clay. The draining of the wetter moors often promotes the development of grass moor. The existence of the thick "mat" of peaty humus lying on the impermeable subsoil has an important bearing on the water-supply of the plants. The underlying subsoil has a moist period (November to May) alternating with a drier period, and the summer rains are mostly absorbed by the humous mat and do not penetrate to the subsoil, though the latter is always moist to the eye[1].

[1] In the case of siliceous grassland a similar condition obtains, but in the autumn the underlying subsoil is in this case very dry.

It is doubtful whether this moisture is available for the plants of the grass moor, and the alternate wetting and drying of the soil may, as Ostenfeld suggests, prevent the accumulation of acidity to so great an extent as in the soil of the Eriophoretum and Callunetum.

The root system of the more abundant associated plants of the grass moor ramify mainly in the humous mat, but the thick "cord-roots" of the dominant plants (*Nardus stricta*, *Molinia cærulea* and *Juncus squarrosus*) also penetrate the underlying subsoil.

Physiognomy, Utilisation, and Relationships.

The physiognomy of a typical grass moor in summer shows an uneven surface of a dingy green colour, not unlike an Eriophoretum or Scirpetum, and contrasting sharply with the brighter green of the siliceous grassland. In winter and spring the tracts of bleached *Nardus* and *Agrostis*, and the masses of windblown shoots of *Molinia* are very characteristic. By the flockmaster and shepherd the grass moor is distinguished as a form of "green land," which has a certain value for grazing, though inferior to that of siliceous, and still more to that of calcareous grassland.

The close relationship of grass moor with siliceous grassland on the one hand, and with the typical moorland associations on the other, will be obvious from what has been said, and grass moor rapidly passes into and is frequently mixed with these kindred associations.

Facies.

In this mixed and varying association two facies (subassociations), a damper and a drier, may be distinguished, corresponding with the dominance of *Molinia cærulea* and *Nardus stricta* respectively, though all transitions between these facies occur.

(1) *Molinietum cæruleæ.*

Molinia cærulea	d	Potentilla palustris	f
Ranunculus Flammula	f	Parnassia palustris	f

Hydrocotyle vulgaris	f	Eriophorum angustifolium	la
Galium palustre *var.* Witheringii	f	E. vaginatum	la
Cnicus palustris	f	Carex echinata	a
Oxycoccus quadripetala	f	C. Goodenowii	a
Erica Tetralix	la	C. dioica	f
Pedicularis palustris	f	C. panicea	a
Myrica Gale	la	C. curta (canescens)	f
Narthecium ossifragum	f	C. flava	f
Juncus effusus	ld	Agrostis alba	a
J. articulatus	ld	Deschampsia cæspitosa	ld
Triglochin palustre	f	Sphagnum *spp.*	a
Scirpus cæspitosus	la	Aulacomnium palustre	a

(2) *Nardetum strictæ.*

Nardus stricta, Agrostis tenuis, Juncus squarrosus } d

Ranunculus acris	f	Luzula multiflora	a
R. repens	la	Carex pulicaris	f
Polygala serpyllacea	f	C. leporina	f
Potentilla erecta	la	C. flacca	f
Galium saxatile	la	C. pilulifera	f
Scabiosa Succisa	f	Anthoxanthum odoratum	a
Antennaria dioica	f	Agrostis canina	la
Vaccinium Myrtillus (dwarf)	f	Deschampsia flexuosa	a
Calluna vulgaris (dwarf)	f	Holcus lanatus	la
Euphrasia officinalis (agg.)	f	Sieglingia decumbens	la
Prunella vulgaris	la	Briza media	la
Rumex Acetosa	f	Polytrichum *spp.*	f
Empetrum nigrum	f	Hypnum *spp.*	f
Orchis maculata (agg.)[1]	f	Hylocomium *spp.*	f

This facies comes very close to the Nardetum of siliceous grassland. In general, however, it may be separated from the latter by the absence in grass moor of species characteristic of the drier grassland with its less peaty soil. Of these the two species of gorse (*Ulex europæus* and *U. Gallii*) and the bracken fern (*Pteris aquilina*) are the most conspicuous and characteristic.

[1] Probably *O. ericetorum*.

To these may be added the ferns *Blechnum Spicant* and *Lastrea montana*, as well as a whole series of flowering plants (*e.g. Linum catharticum, Lotus corniculatus, Galium verum, Hieracium Pilosella, Achillea Millefolium, Campanula rotundifolia, Thymus Serpyllum, Teucrium Scorodonia*).

Distribution.

The distribution of grass moor is extensive, especially in Scotland. The largest areas occur in the north and west of Scotland[1], and extend eastward across the Caledonian canal to the Monadhliath mountains and the districts of Appin, Lochaber and Badenoch. On the hill ranges bordering the Lowland Plain—the Ochils, Pentlands, Moorfoots, etc.—grass moor alternates with siliceous grassland, and similarly it occupies considerable areas in the Southern Uplands. It is well represented in Westmorland and Teesdale, and probably also in Wales and Cornwall[2].

Summary of Moor Associations

The essential identity of the habitats, whether "lowland" or "upland," of the moor formation clearly appears from the recurrence of the same associations on both types of moor, the different associations being determined, not by altitude, but mainly by the water content of the peat. On lowland moors a definite succession from associations characterised by a higher to those characterised by a lower water content can often be traced; this succession is greatly hastened by artificial draining, but is primarily due to the building up of the peat to higher levels. On upland moors such a succession is only locally present, and is then due to increased drainage by the cutting back of streams into the moorland plateaux, or sometimes to artificial drainage. Tree associations do not at present occur on the upland moors.

[1] The "pâturages mouilleux" of Hardy, 1905.

[2] These last areas have not been surveyed.

The following are the associations of the moor formation which have been hitherto recognised in these islands. They are arranged roughly according to decreasing water content, but the sequence is not always exact.

Sphagnetum
Rhynchosporetum albæ (lowland only)
Eriophoretum angustifolii
E. vaginati
Scirpetum cæspitosi (upland only)
Molinietum cæruleæ
Nardetum strictæ (upland only)
Vaccinietum Myrtilli
Callunetum vulgaris
Betuletum tomentosæ (lowland only)
[Pinetum sylvestris] (lowland only)

CHAPTER XIII

ARCTIC-ALPINE VEGETATION

By W. G. Smith

What may be described as arctic-alpine vegetation forms the highest altitudinal zone of British vegetation, and is best represented in the Highlands of Scotland, where the most extensive elevated areas occur[1]. Many of the characteristic plants with a limited distribution have been much sought after and often recorded, but so far little has been done towards describing the natural plant-communities seen in the various habitats belonging to this zone. The main lines of the following account have been discussed with several botanists familiar with the species and localities, and on the whole it appears to include the commoner plant-communities of our higher mountains.

Concept. The recognition of an arctic-alpine zone of vegetation is based on the fact that in ascending the higher mountains one finds that at some altitude the general tone of the vegetation alters, and that new species characteristic of higher altitudes appear, forming plant-communities distinct from the lowland ones. The species also present various growth-forms adapted to

[1] Arctic-alpine vegetation also occurs on the higher mountains of northern England, Wales and Ireland, as well as at lower levels on parts of the Atlantic coast (see below).

conditions ranging from the barrenness of windswept slopes and summits to the comparative shelter and generally favourable environment afforded by rock-crannies amongst precipitous crags and ravines. In the central Highlands of Scotland the contour line of 2000 feet (*c.* 610 m.) indicates roughly the lower limit of this zone: above this the vegetation is characterised by an increase of such plants as saxifrages, dwarf-willows and other low-growing perennials, generally associated with characteristic mosses and lichens. Such woodlands as still remain in the country cease either below or near this limit. On the sub-alpine ericaceous moorland the change is marked by decrease of *Calluna* and increase of *Vaccinium*, while the grasslands become poorer in species belonging to such natural orders as Leguminosae and Umbelliferae, and assume a new physiognomy through increasing frequency of arctic-alpine plants, such as *Alchemilla alpina*, viviparous grasses, and arctic species or varieties of *Carex*, *Juncus*, *Luzula*, *Draba*, *Cerastium*, *Potentilla*, etc.

On the Atlantic side of Scotland and Ireland a similar vegetation occurs at much lower altitudes, almost down to sea-level. These plant-communities include many of the same species grouped in a manner comparable to that of the communities of the higher mountains. It is thus evident that in defining the arctic-alpine vegetation, too much stress should not be laid on zones of altitude expressed numerically; other factors must be sought which may explain the limited distribution of this vegetation.

Distinguishing features.

The arctic-alpine vegetation presents well-marked features both in its plant-communities and in its characteristic growth-forms. It can thus be characterised not only on a floristic, but also on a strictly ecological basis, especially as sub-alpine species occurring in the arctic-alpine zone assume the growth-forms of low-growing cushions, rosettes, or mats, more or less similar to those of the arctic-alpine species proper. It will, however, be more convenient to take as the index of the arctic-alpine vegetation, the presence of those species already so familiar to botanists in Britain as the "Highland" type of H. C. Watson[1].

Watson's "Highland" type.

Of this author's "types"—British, Germanic, English, Scottish, etc.—the "Scottish" is characterised as being most prevalent in Scotland, extending to the hill-districts of England and Wales, and includes *Trollius europeus*, *Viola lutea*, *Geranium sylvaticum*, *Rubus saxatilis*, *Sedum villosum*, *Linnæa borealis*, *Pyrola minor*, *Empetrum nigrum*, and *Listera cordata*, all upland species which occur in certain habitats of the arctic-alpine zone. The "Highland" type is boreal in a more intense degree as regards climate, and is limited to high altitudes in Britain, many occurring only in Scotland, either on mountains, or, under certain conditions, descending to sea-level[2].

Geographical distribution.

If the extra-British geographical distribution of these species be considered, it will be found that the "Highland" type occurs almost exclusively in high northern latitudes in arctic or sub-arctic Europe and Asia; further south a considerable proportion, but not all, are found at high altitudes in the European Alps or other lofty hill-masses. On this

[1] H. C. Watson. The original number of Watson's "Highland" species was somewhat over 100, but is now considerably greater owing to subdivision: when cited in this chapter "Highland" species are indicated.

[2] The usefulness of the "Highland" type as an indicator is illustrated by G. Dickie (*Botanists' Guide to Aberdeenshire*, 1860). On a hill in Aberdeenshire 1700 feet (*c.* 518 m.) high, the summit flora consisted of 14 "British" species; on another summit 2000 feet (*c.* 610 m.) there were 19 species; viz. 12 "British," 2 "Scottish," and 5 "Highland"; on Mount Keen (3125 feet, *c.* 952 m.) 3 "British" and 5 "Highland"; on Lochnagar (3800 feet, *c.* 1158 m.) 1 "British," 1 "Scottish," and 6 "Highland"; and on Ben Muic Dhui (4300 feet, *c.* 1310 m.) "Highland" species alone were represented, viz. *Silene acaulis*, *Saxifraga stellaris*, *Salix herbacea*, *Luzula spicata*, *L. arcuata*, *Carex rigida*, and *Festuca ovina* f. *vivipara*.

account the designation "arctic" or "boreal" seems more appropriate in Britain than "alpine" alone, which indicates relationship with the Alps. The term "arctic-alpine" is used here to designate the zone indicated by Humboldt and other writers on phytogeography. Thus Schimper[1] distinguishes lowland, montane, and alpine stages of vegetation, the latter possessing a characteristic aspect or physiognomy influenced by the collective climatic factors of elevated regions. He also points out that while his three regions or zones can be distinguished generally without much difficulty, they are more definite on mountains lying between the tropics, and less easy to distinguish in circumpolar lands, because of the mixture with arctic species as the more northerly latitudes are reached. This latter difficulty applies in Britain, so that the lower limits of the arctic-alpine zone cannot be defined except for each locality or district.

Distribution in Scotland.

Considerable fluctuations exist in the variety and frequency of typical arctic-alpine species in Britain. Any high summit may be expected to bear some species of this type (*e.g. Rubus Chamæmorus, Lycopodium Selago, Festuca ovina* f. *vivipara, Alchemilla alpina,* etc.), but as regards floristic variety most of the mountains may be generally described as "poor." The term "rich" is limited by botanists to comparatively few hill-groups. Breadalbane in Perthshire is one of the richer districts, Ben Lawers being familiar to many botanists, but the district includes other mountains lying to the west[2]. Quite detached from Breadalbane, and almost as rich, is the district lying on the watershed between Aberdeenshire and Forfarshire. This area was the scene of many of George Don's explorations about 100 years ago, and he first made known the

[1] A. F. W. Schimper, *Plant Geography on a physiological basis* (Eng. ed. 1901).

[2] F. Buchanan White, *Flora of Perthshire*, 1897.

floristic wealth of such localities as Clova, Glen Doll, Canlochan, Lochnagar and Glen Callater[1]. The Cairngorm Mountains, between the basins of the Dee and the Spey, come next; here, on the greatest continuous tract of high mountain plateau in Britain, the granitic hills are not very rich as a whole, but there are many places where arctic-alpine species find a suitable habitat[2]. Away from these three centres there are few really rich hills in Scotland, although Ben Nevis and other Bens near the tourist routes may bear a considerable proportion of the more widely distributed species[3].

The Vegetation of Ben Lawers

General features.

The ecological conditions which limit the arctic-alpine vegetation to higher altitudes are best realised if attention is first directed to some definite area. Ben Lawers is a suitable example known to most botanists. This mountain, one of the Breadalbane range, rises to almost 4000 ft (*c.* 1220 m.), from the northern side of Loch Tay, one of the larger inland lakes of Scotland, and among other features presents to the south a broad open grass-covered flank. In this respect Lawers and its neighbours differ from the majority of Scottish hills, which in general carry a darker-hued ericaceous moorland or heath. Generally speaking, mountains with the most varied arctic-alpine vegetation are grassy on their lower slopes, whereas the darker mountains are floristically poorer. On the southern side of Ben Lawers lies Loch Tay, a lake about 500 feet deep,

[1] W. Gardiner, *Flora of Forfarshire*, 1847.

[2] G. Dickie, *Botanists' Guide*, 1860; also numerous recent papers by J. W. H. Trail in *Ann. of Scottish Natural History*, etc.

[3] Ed. Moir (*Scottish Naturalist*, VI. p. 306, 1881) in an analysis of the distribution of the "native alpine flora" gives 91 species, 88 of which occur in Scotland, 73 being found in the Breadalbane and Clova districts, many of them endemic there.

with a surface level at about 350 feet (*c.* 107 m.) above the sea; Glen Lyon, the valley on the north, is about 600 feet above sea level.

It is noteworthy that as a rule the floors of the larger Scottish valleys lie at no great altitude, cutting deeply into the mountain masses, and that glacial deposits have accumulated up to considerable altitudes on the valley slopes. The valleys are also comparatively sheltered and suitable for cultivation, so that almost every large valley in the hilly districts includes a tract of farmland on its lower slopes. The vegetation on the valley slopes merits special attention because it has an important influence on the present limited distribution of the arctic-alpine vegetation.

1. *Zone of Cultivation and Pasture.*

Cultivation of lower lands.

Although recent economic changes have led to extensive depopulation in hill-districts, the influence of cultivation still remains as an important factor in plant-distribution. The main support of the hill-farm is sheep, but, especially in the earlier days, it was necessary to raise crops for the human population. Thus there was in most valleys a certain amount of ploughland and a considerable tract of enclosed grassland similar to that which still exists round the village of Lawers. Here the grassland probably occurred as a semi-natural vegetation and only needed enclosure, but in most districts the prevailing ericaceous heath or moor had to be reclaimed by draining, ploughing, and liming. Much of the grassland seen from the various railways shows by the ridges that it has been ploughed at some time.

The enclosure of the farmlands has thus played a great part in modifying the previous vegetation. The woodlands have been greatly reduced, for it was from them that enclosures were largely made and supplies of timber and fuel were taken.

Influence of arable land.

The ploughed land of the farms has also favoured the increase of certain species, some derived from the original vegetation, others introduced with seeds or in other ways. Thus many years after land has been ploughed, it is not unusual to find introduced grasses and clovers (*Phleum pratense*, *Trifolium hybridum*, etc.) still existing in the pastures. Gorse (*Ulex*) and broom (*Cytisus scoparius*) have also probably increased their area of distribution, for they were formerly grown as fodder plants. The general rule in pastures formed on ploughland is that they consist largely of a few species of indigenous plants which have assumed dominance in the early years of the grassland period, *e.g. Agrostis tenuis*, *A. alba*, *Anthoxanthum odoratum*, *Holcus lanatus*, *Cynosurus cristatus*, *Trifolium repens*, *Lotus corniculatus*, *Leontodon autumnale*, *Hypochæris radicata*, *Bellis perennis*, *Prunella vulgaris*, *Euphrasia officinalis*, etc. Many species of the original vegetation are almost suppressed or are limited to untilled places such as the banks of roads or streams.

Influence of pasturage.

Passing on now to pastures which have never been ploughed, there is a more natural, but still somewhat modified, vegetation. The introduction of pasturage of sheep on a large scale in the Highlands took place some 150 years ago. The operation which has had the greatest influence on vegetation is the drainage of wet land by open "sheep drains" or "grips" cut in the surface to carry off surplus water; these reduce the amount of wet moor or bog, and tend to promote drier types of vegetation. In a small pasture enclosed with walls, the influence of sheep is probably great, and it may frequently be observed that the pasture side of a wall is very different in vegetation from the moorland side. In the case of large open pastures the effect of grazing may, we think, be over-estimated. Many parts of a pasture are only grazed for short periods, and it is only on the few places where sheep lie[1] that excessive manuring occurs, thus favouring a "Lager-platz" or "lair"-flora consisting of species encouraged by high manuring (*e.g. Urtica dioica, Rumex Acetosa*). Where the enclosed pastures in the grassland areas are large, a considerable number of species of plants may be expected, and they give an indication of the unmodified natural vegetation. Thus on Lawers there is a well-marked belt of dwarfed *Calluna* traversing a zone of drier soils in the larger walled pastures on the slopes. Bracken also persists and on the lower slopes appears to become stronger as time goes on.

Influence of streams.

Within the pasture zone, the greatest variety of species occurs where the habitat favours open plant associations, and it is in such places that one may expect to find stragglers from the higher arctic-alpine zone, as well as the higher stations of the sub-alpine species. The wet spongy banks of streams, or undrained places suitable for Junceta, are specially favourable to many species, *e.g. Caltha palustris, Ranunculus Flammula, Cochlearia officinalis, Viola palustris, Stellaria uliginosa, Montia fontana, Lychnis Flos-cuculi, Potentilla palustris, Parnassia palustris, Sedum villosum, Galium palustre, Crepis paludosa, Pinguicula vulgaris, Narthecium ossifragum, Triglochin palustre, Habenaria virescens*, etc. The drier banks of the streams are also favourable for other species which, though for the most part not extending above the sub-alpine zone, do occur at higher elevations: *e.g. Trollius europæus, Geranium sylvaticum, Rubus saxatilis, Angelica sylvestris, Galium boreale, Solidago Virgaurea, Cnicus heterophyllus, Thymus Serpyllum, Salix aurita, Lastrea aristata, L. montana*,

[1] The sheep lie during the day or by night on a few favourite knolls or slopes, returning there after each period of feeding.

etc. Conversely, by the same track, plants from the arctic-alpine zone may descend to low altitudes, utilising the places where competition is not excessive; thus on Ben Lawers the following may be met with on the stream sides down to Loch Tay: *Saxifraga aizoides, S. stellaris, Tofieldia palustris, Alchemilla alpina, Polygonum viviparum, Oxyria digyna*, etc. These and other arctic-alpine species have been recorded at low altitudes on the shingle banks and margins below the flood level of most of the larger rivers.

2. *Zone of closed Moorland Associations.*

Glacial terraces.

Between the highest enclosed pastures and the arctic-alpine zone, there intervenes on most hillsides a zone of ericaceous moor or heath, or of some type of grassland. These are closed sub-alpine associations such as have been described in earlier chapters, and they generally follow the terrace-like glacial deposits. This common feature throughout the Highlands is due to lateral moraines forming a series of terraces one above the other, and marking phases in the gradual retreat of the ice sheet. The flat part of each terrace is badly drained, while the margin is steeper and well drained. It thus follows that during an ascent, one passes over flats, with vegetation of a wet type, alternating with drier belts on the slopes between the terraces. The knoll Meall Odhar (1794 feet, *c.* 547 m.), a conspicuous landmark between Lawers village and the summit of Ben Lawers, marks the position of the highest and most extensive terrace. The trees on East Mealour (Scots pine and larch) are planted, but they suggest a former extension of the woodland area to at least this altitude. The terrace referred to extends from behind Meall Odhar eastwards to Lawers Burn, and is of considerable breadth (a mile or more). It is almost always wet under foot, and carries a grass moor (see p. 282) with *Nardus stricta*, *Juncus squarrosus* and *Scirpus cæspitosus* as the dominant plants; *Deschampsia flexuosa*, *Anthoxanthum odoratum*, *Molinia cærulea*, *Eriophorum angustifolium*; species of *Sphagnum* and *Carex* are abundant. In places the underlying black slimy peat is exposed.

Grass moor.

The grass moor extends upwards from the terrace on to the comparatively steep slopes above, but here the humus is drier and more fibrous and there is much less *Scirpus cæspitosus*. Instead of *Sphagnum*, species of *Hypnum* are present. This drier grass moor passes up into the arctic-alpine grassland described in the next section.

The eastern part of this zone carries the only extensive peat deposit on the southern side of Ben Lawers. This extends over the eastern flanks to Lochan-'a-Chat, a small lake surrounded by deep channelled peat in a depression north of the summit, but nearly 1500 feet (*c.* 450 m.) below it. The vegetation here is Vaccinium-Calluna moor with *Vaccinium Myrtillus*, *V. Vitis-Idaea*, *Calluna vulgaris*, *Empetrum nigrum*, *Eriophorum vaginatum*, and a few characteristic "Highland" species, such as *Rubus Chamæmorus*, *Arctostaphylos Uva-ursi*, *Lycopodium Selago* and *Cornus suecica*. The remains of *Betula* are found embedded in this peat. This type of moorland is common in many parts of the Highlands, and is extensively developed on the plateau of the Clova district up to nearly 3,000 feet (*c.* 914 m.).

Calluno-Vaccinietum.

Thus far the ascent has revealed only such features as are found on the lower hill ranges. Few arctic-alpine species occur, and it is noteworthy that these are limited mainly to stations where the excessive competition of the closed moorland associations is absent, as on stream-banks or on a few of the exposed knolls. The existence of continuous belts of sub-alpine moor or heath round the centres of arctic-alpine vegetation would thus appear to impose an important restriction on the spread of the arctic-alpine species, and the factor of most effect is that this surrounding vegetation forms closed associations.

3. *Zone of Arctic-alpine Vegetation.*

Arctic-alpine Grassland Formation

(Plate XXVII a)

Arctic-alpine grassland.

The next altitudinal zone of vegetation is generally speaking developed where the individual summits begin to be differentiated from the more continuous undulating slopes of the different valleys. On Ben Lawers the grass moor just described gives place on the southern and eastern slopes at an altitude of over 2500 feet (*c.* 610 m.) to another zone of grassy vegetation. There is, however, a strong contrast between the dingy brownish green of the grass moor and the fresh bright green of the higher zone. The change is evidently associated with increased drainage. Grasses are prominent but mosses frequently occupy considerable tracts, and by their growth and decay furnish a thick surface layer of humus. The habitat is kept moist throughout the greater part of the year by numerous springs emerging at various levels, many of them marked by swelling cushions of bright green mosses. This water is mainly derived from schistose rocks comparatively rich in calcium, magnesium and potassium, and owing to the steepness of the slopes it drains away and does not become stagnant and acid. An essential condition of the habitat is that the soil is stable and little disturbed by surface erosion. On Ben Lawers the substratum is mainly rock disintegrated more or less *in situ*, mixed with some glacial débris, and the whole southern and eastern slopes form a regular and continuous surface with comparatively few emergent rocks[1].

Altogether the habitat may be considered as sufficiently distinctive to warrant the treatment of the group of plant-communities of which this vegetation is composed as a separate plant-formation.

This arctic-alpine grassland is of considerable interest amongst plant associations in Britain, because it lies above the limits of former woodland, and is not grazed to any considerable extent by sheep or deer, nor interfered with by artificial drainage. On Ben Lawers the zone attains an altitude of 3000 feet (*c.* 910 m.) and thus rises above many of the stations for well-marked arctic-alpine species which occur on adjoining rocky crags. It was recognised by R. Smith as a characteristic type of vegetation—"the alpine pasture"—of many of the botanically richer hills in the Highlands. It differs mainly from similar types of grassland in the lower hilly districts in possessing a distinct representation of "highland" species. The zone may thus be regarded as transitional between the sub-alpine and the arctic-alpine zones, in that it consists, like the sub-alpine grasslands, of closed associations, but is developed under conditions of stable and moist yet well-drained soil with a favourable climatic exposure. On the northern slopes of Lawers it is replaced by a meagre vegetation with a very different aspect, described later under the formation of mountain-top detritus. This limitation of the lower arctic-alpine grassland to southern slopes is recorded for several mountains in Hardy's memoir[2].

The plants occurring in this vegetation are numerous.

Variations.

The habitat is a favourable one for the hardier species, and the frequent local variations of the water-supply due to the numerous springs and

[1] The "Grass-Slope" of the Faröes (C. H. Ostenfeld) is a comparable community. (*Botany of the Faröes*, p. 962.)

[2] M. Hardy, 1905, p. 95.

streamlets offer great variety in the substratum. This arctic-alpine grassland on Ben Lawers has frequently been examined floristically, but the grouping of the species according to newer methods still remains to be carried out. Certain species are evidently grouped round the drier knolls, others round the mossy springs, and others again along the stream banks, which generally form spongy moss-carpets broken here and there by emerging rocks.

A feature of the arctic-alpine grassland is the abundance of *Alchemilla alpina* (Plate XXVII a). This species is dominant over considerable areas. The vegetation as a whole has not been sufficiently studied to admit of the recognition of different associations.

The following lists[1] indicate the plants of the drier, more typical, grassland; plants limited to the wet stream-banks are omitted.

A. "Highland" species of the lower arctic-alpine grassland.

Alchemilla alpina Festuca ovina *f.* vivipara	} characteristic and abundant, ld
Luzula spicata	Cerastium alpinum (*forma*)
Carex rigida	Sagina saginoides
C. capillaris	Potentilla Crantzii
Poa alpina (*forma*)	P. Sibbaldi
Phleum alpinum (wet places)	Selaginella selaginoides
Lycopodium Selago	Lycopodium alpinum

B. Sub-alpine species ascending over 2000 feet (*c.* 610 m.) on the lower arctic-alpine grassland; many of them occur as arctic-alpine varieties or forms.

Festuca ovina (agg.) Agrostis tenuis Anthoxanthum odoratum Deschampsia flexuosa	} General and abundant
Nardus stricta Molinia cærulea Carex Goodenowii (*forma*)	} Locally dominant.

[1] Plants mainly recorded by R. Smith, 1900.

Phot. A. G. Tansley

a. Arctic-alpine grassland of Ben Lawers (alt. 2700 feet): *Alchemilla alpina*, *Festuca ovina*, etc.

Phot. F. F. Laidlaw

b. Hydrophilous chomophyte formation. Side of streamlet. *Webera albicans*, var. *glacialis*, *Chrysosplenium alternifolium*. Ben Lawers.

Arctic-alpine vegetation.

Ranunculus acris
Anemone nemorosa
Viola lutea (*f.* amœna)
Polygala serpyllacea
Cerastium vulgatum
Sagina procumbens
Hypericum pulchrum
Linum catharticum
Oxalis Acetosella
Trifolium repens
Lotus corniculatus
Lathyrus montanus
Potentilla erecta
Alchemilla vulgaris
Heracleum Sphondylium
Galium saxatile
Scabiosa Succisa
Bellis perennis
Antennaria dioica
Achillea Millefolium
Taraxacum officinale
Leontodon autumnale
Vaccinium Myrtillus (sparse and dwarf)
Campanula rotundifolia
Veronica serpyllifolia
Euphrasia officinalis (agg.)
Melampyrum pratense *var.* montanum
Thymus Serpyllum
Plantago lanceolata
Rumex Acetosa
R. Acetosella
Orchis maculata
Juncus squarrosus
Luzula multiflora *var.* congesta
Scirpus cæspitosus
Carex echinata
C. pilulifera
C. flava (*forma*)
C. binervis
Avena pratensis (*forma*)
Deschampsia cæspitosa
Botrychium Lunaria
Blechnum Spicant
Lycopodium annotinum

Other types of association at the same level.

The arctic-alpine grassland formation is limited to certain mountain groups, and was recognised by R. Smith as one of three main types of association developed about this level. Of the others, he writes: "The second type of mountain is also largely covered with grasses, but with the blaeberry (*Vaccinium Myrtillus*) very much mixed with them, and still little heather (*Calluna*)." "The third type is still poorer. Grasses are here relatively rare, and peat plants in contrast dominate. Heather, although abundant, is nowhere in good condition, and is partly superseded by *Vaccinium Myrtillus*[1]." These two last types of association extend upwards in places as high as 3000 feet (*c.* 914 m.) in the

[1] R. Smith, 1900, p. 462.

form of a closed association which still retains the dominant species of the lower associations[1]. In spite of their high altitudinal range no fundamental distinction has yet been pointed out between the Calluno-vaccinietum and the Gramino-vaccinietum at low and high elevations; they are here regarded as upward extensions of the sub-alpine moorlands over slopes and plateaux of such a character that the plant-covering remains close enough to prevent denudation[2]. The existence of numerous centres of arctic-alpine vegetation on knolls and other places with imperfect plant-covering indicates a transitional stage, which, with denudation of the closed associations, would lead to extension of the arctic-alpines or, *vice versa*, the latter might be exterminated altogether.

Upper Arctic-alpine Formations

It is a feature of most British mountains that the summits may be reached over continuous slopes with few outstanding rock masses. Less commonly crag and scree, precipice and ravine break up the surface. On Ben Lawers the ascent from 2000 to 3000 feet (*c.* 900 m.) on the south-eastern and eastern slopes is almost entirely over the continuous turf of the grassland. The northern face also presents a continuous slope to Lochan-'a-Chat, very steep, but mainly a slope of disintegrated rock, strewn with boulders and rarely broken by ranges of crags. Amongst the rock-detritus, fine soil and moss-humus have lodged and bear a scanty vegetation, mainly consisting of mosses and lichens with comparatively few flowering plants, which is described below as the moss-lichen open association

Mountain top detritus.

[1] See "Moss-lichen moor association," p. 314.

[2] The elevation of phytogeographical zones is a recognised feature on lofty and complex mountain masses such as exist in the Central Highlands.

(p. 311). From the summit towards the north-west, considerable tracts of this stony waste occur, and on almost all the higher mountains it is a characteristic feature; the Cairngorms in particular include many square miles of this inhospitable rock débris.

The corries. If, however, the ascent of Lawers is made on the south-western flanks by ascending the Carie Burn (two miles west of Lawers village) a very different aspect presents itself. After traversing the pastures, the grass-moor and the lower arctic-alpine grassland, at about 3000 feet one enters the "corries." These are steep ravines cut into the south and south-west flanks of the summit. The corries are filled with fallen blocks and finer rock material, and show numerous ledges and fissures, with many typical arctic-alpine species. Ascending the corries to their steep heads some distance below the summit, one passes out on to the mountain top detritus, with its meagre flora in strong contrast to the wealth of the corries. Looking northwards from the summit many similar corries and crag terraces may be seen, separated by stretches of stony waste.

The two formations. The two groups of habitats just outlined are characteristic of the arctic-alpine vegetation wherever it occurs in Britain, and they give rise to the two plant formations of the upper arctic-alpine zone. Within these major habitats, there are subdivisions according to ecological conditions, amongst which the water supply is an important differentiating factor. A series of habitats where water is a prime factor, as around springs and streams, also occurs, but these are here regarded as subordinate to the formations described. The natural grouping of the vegetation, taking into consideration both habitat and floristic composition, is therefore:

I. The arctic-alpine formation of mountain top detritus.

II. The arctic-alpine chomophyte[1] formation of crags and corries.

The formation of mountain top detritus has by far the wider distribution, but includes the smaller number of species, and at no stage in its succession are the conditions favourable for the growth of flowering plants. The chomophyte formation is limited as regards distribution, but in its various phases it includes the majority of the "Highland" species. More or less analogous plant-communities exist as associations of subalpine plant formations, but they differ through the absence of characteristic species and growth forms. The arctic-alpine plant communities owe their distinctive features to their development in relation to their very distinctive habitat, and must certainly be considered as separate formations.

The question of the origin of the flora need not be discussed here; it may be the relics of a preglacial flora, or it may represent a phase of the vegetation of Britain which came after the Ice Age, or the plants may be immigrants at some later period[2]. The present distribution is limited because of the successful competition of closed plant associations and the work of man, but within the arctic-alpine zone it is determined by the habitats available. Some features of these will now be considered.

Origin of mountain top detritus.

The close topographical relations between the rock débris of the mountain plateau and certain plant associations, such as the *Rhacomitrium*-association to be subsequently described, was first noted by Robert Smith[3]. The recognition of the main geological features and of the probable origin of the mountain top detritus in Scotland is due to Messrs Clough and Crampton of the Geological Survey, and its distribution in relation to the higher mountain plateaux

[1] See below, p. 307.

[2] Clement Reid, *Origin of the British Flora*, 1899.

[3] R. Smith, 1900, p. 463.

in various areas of the Northern Highlands is dealt with in forthcoming Geological Survey memoirs on sheets 93, 82, and the memoir on Caithness. The material is considered to have accumulated from the continued action of frost on the solid rock of the higher glaciated mountain plateaux which early emerged from the "mer de glace," and its main features are due to its formation *in situ* under the influence of frost and wind, its frequent extreme porosity and great depth, and its position safeguarding it from all disturbance except by wind erosion and gradual earth creep. Dr Crampton deals with the influence of the great porosity of this mountain top débris on the plant associations in his forthcoming memoir on the plant ecology of Caithness[1].

Influence of rock exposures.

The occurrence of exposed rock-faces, and of ledges and corries, has been recognised by many British botanists as an essential factor in the development of a rich arctic-alpine flora[2]. This applies more particularly on the richer mountains of Breadalbane and Clova to the outcrop of sericite schist mapped by the Geological Survey as the Ben Lawers and Canlochan Schist. At higher altitudes, above the limits of the deeper glacial deposits, this outcrop weathers into crags with numerous shelves, fissures and corries, with a considerable amount of rock detritus and fine soil, which is more or less calcareous and provided with abundant moisture.

The influence of rock-outcrops is not however limited to the arctic-alpine vegetation, nor to this particular

[1] On the pavements of the mountain (carboniferous) limestone of northern England and Ireland (see p. 160) the process of weathering results in a different type of habitat. The deep crevices of the pavement afford shelter to the vegetation, which includes many sub-alpine but few arctic-alpine species.

[2] P. Macnair, "The Geological Factors in the distribution of the Alpine plants of Britain." *Trans. Perthshire Soc. of Nat. Science*, II. p. 240, 1898.

series of strata. R. Smith has referred to the influence of rock exposures, especially on the basaltic hills of the Lowlands round Edinburgh. These rocks frequently become exposed as crags and in weathering give rise to almost pure rock-soils; a distinct type of vegetation results, partly in consequence of edaphic factors, partly from local conditions as regards exposure and shelter. In Aberdeenshire the floristically rich localities coincide with outcrops of limestone and other rocks, and the "serpentine tract" is frequently[2] referred to as a rich one. In Teesdale, North Yorkshire, the rich and well-known flora is closely related with exposures of "Whin sill," a basaltic mass intruded into the Mountain Limestone in such a way that the intrusive lavas have altered the limestones contiguous to them; a more easily weathered form of rock, "the sugar-loaf limestone," results, furnishing a soil which in texture and in water-supply is favourable to characteristic species.

Conditions of rock habitats.

The effect of rock exposure is that new habitats are provided for colonisation by plants. As indicated by Oettli, the principal conditions are as follows[3]. The habitat is unstable in consequence of frequent falls from the rock faces which are being eroded; these destroy or displace existing plants and the station is open for re-colonisation; at the same time the fallen débris becomes lodged and will be available sooner or later for plant growth. From the nature of the habitat, rock-exposures remain for a long time loosely covered with vegetation, because on them the higher plants only grow where there is root-hold, that is in fissures or crevices, and on ledges where rock-débris has accumulated. For such plants Oettli proposes the

[1] R. Smith, 1900, p. 407; W. G. Smith, 1904–5, p. 18.

[2] In publications, *e.g.*, by G. Dickie and J. W. H. Trail.

[3] Oettli, M., *Beiträge z. Ökologie der Felsflora*, Zurich, 1905; see also Ostenfeld, *Botany of the Faröes*, p. 971.

name "Chomophytes[1]" and this term is here adopted to designate the formation. Amongst the rocks and fissures there is great variation in moisture, both as regards water-content of the soil and atmospheric humidity, so that moist places may be quite close to dry places. Atmospheric factors generally—wind, light and heat—have a marked influence on plant-growth in consequence of the steepness of the surface and the proximity of exposed rock. Crag vegetation is also protected from grazing and trampling, since it is inaccessible to animals.

These conditions favour colonisation by chomophytes at any altitude where rock is exposed. The Lawers-Canlochan schists are particularly favourable because they weather rapidly and furnish a very jagged rock-face presenting many ledges and crevices with much fine soil. On harder rocks which weather slowly, the conditions for higher plants are less favourable. The extreme case is a basaltic rock which weathers so that its face presents few fissures, while the blocks dislodged by erosion form a scree or talus remaining uncovered by plants for a long time.

Influence of water supply.

Passing on to other features of the habitats presented by mountain tops and rugged crags respectively, the conditions of water-supply claim attention. The formation of mountain top detritus is strongly xerophytic in its growth-forms; it is exposed to drying influences arising partly from insolation and consequent drought during the drier periods of the summer, but much more from wind action throughout the greater part of the year. The substratum is very porous, and although wet places occur round springs within this formation, a large and constant supply of telluric water would result in a ravine, thus altering the topographical conditions entirely. The chomophyte formation on the whole represents moister conditions, bu

[1] Gk χῶμα = earth thrown up. Oettli, *l. c.*, also Warming, *Oecology of Plants*, p. 240.

with a wide range, from a habitat on bare rock to one in running water (see p. 325). As regards precipitation, there can be no great difference between the two habitats, since they are frequently adjacent, but on the detritus the moisture is readily dissipated, whereas in the sheltered corries the atmosphere remains more humid. There is generally an abundant precipitation in the arctic-alpine zone. Mists are the rule by night and are frequent by day, and prolonged periods without rain are comparatively few[1]; when they do occur, the retarding effect on arctic-alpine species is very evident.

Influence of snowfall.

Snowfall is probably a differentiating factor between the two formations, although accurate observations have not been made in Britain. In the deeper corries the first snows in October or November will lie and be augmented by drifting from adjacent slopes. With each successive storm this snow is added to and is not reduced till spring, so that even in May the depths of the corries are embedded in ice-bound snow. Later this melts and by July snow is rare on Lawers, but on the Cairngorms snow-patches persist until September, especially in corries facing northwards. In some few localities the corrie-snow rarely melts completely and in the northern corries of Ben Nevis it is said to be permanent. On the open crags, slopes and summits, the first snows are drifted off by wind and collect as "wreaths" or in sheltered places. The formation of mountain top detritus is thus exposed to a period of very low temperatures and strong winds before a snow-covering is established. The snowfall in the early winter months

[1] Rainfall statistics are mainly compiled from observations at low levels and give little indication of arctic-alpine conditions; the actual precipitation at the observatory on the summit of Ben Nevis (now abandoned for lack of State aid) was 151 inches (3835 mm.) *per annum*, for a period of 15 years (Bartholomew's *Physical Atlas*, Part IV. *Meteorology*, 1899, p. 23).

is generally low (see Ben Nevis records[1]), but from December onwards the area may be regarded as mostly under snow. These conditions in early winter must have considerable effect in determining the death or survival of plants; and again in early summer (April—June), the habitat is free from snow several weeks before the corries, during a season when low temperatures are common. In the corries there is also a supply of telluric water which flows under the snow, whereas on the mountain top detritus any water present will drain into the substratum. The relation between a lasting snow-covering in the corries and a rich arctic-alpine flora has been recorded by several writers, and the situation of the more mesophilous species suggests shelter and moisture before the flowering season. The greater accumulation of soil in the lower parts of the corries also indicates a snow-covering during the severer months. The open exposed rocks and crags are less protected by snow and bear a less luxuriant type of vegetation.

Influence of sunshine.

Sunshine may also be a differentiating factor in the habitats, since the open formation of mountain top detritus will be more exposed to insolation than the chomophyte formation on

[1] Depth of Snow on Ben Nevis (*Trans. Roy. Soc. Edinburgh*, 43, 1905). The observations were made on the lying snow every two weeks from October to July for 20 years (1883–1903) on the flat summit, a typical area of mountain top detritus with little or no fine soil; the observation station was located so as to be free from wreaths or drift; lying snow first recorded before October 1st, 2 years; November 1st, 5 years; December 1st, 9 years; January 1st, 4 years. Last records: May 1st, 2 years; June 1st, 10 years; July 1st, 7 years; July 15th, 1885, 7 inches (17·5 cm.) still remained. The amount of snow (20 years mean) is much less in earlier months, November 1st, 5·2 inches; December 1st, 14·5 inches; January 1st, 33 inches; there is a marked increase between this and February 1st, 51·3 inches; March 1st, 62·3 inches; April 1st, 72·4 inches; May 1st, 74·3 inches, then the amounts rapidly diminish. The lowest maximum of snow recorded in

crags. On the whole this is not an important factor in the arctic-alpine zone, because cloud and mist are so frequent[1]. It is, however, significant that the grasslands of the lower arctic-alpine zone ascend to greater altitudes on the southern slopes, while the moss-lichen association is more strongly developed on the northern slopes. During the winter half of the year, northern slopes receive no direct sunshine because of the low angle of elevation of the sun[2].

I. Formation of Mountain Top Detritus

Succession. In this are included the plant associations on continuous slopes or summits where the substratum is mainly rock débris. In the earlier phases, lichens form crusts and xerophytic mosses gradually cover the rock detritus and bridge the gaps, till, by their growth and decay, cushions of humus become available for the growth of flowering plants. At the same time fine soil collects in the pockets amongst the stones.

The formation includes two extreme phases. There is an open plant-community on a substratum where stones and boulders are still visible, but are covered with crusts of lichens and a partially discontinuous carpet of moss and lichen (moss-lichen open association). This leads by gradations to the *Rhacomitrium*-heath association in which the substratum is often completely hidden by masses of woolly fringe-moss (*Rhacomitrium*) which frequently form

any winter is 54 inches (April 11th, 1895), the highest is 142 inches (April 3rd, 1885); the maximum depths in all the years occurred in March, April, or May.

[1] The Ben Nevis records (20 years mean) give 17 per cent. of total possible sunshine, while Edinburgh has 31 per cent. and Jersey 44 per cent.

[2] At Edinburgh the sun's angle of elevation on Dec. 26th is 12°, on Jan. 24th, 15°; this means that the north side of a moderately steep hill 100 to 150 feet high, 350 to 500 feet above sea level, is sunless for 3 months (November—February).

a thick layer of humus. This latter is the final or closed stage of the formation, but at the lower altitudes it may advance a stage further and become an association of ericaceous shrubs. The final stage has however a precarious existence on the summit ridges, as the moss carpet may frequently be observed completely torn up by wind so that the rocky floor is exposed, and the succession begins again. No hard and fast line can be drawn between the extreme associations, and all transitional stages between them may frequently be seen within a limited area.

The two associations.

The formation coincides topographically with the "alpine plateau" of Robert Smith. The two most marked associations are described in the Faröes by Ostenfeld as the "alpine formation on the rocky flat" and the "*Grimmia* (*Rhacomitrium*)-heath"; the species given by Ostenfeld correspond closely with those recorded on the Scottish mountains. The *Rhacomitrium* association was recognised by R. Smith[1]:—"above 3,000 feet (*c.* 910 metres), and even on bare exposed places at lower elevations, such as the summits of hills, the ground is usually stony and sparsely covered with vegetation. Mosses and lichens dominate; in particular, the woolly fringe-moss (*Rhacomitrium lanuginosum*) which on many of the steep mountain rubbles forms the peat on which the alpine humus plants develop[2]."

Status.

The arctic-alpine formation of mountain top detritus is primarily climatic and represents the extreme limit of plant-life on an extremely

[1] R. Smith, 1900.

[2] Pethybridge and Praeger (1904, p. 168) record *Rhacomitrium* associations from Killakee and other summits in Ireland below 2000 feet; two forms of this association are recognised by them, one in which this moss forms high bosses or cushions on wet boggy moorland, another in which *Rhacomitrium* forms a flatter carpet over granite débris. In these, however, the arctic-alpine species are almost absent.

porous substratum in places greatly exposed to wind, subject to extremes of temperature for the greater part of the year and probably under snow for a considerable period. On the one hand, it may be regarded as an outpost of vegetation represented in the extreme case by the first lithophytes on bare rock, and progressing towards a condition of permanent covering. On the other hand, passing upwards from below it appears as the remnant of the grass slope and the ericaceous moorland of the lower arctic-alpine zone, with all species eliminated which cannot adapt themselves successfully to the more extreme ecological conditions. At the same time the vegetation is recruited as regards "highland" species, *e.g.* *Silene acaulis*, *Arenaria sedoides*, *Saxifraga oppositifolia* on Lawers, from the more sheltered corries where these species find their most favourable habitat.

Conditions of life.

The surface is dry and well-drained either because of its porous character or from steepness of slope, but the frequency of rain, snow, or mist makes it periodically wet. There is little or no soil-layer, so that the competition of the closed moorland associations is excluded. The plant covering is distinctly xerophilous in response to frequent dry periods as a result of wind action, less frequently to insolation and lack of precipitation. The growth forms are those indicated by Warming as characteristic of the Arctic mat-vegetation and "Felsenflur," the moss tundra or heath, and the lichen tundra or heath. Most of the species of flowering plants are deeply rooted amongst the stones and boulders, in pockets of soil which are deeper than a superficial glance might suggest. Advantage is taken of any shelter afforded by boulders, so that depressions are more rapidly occupied than the more exposed ridges. An observation (supplied by Mr W. E. Evans) illustrates this; while ascending the

Cairngorm, the stony waste appeared devoid of green plants, but on looking down the slope from above there was a distinct green tint, due mainly to *Juncus trifidus* growing in the shelter afforded by the slightly raised edges of a series of terraces of rock débris.

Growth forms.

The more characteristic growth-forms on the drier substrata are as follows: cushion plants (*Silene acaulis, Arenaria sedoides*), low mats of decumbent intertwined branches (*Azalea procumbens, Potentilla Sibbaldi, Empetrum nigrum, Saxifraga oppositifolia*, which has also a cushion form), mats formed by rhizomes and shoots intertwined just below the surface (*Luzula spicata, Carex rigida*), mats with rhizomes deeply buried (dwarf *Vaccinium Myrtillus, Salix herbacea*), rosettes (dwarf *Ranunculus acris, Festuca ovina* f. *vivipara*). These growth-forms give much protection, and this is further increased by hairs and other protective adaptations. Most of the species are well adapted for vegetative propagation and several are viviparous.

The lists of species available at present do not warrant an attempt to draw up complete lists; as to the lower plants especially, there is little information. Where the moss-lichen association is adjacent to exposed rock-faces and corries with a rich flora (as on Ben Lawers), the list of species is considerably increased. This indicates a great wastage of plant-life under the extreme conditions of this formation, a loss which can only be replaced from habitats more favourably situated. While the closed *Rhacomitrium* association would appear to increase the shelter for other species, it is noteworthy that the proportion of "highland" species is generally less in it than on the more open stony waste. S. M. Macvicar[1]

[1] Distribution of Hepaticae in Scotland, *Trans. Botan. Soc. Edinburgh*, xxv., 1910.

states that Hepaticae are almost absent from the Rhacomitrium association. The following are characteristic species[1].

Moss-lichen open association.

Ranunculus acris (dwarf)	Azalea procumbens (H)
Arabis petræa (H)	Polygonum viviparum (H)
Draba incana (H)	Salix herbacea (H)
Silene acaulis (H)	Empetrum nigrum
Lychnis alpina (H) (Clova)	Juncus trifidus (dwarf) (H)
Arenaria sedoides (H)	Luzula arcuata (H)
Potentilla Sibbaldi (H)	L. spicata (dwarf) (H)
Alchemilla alpina (H)	Deschampsia flexuosa (dwarf)
Saxifraga oppositifolia (H)	Festuca ovina *f.* vivipara
S. stellaris (dwarf) (H)	Nardus stricta
Solidago Virgaurea	Lycopodium Selago (H)
Gnaphalium supinum (H)	

Rhacomitrium heath association.

Astragalus alpinus (rare) (H)	Arctostaphylos alpina (H)
Potentilla erecta	Calluna vulgaris
Cornus suecica (H)	Euphrasia officinalis (agg.
Cerastium arcticum (H)	Rumex Acetosa
C. alpinum (H)	Carex rigida (H)
Galium saxatile	Lycopodium alpinum (H)
Campanula rotundifolia	Rhacomitrium lanuginosum
Vaccinium Vitis-Idæa	R. ericoides
V. uliginosum (H)	Cetraria islandica
V. Myrtillus (dwarf)	Cladonia rangiferina, etc.
Arctostaphylos Uva-ursi (H)	Peltigera canina

Moss-lichen moor association.

Under certain topographical conditions which favour accumulation of peat, moss-lichen associations occur as a part of the moor formation. This is a characteristic feature of the extensive Clova-Canlochan plateau at 2500—3000 feet (*c.* 760—915 m.), where on a flat and undulating surface with impeded drainage, peat has accumulated and covers large areas of the flat watersheds. Although the altitude is higher than the stations for many of the

[1] Taken from field-notes of R. Smith, indicated in local floras, or observed by the author.

"Highland" species of this rich district, the plateau vegetation is mainly made up of species common on wet acid peat at lower levels. The prevailing vegetation has been distinguished as a Vaccinio-Callunetum[1]. Many square miles of peat-hags are characterised by species of *Vaccinium*, *Scirpus cæspitosus*, *Eriophorum*, Carices, etc., but with a marked reduction of *Calluna vulgaris* and *Erica Tetralix* when compared with typical sub-alpine moors. On somewhat bare places amongst the peat *Cornus suecica* and *Betula nana* associated with mosses and lichens introduce an arctic element. On the higher knolls *Rhacomitrium* becomes dominant and the Vaccinia are thin and dwarfed. This association includes several well-marked arctic-alpine species and becomes more conspicuous on the more isolated higher knolls, evidently where competition from the more gregarious moor plants becomes reduced by the conditions of the habitat. Further observations on the conditions of the habitat are required, notably the part played by snow, which covers this area for several months; but the following example will indicate the distribution of the plants. On the plateau forming the summit of the "Snub" near Clova, altitude between 2500 feet (*c.* 760 m.) and 2750 feet (*c.* 837 m.), the vegetation is Vaccinio-Callunetum on the lower parts and Rhacomitrio-Vaccinietum on the higher knolls:

Vaccinio-Callunetum.

Vaccinium Myrtillus (dominant)		Rubus Chamæmorus
V. Vitis-Idæa	sub-dom.	Vaccinium uliginosum
Eriophorum vaginatum	sub-dom.	Trientalis europæa
Scirpus cæspitosus	sub-dom.	Galium saxatile
Calluna vulgaris		Anthoxanthum odoratum
Empetrum nigrum		

[1] W. G. Smith, 1905, p. 63. This vegetation is best regarded as an example of a particularly elevated sub-alpine moorland association, while the higher knolls where *Rhacomitrium* becomes dominant and the arctic-alpine species more prominent are locally developed arctic-alpine moss-lichen associations occurring in the moor formation.

Rhacomitrio-Vaccinietum.

Rhacomitrium lanuginosum	Vaccinium *spp.*
Cladonia rangiferina, etc.	Salix herbacea
Carex rigida	Gnaphalium supinum
Alchemilla alpina	Lycopodium Selago

Snow-patch vegetation.

Although very few observers have recorded the influence of snow lying over certain tracts till late in summer, there is reason to believe that this is a differentiating factor within the formation of mountain top detritus. Swiss botanists distinguish the vegetation of basin-shaped depressions where patches of snow lie long ("Schneeflecken") and snow-troughs the tracks of streamlets of snow-water ("Schneetälchen"). The chief factors are patches of fine silty soil derived from dust accumulated in the snow, and an abundant supply of cold snow-water by which the soil is saturated; then as drying proceeds there results a slippery slime and finally in summer a dry cracked surface[1]. On such patches of soil, special conditions are introduced leading to the elimination of some species and the favouring of others. Two examples may indicate this vegetation. (*a*) On Ben-y-Gloe (near Blair Atholl, Perthshire), altitude nearly 3000 feet (*c.* 900 m.) on quartzite rock, on patches whence snow had recently melted, R. Smith records (MS. fieldnotes, Aug. 13, 1898) *Salix herbacea*, *Gnaphalium supinum*, *Azalea procumbens*, *Alchemilla alpina*, and *Galium saxatile*. (*b*) On Ben Lawers, altitude 3100 feet (*c.* 950 m.) on mica-schist, on wet recently exposed surfaces where snow lies long, Dr F. F. Laidlaw records (*in litt.*) *Cochlearia micacea*, *Juncus triglumis* (dwarf), *J. biglumis* (dwarf), *Phleum alpinum* (dwarf), and *Saxifraga stellaris*.

[1] See Warming, *Oecology of Plants*, p. 319; Oettli, *l. c.* p. 15; Brockmann-Jerosch, *Flora des Puschlav*, p. 335; L. Schröter, *Taschenflora des Alpenwanderers*, etc.

II. Arctic-alpine Chomophyte Formation

This formation is floristically richer than that of the mountain top detritus, but has a more restricted distribution, because places suited to its development are limited in number and extent. The substrata occurring amongst crags and corries are extremely varied, grading from the surface of recently disintegrated rocks through all degrees of weathered débris, lodged in fissures, on ledges and amongst fallen blocks. Plants with different requirements are thus suited. In a well-stocked corrie several categories of plants exist together. The most interesting group includes the arctic-alpine species native to the richer corries and found nowhere else:—*Myosotis pyrenaica*, *Saxifraga cernua* (limited to Breadalbane), *S. nivalis*, *S. rivularis*, *Erigeron alpinum*, *Saussurea alpina*, *Gentiana nivalis*, *Veronia saxatilis*, *Arenaria rubella*, *Draba rupestris*, *Carex atrata*, *Salix reticulata* and other species, *Polystichum Lonchitis*, etc.; most of these are typical chomophytes on soil lodged in fissures or on ledges. Species from the mountain top detritus also find places in the corrie, especially on the upper more rocky and exposed parts, which in many ways resemble the rocky terrain. The corrie vegetation also includes many species which are not strictly arctic-alpine, since they find a place in various lowland associations. Some are mesophilous species from stream-banks and woodland which find a habitat at high altitudes in the deeper moister soils of the sheltered corries, *e.g.* *Trollius europæus*, *Cardamine flexuosa*, *Lychnis dioica*, *Heracleum Sphondylium*, *Geranium sylvaticum*, *Oxalis Acetosella*, *Rubus saxatilis*, *Galium boreale*, *Adoxa Moschatellina*, *Mercurialis perennis*. Others are immigrants from the closed associations of grassland or heath.

Vegetation of the corries.

Ecological features of the vegetation have already been pointed out by various authors. Schimper expressed a well-marked distinction in using the term "lithophytes" for plants able to live on the surface of rocks or boulders in the earlier stages of erosion, and "chasmophytes" for plants rooted in crevices where fine detritus and humus have accumulated. The more recent memoirs by Oettli[1] and Ostenfeld[2] have added considerably to our knowledge. No attempt can be made here to deal exhaustively with British arctic-alpine rock-plants nor to classify them into associations, because the necessary observations are lacking, but the two memoirs mentioned seem to supply a basis for classification.

Lithophytes and chasmophytes.

The arctic-alpine chomophyte formation reaches its fullest development in the mountain corries of the mica-schist and presents a series of phases in succession. In its more typical condition it is an open vegetation consisting of small communities which are plant societies rather than plant associations. In some cases it is possible to distinguish one or more dominant species to serve as index-plants, but it seems better at present to group the species provisionally according to obvious habitats. It is therefore necessary to examine briefly the habitats and their origin.

Variations of habitat.

So far as the rock itself is concerned, the stations available for plant-growth are of two kinds; the habitat offered by the rock during its stages of disintegration and decomposition, and the habitat prepared by the transported rock fragments lying in a new situation. Weathering rock and the surface of boulders will in course of time offer various habitats, mainly those resulting from the formation of horizontal ledges or shelves,

[1] Oettli, M., *Beiträge z. Ökologie der Felsflora*, Zurich, 1905.

[2] Ostenfeld, C. H., "Land Vegetation of the Faröes" (*Botany of the Faröes*), Copenhagen, 1908.

(Plate XXVIII) and of crevices and fissures which frequently extend up and down vertically. The detached rock-material becomes re-sorted in different ways according as the fragments are large or small. With easily friable rocks like the mica-schists of the richer Scottish hills, a large amount of fine débris is formed and lodges amongst the fissures and ledges of the rock-face, which gradually become filled up and may ultimately be obliterated. The larger fragments of rock fall over the rock-ledges and come to rest as a block-scree at some lower level; at first these screes are unstable and mobile, but in time become more settled as the blocks weather and the gaps are filled with finer material.

Succession.

The succession begins on the bare rock faces with lithophytes, coatings of algae and lichens, followed by typical rock-mosses (*Andreæa*, etc.); flowering plants rarely occur in this phase. As weathering proceeds, and soil accumulates in fissures, the chomophytes become established. Xerophilous species (*Arenaria sedoides*, *Silene acaulis*, *Saxifraga oppositifolia*) here precede the mesophytes, which will ultimately occupy the deeper deposits of fine soil and humus wherever they are accumulated. Later still, as the ledges become more or less obliterated by soil accumulations, and as the crevices become filled up, there is invasion by surface-rooting and mat-forming plants ("exochomophytes" of Oettli) from the closed plant-associations of the grassland or heath of the lower arctic-alpine or sub-alpine zone. This last phase is well seen on many Scottish hills (*e.g.* Ben Lawers), where the lower terraces of the corrie area are covered with closed grassy swards, although they are clearly parts of the same system of rock exposures as the ledges and ravines bearing open associations at a higher level. There is reason to believe that if erosion of the rocks ceased, the whole system of rock-ledges and corries would ultimately become closed plant-associations, and as is the case now on the closed terraces, the characteristic arctic-alpine chomophytes would be largely exterminated. The chomophyte formation is therefore strictly speaking a series of phases, none of them closed associations; but as it is extremely improbable that erosion will be reduced under present conditions (increased erosion is more probable) the habitat of the chomophytes is relatively constant, though each part of it is undergoing fairly rapid change.

Succession in scree vegetation.

The same phases may be traced on screes, beginning with lithophytes and ending with a closed vegetation. Mosses play a larger part in the intermediate phases, they become established on the blocks, and extend over the narrower cavities, and by their growth and decay provide a humus covering, but frequency of drought prevents any but xerophilous flowering plants from becoming established. Within the larger interstices, vegetation is restricted to liverworts, mosses and such species as can live in deficient light, but as soil and humus accumulate the conditions become not unlike those of the chasmophytes or shade chomophytes of fissures in the rock-face.

Relations to water.

Other factors, besides the progressive disintegration of rock, obviously play an important part in modifying the vegetation, especially the supply of moisture, and its conservation for plants through shelter from wind. Atmospheric humidity is probably quite as essential for many species as soil moisture. Lithophytes and other plants on the rock-faces depend mainly on aerial sources for moisture, and are therefore subject to recurrent drought. On the ledges and in crevices there is frequently telluric water available, more constant in supply and richer in food-materials. Sometimes, as on Ben Lawers, this supply is so ample that throughout the summer water continues to run over extensive ledges and forms the sources of numerous streamlets; this is the habitat for the association of hydrophilous chomophytes referred to later. Another aspect of water-supply is the temporary supply of cold water from melting snow, which in early summer can be seen trickling down the rocks in streamlets or spreading over ledges that later become comparatively dry. Here we have a "snow-valley" (Schneetälchen), and with it the occurrence of certain species of plants, *e.g. Saxifraga rivularis*[1], *Veronica alpina, Arabis petræa.*

Classification of Chomophyte vegetation.

A complete grouping of the chomophytes is not possible at present, as there is little information available. The scheme suggested by Ostenfeld[2] has much to recommend it, as it takes account of the natural grouping of the chomophytes in relation to their respective habitats. This author classifies the habitats according to the amount of soil-detritus, moisture, light, and shelter from wind, available on the different parts of a rocky face. Generally speaking the groups proceed from the drier and more xerophilous forms, common to the mountain top detritus and open crags, to more mesophilous forms characteristic of the more sheltered and more congenial parts of the corries[3].

The following is a provisional grouping according to the more obvious habitats.

A. *Open communities on exposed rock-faces.*

The habitat is an open ledge exposed to full illumination and without much soil. The surface supply of water

[1] I. H. Balfour (*Excursion Notes*) several times records this plant as "near melting snow."

[2] C. Ostenfeld, *Botany of the Faroes.*

[3] The placing of the Scottish species has been greatly facilitated by assistance from Messrs T. Anderson, W. Evans, F. F. Laidlaw, and others.

depends on precipitation, but within the fissures the deeply-rooted plants find moisture fairly constant and at no great depth. The growth form is generally that shown by *Sedum roseum* or *Dryas octopetala*, a more or less compact cushion or mat with a single rootstock extending into a fissure, within which a branching root-system anchors the plant firmly. From the situation, the plants must be considerably exposed to wind and insolation, and in early winter there is probably insufficient snow-covering. In Britain the ledges often have a northern exposure. The plants occur isolated or form small communities, and rarely cover much space, except where the more luxuriant cushions or mats hang down over the rock face. The following are representative species, especially of the mica-schists[1] (Plate XXVIII).

Arabis petræa (H)*	Sedum roseum (H)
Draba incana (H)	Erigeron alpinum (H) rare
Cerastium alpinum (H)	Campanula rotundifolia
Arenaria sedoides (H)	Thymus Serpyllum
Silene acaulis (H)	Bartsia alpina (H) rare*
Astragalus alpinus (H) rare*[2]	Juncus trifidus (H)
Oxytropis uralensis (H) rare	Luzula spicata (tall form) (H)
Dryas octopetala (H)*	Salix reticulata etc. (H)
Alchemilla alpina (H)	Woodsia alpina (H)
Potentilla Crantzii (H)	Asplenium viride (H)
P. Sibbaldi (H)	Cystopteris fragilis
Saxifraga oppositifolia (H)	Rhacomitrium lanuginosum
S. nivalis (H)	Polytrichum *spp.*
S. hypnoides (H)	Andreæa *spp.*

* Do not occur on Ben Lawers.

The following have a very restricted distribution on fairly open ledges, and apparently require some special

[1] Oettli (*l.c.* p. 142) includes many of the species located here amongst his cliff plants ("Felsenpflanzen"); he discusses in detail the species in relation to habitat.

[2] "Rare" means rare within the habitat, most of the species occurring in a few stations only, for details of which the local floras must be referred to.

Phot. F. F. Laidlaw

a. Association of dry ledges. *Erigeron alpinum*, *Festuca ovina*, f. *vivipara*.

Phot. F. F. Laidlaw

b. Association of dry ledges. *Rhacomitrium lanuginosum*, *Salix reticulata*.

Arctic-alpine chomophyte formation.

feature in the environment; all are rare species on Lawers.

Saxifraga cernua (H)	Arenaria rubella (H)
S. rivularis (H)	Sagina nivalis (H)
Draba rupestris (H)	

B. *Association of sheltered ledges.*

The habitat is on ledges or terraces with deep moist humous soil which remains firm. The luxuriant vegetation of ledges of this kind is very striking after traversing summits or rock-faces scantily covered with vegetation. The accumulation of deep soil indicates shelter from wind erosion and protection by snow during the winter months. The plants are more massed together than on the open ledges, many having a loose mat habit which, with a covering of mosses, assists in binding the soil. Broad leaves of a mesophilous or partial shade type are often conspicuous, largely due to the high proportion of species from subalpine stream-banks. Transitions may be traced from the open vegetation of exposed ledges to the closed phase where grasses and sedges or heath-plants form closed associations.

The following are regarded as groups of characteristic species:

(*a*) *Plants of dry well-drained rock-ledges with abundance of soil and relative freedom from competition; on southern exposures:*

Veronica saxatilis (H)*	Myosotis pyrenaica (H)
Gentiana nivalis (H)*	Potentilla Crantzii (H)
Botrychium Lunaria*	Viola lutea
Rhinanthus borealis (H)	Linum catharticum

* Three species recorded together in various stations (D. A. Haggart, Killin).

(*b*) *Plants of ledges with deep soil and shelter; approximating to the closed phase:*

Trollius europæus	Galium boreale
Geranium sylvaticum	Hieracium alpinum, etc. (H)
Lychnis dioica	Rumex Acetosa
Rubus saxatilis	Orchis maculata
Alchemilla alpina (H)	Luzula sylvatica
Angelica sylvestris	Carex atrata (H)
Pyrola minor	Poa alpina, etc. (H)
P. secunda	Deschampsia alpina (H)
P. rotundifolia	Lastrea aristata

C. *Association of shade chomophytes.*

Where light is deficient, as in deep fissures hollowed into the rock-face or in hollows amongst block-screes, there is also much shelter, soil-moisture and atmospheric humidity. Amongst the mica-schists such cavities have earthy or rocky walls, moist and bare, except where covered by mats of mosses and liverworts; here and there a few flowering plants grow, more or less pale and etiolated. Or it may be that flowering plants are more abundant on shaded rocks where water oozes down through cushions of algæ and mosses. The deeper fissures are evidently buried in snow early in the winter, but the situation is such that plant-growth is promoted by water trickling under the snow, and in early summer the snow melts close to the rocks while it still lies in the open corries. The following are characteristic plants:

Anemone nemorosa	Taraxacum officinale (*forma*)
Cardamine flexuosa	Lactuca alpina (H)
C. hirsuta	Saussurea alpina (H)
Geranium Robertianum	Oxyria digyna (H)
Oxalis Acetosella	Polygonum viviparum (H)
Saxifraga stellaris (H)	Salix Lapponum (H)
Chrysosplenium alternifolium	Luzula spicata (H)
C. oppositifolium	Carex atrata (H)
Adoxa Moschatellina	Agropyron Donianum (H)

Polystichum Lonchitis (H)
Athyrium alpestre (H)
A. flexile (H)
Cystopteris fragilis (H)
C. montana (H)
Phegopteris Dryopteris

D. *Association of hydrophilous chomophytes.*

The banks of streams in the arctic-alpine zone are to be recognised as the habitat of characteristic species. Springs emerge even at high altitudes, and these, with abundant water from precipitation, drain off by a network of streamlets. The streamlets are generally small, so that on hard ground they have shallow channels. On the mountain top detritus drainage takes place through the rocky substratum, although limited areas of mossy peat occur. Amongst the peat hags the water drains away in deep channels or below the peat, and although originally spring water, it soon becomes mixed with acid peat water. The effect of slow and rapid flow and other conditions of aquatic vegetation, already dealt with to some extent in Chapter VII, need only be referred to here so far as they influence arctic-alpine vegetation in a few examples.

(*a*) *Moss communities.*

Correlated with rock outcrops in the schistose areas there is a characteristic type of vegetation where excess of water from numerous springs flows gently over a succession of terraces. The main habitat on Ben Lawers is towards the lower levels of the corries, but on a smaller scale many ledges at higher levels present the same moist conditions. There are few well-marked stream channels, so that most of the water flows dispersed through a bright green carpet of mosses, hepatics and algae spread along the flat terraces, and it is only on traversing these that the amount of water is appreciated.

The substratum consists mainly of flakes of mica-schist or of other rocks, caught amongst larger stones and boulders, and forming a fairly uniform surface, except at the margins of the terraces where the water trickles over in numerous tiny waterfalls. The water supply is mainly from springs, and on the south-western flanks of Ben Lawers the amount is considerable even in late summer. The constant supply of water from springs and the drainage ensure aeration, so that the vegetation is bright green; only locally is there any approach to boggy conditions. There is evidently no great periodic augmentation of the water, because there is not much erosion such as would indicate great periodic flushing. In early summer melted snow from higher altitudes must come down, also heavy rains, but most of this water escapes by a system of stream channels. On descending the slope, the streamlets become more defined and gather together ultimately to form a hill burn. Along the lower stream-flats drainage is impeded, and the conditions favour mosses, sedges and other plants, which on Lawers make up a dull green stream-bank vegetation.

In abundance and variety of species the hydrophyte moss association presents a striking physiognomy. It is evidently similar to the type described by Ostenfeld in the Faroes as the "Philonotis formation," distinguished by *Philonotis fontana* as the characteristic species; but further observations, especially those of bryologists, are required before this vegetation can be satisfactorily classified. Species of the "Highland" type are fairly abundant in this group of associations, especially at higher levels near the springs, but there is a large proportion of "exochomophytes," especially plants occurring along streams in the subalpine zone.

The following are representative flowering plants:

"Lowland" species.	*"Highland" species.*
Caltha palustris, *var.* minor	Thalictrum alpinum
Cochlearia officinalis (*forma*)	Cerastium cerastioides
Stellaria uliginosa	Saxifraga stellaris
Alchemilla vulgaris (*forma*)	S. aizoides
Montia fontana	Epilobium alsinefolium
Sedum villosum	E. anagallidifolium
Chrysosplenium oppositifolium	Veronica alpina
Veronica serpyllifolia, *var.* tenella	Polygonum viviparum
Pinguicula vulgaris	Oxyria digyna
Narthecium ossifragum	Tofieldia palustris
Carex dioica	Carex vaginata
C. echinata	
C. curta	

(*b*) *Sedge communities.*

Besides the moss communities the two following communities have been recorded by Dr F. F. Laidlaw in similar situations on Ben Lawers.

(i) *Junco-Caricetum.*

Carex saxatilis (H)	Juncus castaneus (H)
C. dioica	J. triglumis (H)
C. Goodenowii	J. biglumis (H)

(ii) *Caricetum curtæ.*

Carex curta covers a considerable area (*c.* 3000 feet altitude) and is the dominant plant, probably from its prostrate matted habit; associated with it are *Carex saxatilis, Pinguicula vulgaris,* etc.

Synopsis of Scottish Arctic-alpine vegetation (mainly based on the vegetation of Ben Lawers)

I. *Zone of cultivation and semi-natural pastures* (formerly woodland?).

II. *Zone of closed moorland associations belonging to heath and moor formations* (formerly woodland?).

- (*a*) Transitional belt of closed moorland associations (as in 2) with local centres of arctic-alpine species.

III. *Zone of arctic-alpine vegetation* (over 2000 feet Central Scottish Highlands).

1. Lower arctic-alpine belt.
   Formation of arctic-alpine grassland.
2. Arctic-alpine belt.
   - (i) Formation of mountain top detritus.
     - A. Moss-lichen open association.
     - B. Rhacomitrium heath association.
     - C. Rhacomitrium moor association.
   - (ii) Chomophyte formation
     - A. Open communities on exposed rock-faces.
     - B. Association of sheltered ledges.
       - (*a*) Sub-association of dry well-drained soil and southern exposure.
       - (*b*) Sub-association of deep damp soil and shelter.
     - C. Association of shade chomophytes.
     - D. Associations of hydrophilous chomophytes
       - (*a*) Moss communities
       - (*b*) Sedge communities.
         - (1) Junco-caricetum.
         - (2) Caricetum curtæ.

Diagram showing the relations of the plant-communities of the Arctic-alpine formations.

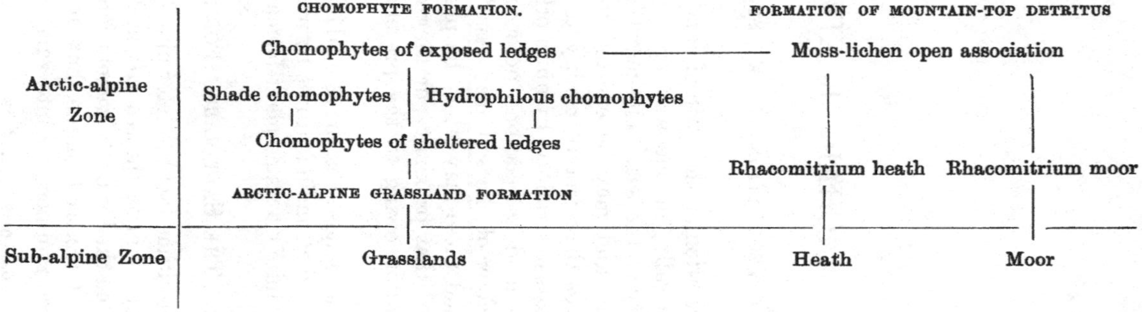

CHAPTER XIV

THE VEGETATION OF THE SEA COAST

THE vegetation of the strip of land bordering on the coast is generally, in one way or another, affected by the proximity of the sea. The maritime influences take various forms and may be exerted in various degrees. In some cases they merely modify to a greater or less extent habitats determined by quite other factors. Thus various grassland associations, determined in each case by the character of the soil, may exist in close proximity to the sea, and these may either be practically unaltered by the fact of that proximity, or they may show more or less modification owing to the presence of "maritime" species.

In other cases, however, the coastal habitats are actually determined in their main features by maritime factors and thus bear definite maritime plant-formations.

I. THE SALT-MARSH FORMATION

Of the maritime factors determining specific habitats, the most important is the *sea-salt,* and the effect of this is seen most markedly in the vegetation of tracts of land, occupied by terrestrial plants, but actually covered by the tides at certain periodic intervals[1]. Sand and shingle

[1] Habitats permanently covered by the sea and those only exposed for short periods, belong to the aquatic formation of salt-water, and are

beaches between tide-marks as well as mud exposed to strong tidal scour are practically destitute of plants because of the constant movement of the substratum by the waves. Rocks in the same zone are commonly inhabited by those marine algæ which can tolerate periodic emergence from the water and which are fixed to the rock substratum by holdfasts. Tracts of tidal mud, or mixtures of mud and sand, laid down on the flat shores of sheltered bays or estuaries, are however the characteristic habitat of a vegetation which is mainly phanerogamic. This type of habitat is generally known as salt-marsh, and, since it is fundamentally distinct both in conditions and in flora from other habitats, its vegetation may be properly called the *salt-marsh formation.*

Habitat.

The plants of the salt-marsh formation are periodically immersed by the tide, but only during high-water of the spring tides, *i.e.* during a certain number of the highest tides occurring once a fortnight. The highest regions of a salt-marsh are not reached by all spring tides, and in extreme cases are immersed only once or twice a year. Besides having to undergo periodic immersions in salt water the plants of the salt marsh have a ground water which is normally salt, though the percentage of salt it contains may vary from a percentage higher than that of sea water to quite a low percentage—approximating to that of fresh water—during considerable periods. The variation in the amount of salt in the ground water is due to the interaction of various factors, immersion by the tide, the rate of evaporation, and the washing out of the salt by rain-water during the periods of exposure.

Conditions of life.

not dealt with in this book. Vegetation exposed to the air most of the time and only occasionally covered is to be reckoned with maritime terrestrial vegetation. This distinction roughly corresponds with the difference between marine algal vegetation and salt marsh vegetation.

Salt-marsh plants form the typical *halophilous vegetation.* The peculiarities of halophytes, even the cause of their most general and most striking character, succulence, are by no means fully understood; but there can be no doubt that the factor which leads to the striking anatomical and physiological peculiarities of halophytes, of which succulence is the most general and striking, is the salt ground water.

The associations of the salt-marsh formation, like all associations determined by a gradual spatial change in the incidence of a well-defined factor, show a marked zonation, in this case in relation to the tidal zone.

1. **Glasswort association** (*Salicornietum europææ*).

The first or pioneer association of the salt marsh, developed on the seaward side, is an open association of one or more of the herbaceous species of *Salicornia*, typically of *S. europæa*.

On the mud flats of estuaries or protected bays this association extends as far as the upper limit of neap tides; on its outer fringe the plants are few and far between, further inshore they become closer set though they never form a closed association. The association is typically pure, though *Zostera marina*[1] sometimes occurs.

The first plants to become associated with *Salicornia* on the landward side vary according to the nature of the soil. In muddy salt-marshes they may be *Glyceria maritima* and *Triglochin maritimum*[2] or *Aster Tripolium* (Plate XXIX a). Where sand predominates in the silt *Salicornia radicans* and *Atriplex portulacoides* may play an important part, with *Glyceria maritima* and other

[1] On certain coasts this plant forms a well-defined association of the aquatic marine formation, beyond the *Salicornia*-zone, and only exposed to the air at low water of the spring tides.

[2] Moss, 1906, p. 19.

Phot. R. H. Yapp

a. Glasswort open association (*Salicornietum europææ*) passing into General salt-marsh association. *Salicornia europæa*, *Glyceria maritima* (very luxuriant) and *Aster Tripolium*. Coast of the Wash.

Phot. R. H. Compton

b. General salt-marsh association. Matted *Glyceria maritima*, *Atriplex portulacoides*, *Statice vulgaris* (left), *Limonium reticulatum* (right), *Plantago maritima* (centre), *Spergularia marginata* (top). Holme-by-the-Sea, Norfolk.

Salt-marsh formation.

Phot. S. Hastings

a. Channel with raised banks fringed with *Atriplex portulacoides*. Community of *Salicornia europæa* and *Glyceria maritima* occupying flat ground on the left. Bouche d'Erquy, Brittany.

Phot. F. F. Blackman

b. Glycerietum maritimæ with *Suæda maritima*, and scattered *Atriplex portulacoides*. Salicornietum occupying the channels. *Juncetum maritimi* in background. Bouche d'Erquy, Brittany.

Salt-marsh formation.

species following[1]. These varying phases lead up to the establishment of the general salt-marsh association.

1 *a*. **Cord-grass association** (*Spartinetum*).

This association, dominated by *Spartina stricta* or other forms of *Spartina*, occurs locally on the south and east coasts, in place of the Salicornietum (see p. 337).

2. **General salt-marsh association.**

This association is developed at higher levels and is covered only by medium tides. It is not dominated by any single species, except locally, and varies from place to place according to the local conditions and to the accidents of colonisation by different species (Plate XXIX b).

The following species are commonly represented:

Spergularia marginata	Atriplex portulacoides
S. salina	Salicornia europæa (and other herbaceous species)
Aster Tripolium	S. radicans
Artemisia maritima	Suæda maritima
Limonium vulgare	Triglochin maritimum
Statice (Armeria) maritima	Glyceria maritima
Plantago maritima	

Atriplex portulacoides is typically dominant on the banks of creeks, especially when these banks, as is usually the case, are raised above the general level of the muddy flats drained by the creeks (Plate XXX a). *Aster Tripolium* grows particularly luxuriantly in the wet mud beside sheltered creeks.

3. **Glycerietum maritimæ.**

The general salt-marsh association gradually accumulates mud and raises the level of the surface of the salt

[1] These data are taken from observations on a Breton estuarine salt-marsh whose soil is largely sand. Cf. Oliver, "The Bouche d'Erquy in 1907," *New Phytologist*, vol. VI. pp. 249—251. The newly described species *Salicornia Oliveri* Moss forms a pioneer association in this locality. Cf. Moss, *Journal of Botany*, 1911.

marsh. An association dominated by *Glyceria maritima*, which forms a closed turf, and is only covered by the higher tides, is gradually developed on these higher levels. In the *Glyceria*-turf most of the species of the general salt-marsh association may occur. Such species as the sea-lavender (*Limonium vulgare*) and the thrift (*Statice maritima*) sometimes occur very abundantly, and in summer the turf, studded with innumerable lavender or rose-coloured flowers, then presents a glorious spectacle.

The Glycerietum is commonly used for pasturage, and the salty herbage is said to impart a distinct flavour to the mutton fed on it.

A very distinct sub-association of *Suæda maritima* with *Glyceria maritima* is sometimes found occupying sandy slopes and hummocks[1], while *Salicornia europæa*, associated with the same grass, occupies lower lying flats (Plate XXX).

4. **Sea-rush association** (*Juncetum maritimi*).

The highest zone of the salt marsh, reached only by the highest spring tides, is frequently occupied by an association dominated by the sea-rush (*Juncus maritimus*) (Plate XXXI).

The lower belts of the Juncetum may contain many species of the general salt marsh association, but the highest belt, which is very rarely covered by the tide and is least affected by the salt water, shows quite a low percentage of salt in the soil water, and typically contains the following plants:

Frankenia lævis (sandy soil)	Plantago maritima
Œnanthe Lachenalii	P. Coronopus
Glaux maritima	Agropyron pungens

[1] The development of these hummocks has been carefully traced in the Bouche d'Erquy, and has been found to be due to initial accumulations of sand by the side of stream channels through the agency of plants of *Salicornia radicans* and other species. See Oliver, *l. c.*; also T. G. Hill, *New Phyt.*, vol. VIII., p. 97, and figs. 4 and 5.

Phot. F. F. Blackman

Sea-rush association, with Glycerietum maritimæ in intervals. *Salicornia europæa* and *Suæda maritima* in foreground. Bouche d'Erquy, Brittany.

Salt-marsh formation.

With these may be associated various non-halophilous plants which can tolerate a certain amount of salt in the soil, *e.g.* such grasses as *Agrostis palustris* and *Festuca rubra*. A similarly composed association is commonly found on the dykes thrown up for the reclamation of salt marshes.

When a salt marsh has reached a certain stage of development the sea is frequently fenced out by dykes and the land is drained. The salt is washed out comparatively rapidly by rain, and the ground is quickly colonised by non-halophilous species, among which a certain number of the less extreme halophytes may linger. Such land makes excellent pasture, and great tracts of such pasture exist on the sites of old estuaries. The vegetation has not been sufficiently closely studied to permit of an adequate description.

Association of spray-washed cliffs and rocks.

Soil-covered rocks or cliffs lying above high-tide mark, but constantly subject to the influence of spray from the waves, have their soil impregnated with salt and bear halophilous species, *e.g. Statice maritima, Limonium vulgare, Plantago Coronopus* and others. The samphire (*Crithmum maritimum*) is particularly characteristic of this habitat.

Brackish water associations.

Other coastal habitats may have a more or less salt (brackish) ground water, and these show features in common with the salt marsh. Brackish marshes possessing characteristic halophilous and brackish-water species, besides inland freshwater marsh species, often occur by the sides of the upper reaches of tidal estuaries where the sea water is much diluted[1], and also where there is an underground salt or brackish water "table" close to the surface, *e.g.* on flat lowlying coasts to which the actual

[1] W. G. Smith, 1904–5, pp. 73, 74.

access of the sea is prevented by sand dunes or shingle beaches or by artificial dykes[1]. There are certainly several distinct associations belonging to such habitats, but they have been, as yet, very insufficiently studied.

THE SALT-MARSH FORMATION OF THE HAMPSHIRE COAST

By W. M. RANKIN

The coasts of Hampshire and the Isle of Wight, which extend in all for some 130 miles (*c.* 210 km.), exhibit great variety, both of rock and outline. Much of their extent is formed of Weald Clay, Eocene clays and Oligocene marls, and the shores are low and wasting. The several estuaries and the Solent (the strait into which they open) are "drowned" river valleys up which the tides run far inland. The archipelago of low islands between the Gosport and the Selsey peninsulas is similarly a result of marine inundation. The shores of these estuaries and channels are edged with broad flats of mud derived from the many soft rocks of the district and redistributed by the tides.

The tides show the phenomenon of double high water, owing to the division of their flow by the great mass of the Isle of Wight. Part of the water coming up Channel from the west takes the western or Needles passage, attains its maximum flow, and begins to ebb before the other part, taking the longer route round the Island, reaches the Solent and Southampton Water by way of Spithead. The two bodies of water meeting make a second high tide, thus lengthening the period of high water during which deposition of mud occurs. This is followed by a very rapid ebb. At the spring tides the rise

[1] See M. Pallis, "On the cause of the salinity of the broads of the River Thurne," *Geogr. Journal*, 1911, p. 284.

within the Solent is about 13 feet (*c.* 4 m.) and at the neap tides about 10 feet (*c.* 3 m.).

The seawrack association (Zosteretum) is well developed in this area on the flats which are exposed only at the low spring tides. The plants are attached to pebbles or larger rock fragments embedded in sand or mud. *Z. marina* is the commoner form, but *Z. nana* also occurs, especially off Ryde.

Cord-grass association (*Spartinetum*).

The most characteristic and interesting feature of the Hampshire salt-marsh formation is the prominence of the cord-grass association or Spartinetum which largely replaces the Salicornietum on these coasts.

Three forms of *Spartina* occur on the salt-marshes of Hampshire, *S. stricta, S. alterniflora* and *S. Townsendi* H. and J. Groves. *S. stricta* has long been known as occurring in occasional patches on the salt mud of this district, as well as locally along the south coast as far west as Devon, and along the east coast as far north as Lincolnshire. *S. alterniflora* has also been known for some considerable period as covering extensive tracts of salt marsh on the shores of Southampton Water, to which it is limited in this country. *S. Townsendi*[1], on the contrary, was first noticed in the Southampton Water salt marshes in 1870, and was described under this name in 1881. Since then it has spread to a remarkable extent, not only covering broad expanses of mud in Southampton Water itself, but also many of the creeks that open into the Solent. It is also beginning to occupy the muddy shores of Poole Harbour (Dorset) and of the channels and inlets between Fareham and Chichester (Sussex). Already between 6000 and 8000 acres (*c.* 2400 to 3200 hectares) of

[1] Considered by Dr Otto Stapf, following Foucaud, to be a hybrid (*S. stricta* × *S. alterniflora*) (*Gardeners' Chronicle*, vol. XLIII, Third Series, p. 33, 1908).

mud which was formerly bare are covered with this vegetation.

The Spartinæ, especially *S. alterniflora* and *S. Townsendi*, form strong tussocks. There is a stout stem with stiff erect leaves; while strong fibrous roots, and thick stolons radiating horizontally in all directions, firmly bind the soft mud. The leaves offer broad surfaces for the deposition of silt from the tidal waters, and their points catch and hold fragments of seaweed and other flotsam. The thick forest of stems and leaves breaks up the tidal eddies and so prevents the removal of mud which has once settled on the marsh. These plants thus play a very important part in the fixation of large areas of mud, raising the level of the flats and preparing them for eventual reclamation from the sea. These consolidated mud flats also prevent the erosion of still untouched coast lines, and play an important part in determining the courses of tidal currents and channels. There is a striking parallelism between the habit of growth and consequent *rôle* of Spartina on these tidal flats and that of *Ammophila* on the sand dunes.

The associations of *Spartina Townsendi* and *S. alterniflora* are remarkably pure. *Aster Tripolium, Spergularia marginata* and *Salicornia europæa* form occasional societies.

The Salicornietum, except round Hayling Island and at a few other places, and the general salt-marsh association are not extensively developed in this region, the Spartinetum largely occupying their places.

Glycerietum maritimæ.

This association occupies the slightly higher and therefore less frequently inundated zone behind the Spartinetum. On the finer muds *Glyceria maritima* attains a luxuriance of growth suggesting meadow vegetation. Associated are

Cochlearia officinalis
C. anglica
Spergularia marginata
Aster Tripolium
Statice maritima
Limonium vulgare
and other species
Plantago maritima
Salicornia europæa
Atriplex portulacoides
Triglochin maritimum
Spartina Townsendi

Aster, *Limonium*, *Spartina* and *Salicornia* often form strong local societies.

The sandy or otherwise drier Glycerietum shows a more varied flora, including, besides the above, *Salicornia radicans*, *S. ramosissima*, *Suæda maritima*, *Juncus maritimus*, *J. Gerardi* and other species.

These Glycerieta pass into pasture land containing non-halophilous grasses and other plants, but the phases of the later stages of succession of the salt marsh are not specially well seen on the Hampshire coast.

II. The Sand Dune Formation

The second factor determining a specific type of coastal vegetation is *blown sand*. The accumulations of blown sand known as sand dunes are not peculiar to the sea coast; they also occur on the margins of great inland lakes and in desert regions, and the vegetation of all these kinds of sand dunes appears to be fundamentally similar. But in this country typical sand dunes are confined to the sea coast, and their vegetation may be regarded as a maritime vegetation constituting a distinct plant formation—the *sand dune formation*. While a sandy beach between tide marks is very bare of plant life, the sand which is blown up beyond the reach of the tide at once becomes colonised by vegetation, and the characteristic sand hills that occur on coasts where there is a constant supply of such blown sand are the result of the combined action of this vegetation and of the wind on the loose mobile sand. Away from the sea the dune vegetation tends to become stabilised, and fixed dunes are formed which contrast with the mobile dunes nearer the sea. Fixed dunes may eventually pass into other formations.

The sand dune habitat continually advances seaward where there is a constant and abundant supply of sand thrown up by the waves, and thus made available for the construction of new dunes by the wind and the sand dune vegetation.

Association of strand plants.

When the supply of sand is arrested the seaward growth of the front mobile dunes is also arrested. In such cases an *association of strand plants* develops on the beach in front of the dunes and between them and the high-water mark of ordinary spring tides. This association is probably determined by the existence of salt ground-water in immediate connexion with the sea. It consists of characteristic species and these show the peculiar features of halophytes. The association does not therefore properly speaking belong to the sand dune formation, but should rather be placed under a subformation of halophilous vegetation. On account of its topographical relations it is however convenient to treat the association in connexion with the sand dune formation.

The following species are the most characteristic of this association:

Cakile maritima	f	A. hastata	f
Crambe maritima	r	A. Babingtonii	f
Arenaria peploides	f	Salsola Kali	f
Atriplex patula	f	Polygonum Raii	r

These plants frequently form miniature dunes. In some cases *Agropyron junceum* is mixed with the strand plants and scarcely forms a separate association (Plate XXXII a).

A certain amount of salt is found in the sand of some mobile dunes, but this is by no means constant; the master factor of the habitat is the mobile sand, and the vegetation is not halophilous. The dominant plant of mobile dunes in this country is the marram grass (*Ammophila arenaria*) which almost invariably colonises blown sea sand on our coasts. By means of its long underground stems and roots and its power of growing up to the surface when buried in sand by the wind, this grass brings about the formation of the typical coastal dunes.

The character of the associations which follow the *Ammophila* association in succession and occupy the fixed dunes, varies with the nature of the blown sand, and particularly with the nature of the neighbouring inland vegetation, which largely determines what plants shall colonise the fixed dunes. In many cases there is a mixed "pasture" association, in which such plants as *Carex arenaria*, and *Festuca rubra* var. *arenaria*, are dominant, while *Ononis repens*, *Erodium cicutarium*, *Lotus corniculatus*, *Phleum arenarium* and others are abundant[1]. Sometimes the dunes are colonised by numerous bushes and come to bear dense scrub; in other cases by the ling and its associates and pass into typical heath[2]. In this country natural woodland associations do not occur on sand dunes, probably because of the general absence of trees, which could supply a sufficiency of easily distributed seed, in the immediate neighbourhood of the coast; in other parts of the world however woodland associations frequently colonise coastal dunes. In the hollows of sand dunes, where the ground water, held up by a relatively impermeable substratum, approaches or reaches the surface, dune marshes are developed; the ground water may be brackish or fresh and the vegetation of the dune marsh shows corresponding characters.

[1] W. G. Smith, 1904–5, pp. 69, 70; Moss, 1906, pp. 13, 14.
[2] W. G. Smith, *l. c.* p. 71.

1. **Sea couch-grass association** (*Agropyretum juncei*).

This association is not found on all sand dunes, but in some cases (*e.g.* on the Somerset and on the Lancashire coasts) it brings about the existence of a characteristic line of low dunes in front of the higher dunes formed by the marram grass (*Ammophila arenaria*). The dominant plant, *Agropyron junceum*, like its more widespread congener *Ammophila*, has the power of binding the sand and growing up through the sand with which it is covered by the wind, but probably in a lesser degree. It also seems likely, from the characteristic position of this species, that it can tolerate more frequent drenching with spray than is the case with *Ammophila*.

Moss[1] gives the following list of the species of this association on the Somerset dunes:

Dominant, Agropyron junceum.
Sub-dominant, Carex arenaria.

Glaucium flavum	Atriplex hastata
Cakile maritima	A. deltoidea
Arenaria peploides	A. Babingtonii
Eryngium maritimum	Rumex crispus
Caucalis arvensis	Euphorbia Paralias

2. **Marram grass association** (*Ammophiletum arenariæ*).

This is the most characteristic association of mobile sand dunes, and occurs on practically every patch of blown sand on the British coasts. Where the Agropyretum is absent, the Ammophiletum is the most seaward of the dune associations, occupying the principal front range of mobile dunes, whose accumulation and primary fixation it determines (Plate XXXII b).

Ammophila arenaria (marram grass, star-grass, "bent") colonises blown sand on the coast with remarkable rapidity, and by its well-known habit and powers of growth leads to the formation of the characteristic mobile dunes

[1] Moss, 1906, pp. 10—11.

Phot. R. H. Compton

a. Association of strand plants. *Arenaria peploides* mixed with *Agropyron junceum* forming miniature dunes. Holme-by-the-Sea, Norfolk.

Phot. A. G. Tansley

b. Progressive dune formation. The low dunes on the right are formed by *Elymus arenarius*, on the left by *Ammophila arenaria*. Main dune range behind. Hemsby, Norfolk.

Sand Dune formation.

("shifting," "travelling," or "white" dunes). In most cases it succeeds in dominating the blowing sand, continually growing up to the surface, when it is covered by fresh supplies of sand. The sand accumulated on the seaward side of the mobile dunes is gradually occupied by the dune grasses; where *Agropyron junceum* is present, first by this plant, and later by *Ammophila*, so that the complex of mobile dunes gradually grows out seawards and new land is gained. But in some cases, *e.g.* on parts of the Irish coast, the dune grasses are overwhelmed, at least temporarily, by vast sand accumulations, and the blowing of the sand inland is then quite unchecked. Marram grass is often regularly planted to check the advance of blowing sand in places where the plant has not established itself naturally.

"Blow-outs"

Very frequently violent winds blow away portions of the *Ammophila* dunes, and on the sides of these excavations, or "blow-outs," the underground portions of the marram grass can be seen extending to the base level of the dunes, thus exhibiting a record of the gradual building up of the dune through the agency of the plant.

The process of fixation

Marram grass is the primary agent in fixing the principal mass of sand in any given sand dune area. The fixation however is by no means complete; a good deal of sand is blown through and between the subaerial portions of the plants and is deposited further inland. In this way lowlying country on the landward side of the dunes is often covered with a thin layer of sand. Occasionally, where the supply of sand thrown up by the sea is very constant and abundant, the mobile dunes "travel" inland over flat country. In most cases however the partial fixation of the sand through the agency of *Ammophila* is completed by other plants which colonise the comparatively protected sand on the leeward side of the first range of dunes, and "fixed dunes," covered by a plant carpet which almost completely withdraws them from the disturbing action of the wind, are formed. Even these however may be excavated by an occasional violent wind storm giving rise to a "blow-out," and the process of colonisation by vegetation begins afresh.

Static and retrogressive phases.

On the other hand when the supply of sand thrown up by the waves ceases, the formation of new mobile dunes is checked, and the dunes eventually become fixed, *i.e.* covered by a plant carpet right up to their front limit. This static phase in the life of the dune formation may be succeeded by a phase of retrogression, if, owing to changes in tidal currents, the sea begins to gain on the land, and the seaward face of the dunes is eaten away by the waves. All these varying relations of sea and dunes, from active growth of the dunes seaward, through the condition of equilibrium to that of active erosion, may be observed on different parts of our coasts.

The steep seaward face of the *Ammophila* dunes is often quite bare, the marram grass itself occupying the crest of the line of dunes, occasionally accompanied by its more local congener the sea lyme grass (*Elymus arenarius*), whose mode of growth is similiar to that of *Ammophila* itself (Plate XXXII b). On the crest, under the shelter of the marram grass and particularly on the leeward side, a number of other plants are able to find a footing. The more conspicuous and abundant of these are:

Arenaria peploides
Eryngium maritimum
Senecio Jacobæa
Taraxacum erythrospermum
Hieracium umbellatum
Hypochæris radicata
Calystegia Soldanella
Euphorbia Paralias
Hippophaë rhamnoides
(native on east coast)

Carex arenaria
Phleum arenarium
Festuca rubra
var. arenaria
F. uniglumis
Corynephorus canescens (very local, but abundant in Norfolk)
Agropyron junceum
Peltigera canina

Phot. A. G. Tansley

a. Dune valley behind main dune ridge with scrub of *Hippophaë rhamnoides* (*Rubus discolor* agg., etc., associated). Dune grassland association in which *Ammophila arenaria* is conspicuous in the intervals. Looking seaward; the horizon is formed by the main dune crest. Hemsby, Norfolk.

Phot. A. G. Tansley

b. Detail of dune scrub. *Hippophaë rhamnoides* (right), *Rubus discolor* agg. (left), *Lonicera Periclymenum*, *Polypodium vulgare* (centre), *Ammophila arenaria*, *Festuca rubra*, *Holcus lanatus*, etc.

Sand Dune formation.

Fixed dune associations.

The carpet of vegetation which gradually establishes itself on the leeward side of the mobile dunes passes over into the fixed dune associations. These, as already stated, vary according to the different conditions which may obtain, and no detailed treatment will be given here. A feature of the initial stages of the passage to fixed dune is the existence of locally developed societies of gregarious plants. Thus a dense scrub of *Hippophae rhamnoides*, with elder (*Sambucus nigra*) and bramble (*Rubus discolor* agg.) sometimes occurs, as on the coast of Norfolk (Plate XXXIII); *Salix repens* may play an important part (see p. 350); or widespread societies of the sand sedge (*Carex arenaria*) may occupy considerable tracts of ground. A large number of species occur, as is usual in these intermediate associations[1], and among them are a number of easily distributed species of light soil which readily colonise sand dunes, where they find comparatively sparsely occupied ground of a suitable kind. In agricultural districts, too, weeds of light arable land are strongly represented on the sand dunes.

As the colonisation of the soil proceeds and the vegetation carpet becomes closed, the "maritime" species tend to disappear, but *Ammophila* often maintains itself in the grassland for a considerable time, though its growth is much less luxuriant than on the mobile sand, and its leaves lose their fresh green appearance.

Some of the more widespread species of dune grassland, which is the commonest association of fixed dunes in this country, are the following:

Erophila verna	Lotus corniculatus
Cerastium semidecandrum	Trifolium repens
Viola ericetorum	T. pratense
Erodium cicutarium	T. arvense
Ononis repens var. horrida	Rubus discolor (agg.)

[1] Moss, 1906, p. 13.

Rosa spinosissima	L. autumnale
R. Eglanteria	Carlina vulgaris
Potentilla reptans	Taraxacum erythrospermum
P. anserina	Carduus pycnocephalus *var.* tenuiflorus
Sedum acre	Cnicus arvensis
Sambucus nigra	Centaurium vulgare (Erythræa littoralis)
Galium verum	Myosotis collina
G. saxatile	Thymus Serpyllum
Senecio Jacobæa	Carex arenaria
Crepis virens	Festuca rubra
Hieracium Pilosella	Holcus mollis
Hypochæris radicata	
Leontodon nudicaule	

Associations of dune marshes.

In the dune hollows, where an underlying impermeable substratum on which the dunes are built is close to the surface, marsh or even permanent pools occur. The water may be brackish or fresh, and consequently the vegetation varies. The brackish water associations have not been properly described (see p. 335; and for a list of species of the freshwater marshes of the Somerset dunes, see Moss, 1906, p. 16).

THE SAND-DUNE VEGETATION OF THE LANCASHIRE COAST[1]

History of the area

The extensive range of sand dunes which line the Lancashire coast from the estuary of the Mersey to that of the Ribble rests upon a wide plain of glacial clay, which in early post-glacial times extended far out into the Irish Sea, and, according to some authorities, probably joined the Isle of Man to the mainland. The surface of this plain was uneven and held numerous shallow lakes or meres, the remains of some of which still exist behind the

[1] Prof. F. E. Weiss kindly collected most of the information on which this account is based and Mr Wheldon was good enough to supply data as to the Bryophytes. [Editor.]

present coastal deposits. Extensive peat formation took place locally at some later epoch, and the remains of forests of oak, birch and pine occur buried in the peat, in some cases beyond the present shore line. Subsidence and sea-erosion gradually caused the retreat of the coast

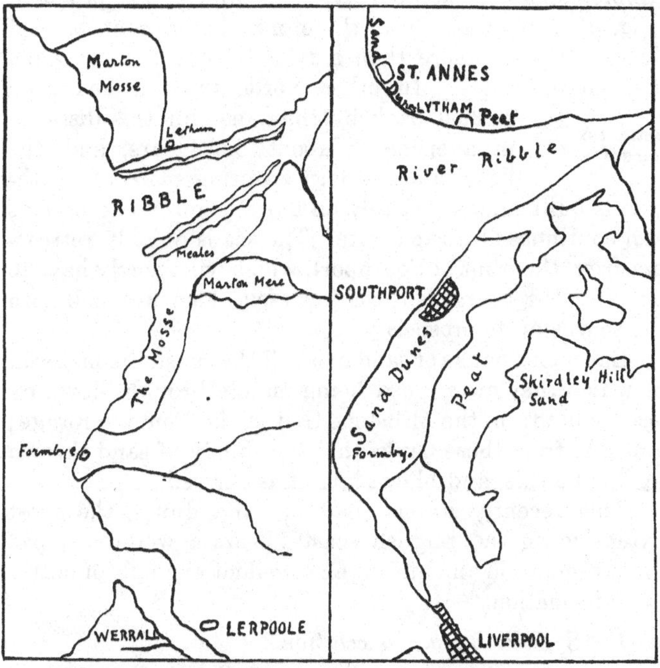

John Speed's Map, 1610. *Present Day.*

Scale of miles 0 1 2 3 4 5 6 7

FIG. 19. Sketch-maps of the southern portion of the Lancashire coast, showing the growth of the sand dunes between the Ribble and Mersey estuaries, and north of the Ribble, during the last 300 years. Southport and St Anne's are built entirely on this recent blown sand.

eastwards, and by about the year 1600 it had probably, between Ainsdale and Southport, reached a line "not far removed from the present Lancashire and Yorkshire Railway[1]," *i.e.* near the *inner* margin of the existing belt of dunes. This old coast line is shown on a map of Lancashire, published by John Speed in 1610, in which the strip of coast from a little north of Formby to the Ribble estuary is marked as "The Mosse," *i.e.* peatland (Fig. 19). At that date, therefore, the present belt of dunes, which is more than a mile (nearly 2 km.) wide, had no existence. In other words the sea has been pushed back by the dunes for this distance in a space of about 300 years, and the flourishing modern watering place of Southport is built almost entirely on blown sand. The process still continues at a rapid rate. The sea is quickly retreating from the front of Southport, which will shortly have to be extended seawards if the town is to keep in touch with the source of its prosperity.

Retreat of the sea

Enormous banks of sand exist off the coast, the material of which they are formed being mainly brought down by the Ribble from the Millstone Grit of the Pennine Range; and it is from these banks that the supply of sand thrown up by the tide, and blown inland, is derived.

This recently formed belt of sand dunes, the most extensive on the English coasts, bears a well-developed dune vegetation and affords an excellent example of active dune formation.

1. *Sea couch-grass association.*

Owing to the rapid advance of the dunes seaward there is no association of strand plants on the beach, but on the outer edge of the main mass of sand hills there is a range of low dunes formed by the sea couch-grass

[1] T. Ashton, *The Battle of Land and Sea on the Lancashire, Cheshire and North Wales coasts, with special reference to the origin of the Lancashire sandhills*, 1909, pp. 97–8. This book contains a collection of valuable historical *data* from which the history of this coast is reconstructed.

Phot. W. Ball

a. Progressive Dune formation. Sea drift and low dunes formed by *Agropyron junceum.* The dunes at the back on the right, somewhat denuded by wind, bear *Ammophila arenaria.* Near Southport, Lancashire.

Phot. W. Ball

b. Dry association of *Salix repens*, in a "slack" (dune valley). Low *Salix*-dunes at the edge. Behind, *Ammophila*-dunes. Near Southport, Lancashire.

Sand Dune formation.

(*Agropyron junceum*) (Plate XXXIV a). A characteristic member of this association is *Senecio vulgaris* var. *radiatus*.

2. *Marram or star-grass association.*

Behind the Agropyrum dunes come the higher mobile dunes with their typical association of *Ammophila arenaria*, locally called "star-grass[1]." The front row of dunes does not exceed 20 or 30 feet in height, but the higher dunes further inland attain a height of "about 60 feet, and here and there between Formby and Ainsdale the hills reach over 80 feet" (*c.* 24 m.) in height[2].

On the exposed seaward side the *Ammophila*-association is nearly pure, but in sheltered places, especially on the landward side, the dominant becomes mixed with other grasses, among which *Festuca rubra* and its variety *arenaria* are conspicuous.

The following is a list of the commoner species of this association, grasses and members of the Compositæ greatly predominating.

Ammophila arenaria d

Eryngium maritimum	Taraxacum erythrospermum
Galium verum	Phleum arenarium
Senecio vulgaris	Aira caryophyllea
S. Jacobæa	A. præcox
Carlina vulgaris	Festuca membranacea
Hieracium umbellatum	F. rubra
Hypochæris glabra	F. rubra *var.* arenaria
Leontodon nudicaule (hirtum)	

Between the first series of *Ammophila*-bearing dunes and the higher sand hills further inland there is a distinct valley. On the eastward side of this valley the sand hills, even on the side exposed to the wind, are covered with a more abundant and more varied vegetation, including in some places a thick carpet of mosses. This seems to be transitional to a fixed dune association (grassland), which,

[1] T. Ashton, *op. cit.* p. 123. [2] *Ibid.* p. 124.

however, does not occur in a typical form in this region. In addition to several of those already mentioned the following species occur in these positions:

Erodium cicutarium	Euphorbia Paralias
Ononis repens	E. portlandica
Rosa spinosissima	Carex arenaria
Calystegia soldanella (very rare near Southport)	Barbula convoluta
Cynoglossum officinale	Tortula ruralis var. arenicola, Braithwaite
Rumex crispus	Ceratodon purpureus

3. *Dry association of Salix repens* (Plate XXXIV b).

The vegetation of the dune valleys is particularly well developed in these Lancashire dunes. The creeping willow (*Salix repens*) is the most characteristic feature of these hollows, occurring in great abundance. Various forms of this plant, *e.g.* f. *fusca*, f. *ascendens*, and f. *argentea*, occur. As described by Massart on the Belgian dunes[1], *Salix repens* forms small dunes in the hollows. These small dunes are formed from sand blowing off the larger sand hills, and can be seen in various stages of development. The willow keeps pace in its growth with the accumulation of sand around it and forms dunes 15 feet high or more. The secondary *Salix*-dunes in all stages of development and denudation by wind are a conspicuous feature of the larger dune hollows, or "slacks" as they are locally termed.

A number of plants become associated with the creeping willow on these smaller secondary dunes. The most conspicuous among them is *Pyrola rotundifolia*, the inflorescences of which project above the prostrate branches of the willow, while in winter the bright green persistent foliage is conspicuous under the leafless twigs. This plant exists as a characteristic form (var. *arenaria* Koch), which does not, however, preserve its distinctive characters

[1] J. Massart, "Les districts littoraux et alluviaux" in *Les aspects de la végétation en Belgique* (Bommer et Massart), 1908.

in cultivation. The following plants occur in this association:

Salix repens	d		
Lotus corniculatus	f	Carex arenaria	a
Blackstonia perfoliata	f	Brachythecium albicans	
Pyrola rotundifolia	a	B. rutabulum	
Monotropa Hypopitys	o	Eurhynchium megapolitanum	

4. *Damp association of Salix repens.*

In damp sandy hollows the *Salix*-association has the following species, including numerous bryophytes.

Parnassia palustris	Bryum Warneum
Carex distans	B. pseudotriquetrum
C. extensa	B. neodamense
Equisetum variegatum var. arenarium	Aneura pinguis var. latifrons
Bryum pendulum	Petalophyllum Ralfsii
B. calophyllum	Pallavicinia hibernica etc.

5. *Dune marsh association.*

In the wettest hollows *Salix repens* disappears and a marsh association is developed including

Parnassia palustris	Carex spp.
Hydrocotyle vulgaris	Hypnum elodes
Anagallis tenella	H. polygamum
Juncus maritimus	

In the same hollows, where there has been no artificial drainage, the marshes may, during the winter months, become actual pools; in some cases these pools persist throughout the summer and then contain *Chara* spp. and various green Algæ.

6. *Grassland association of dune hollows.*

The following species occur in the moderately dry grassland association of dune hollows which represents the nearest approach to a fixed dune association in this region.

Viola Curtisii	C. vulgare
Anthyllis Vulneraria	Gentiana campestris
Lotus corniculatus	G. baltica
Potentilla anserina	Helleborine latifolia
Blackstonia perfoliata	Orchis pyramidalis
Centaurium umbellatum	

The most landward, which are also the oldest, of the higher dunes suffer a considerable amount of denudation from the strong sea winds, and the sand blown inland from them is forming lower dunes which encroach upon the adjacent meadowland.

The Evening Primrose (*Oenothera biennis*) is an interesting alien which has firmly established itself about Formby, near St Anne's, and elsewhere.

Selaginella selaginoides has been found in some of the wet hollows, though it cannot be regarded as a constant associate. It has been transported by natural agencies from some of its stations in the Lancashire hills.

In the same way some interesting mosses have successfully established themselves among the sand hills. *Amblyodon dealbatus*, P. Beans, has its nearest mountain home on the west Lancashire fells; *Meesia trichoides*, Spruce, is an immigrant from boggy ground on the mountains of the north of England; while *Catascopium nigritum* has its nearest mountain station at Widdy Bank in Teesdale.

Several species of *Cladonia* and *Peltigera* occur in the damp Salix-association, and in the hollows of medium age, where there is little herbage, the lichen *Arthropyrenia arenicola* forms extensive greyish-white patches among the carpet of mosses and liverworts.

III. Shingle-beach communities

Another well-marked coastal habitat is the shingle or pebble-beach. Shingle-beaches occur very frequently on our southern and eastern coasts, but, like sandy beaches, they are almost destitute of plant life below the limit reached by ordinary high tides. The strip of shingle carried beyond this limit by exceptional tides is generally narrow, but in some places, owing to various causes, there are large areas of shingle now altogether beyond the reach of the sea.

A thin layer of shingle overlying sand, or a mixture of shingle and sand, bears mainly arenicolous plants, and the vegetation is very closely allied to that of sand dunes. Similarly shingle covering the edge of a salt marsh is partly, at least, occupied by halophilous species whose roots presumably penetrate to the salt bottom-water (see p. 360). But the vegetation of thick banks or expanses of shingle is different from either of these and bears many species which are neither halophilous nor arenicolous. The substratum is quite special, consisting of rounded waterworn stones often of considerable size, in the interstices of which a fine black soil collects. This is derived partly from sea drift, air-dried and pulverised, and partly perhaps from the disintegration of the surfaces of the stones by the lichens which frequently cover them. Many square miles of the flat headland of Dungeness in Kent, and of the low coastland to the west, are covered by this kind of shingle-beach.

The composition and ecology of this vegetation has, as yet, scarcely been studied. Certain species occur very frequently, *e.g. Rumex crispus, Echium vulgare, Sambucus nigra, Glaucium flavum, Silene maritima, Lathyrus maritimus, Solanum Dulcamara* var. *marinum, Geranium Robertianum* var. *purpureum.* Some of these are "maritime" species or varieties affected presumably by "maritime" factors, and others are plants of waste places. Eventually the shingle-beach may become overgrown by grasses, but whether this is a normal and regular succession is unknown. Until the vegetation and the edaphic conditions of shingle-beaches have been properly investigated, the beaches whose vegetation depends on a salt-water factor separated from those in which sand is mixed with the shingle, and both from those into which neither of these factors enters, it is impossible to write anything very definite about the *status* of the shingle-beach communities. The last type however may be provisionally called the shingle-beach association proper: the others are to be reckoned as belonging to the salt marsh and sand dune formations respectively.

THE MARITIME FORMATIONS OF BLAKENEY HARBOUR
By F. W. OLIVER

I. TOPOGRAPHICAL STRUCTURE OF THE AREA

The northern coast-line of Norfolk from Weybourne, at the western end of the low cliffs which line the coast on either side of Cromer, westwards to the Wash, is famous for the extensive stretches of maritime formations that fringe its shores. Though mainly salt-marsh (or "saltings" as these creek-traversed flats are locally termed) the marshes are at many points associated with "meols" or sand hills, which reach above tidal limits and bear a sand-dune vegetation.

The Weybourne-Blakeney shingle-bank.

The eastern portion of this strip of coast is dominated, topographically speaking, by the Weybourne-Blakeney shingle-bank, which runs parallel to the mainland for a distance of 8 miles (*c.* 13 km.) (Fig. 20). The area behind this shingle bank, open to the sea at the western or Blakeney end, has silted up to such an extent that it has been possible to dyke off successive sections and reclaim them from tidal invasion. As another consequence of the silting up, the once flourishing seaport of Cley-next-the-Sea, situated on what was formerly the mouth of the River Glaven, has become inaccessible to all but the

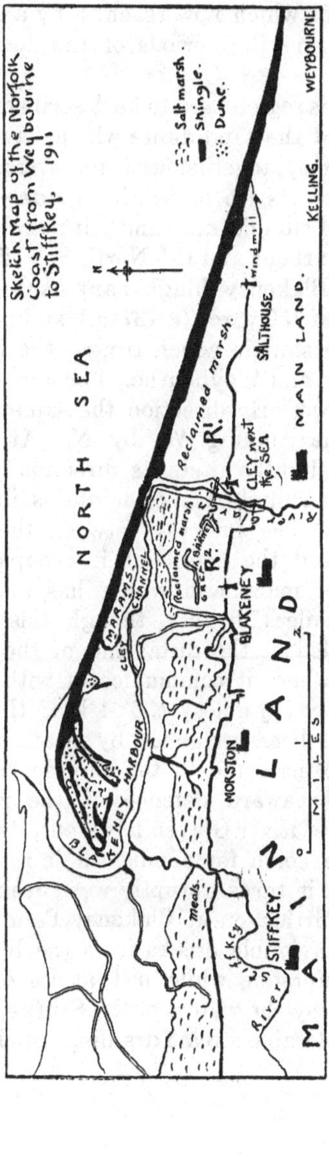

Fig. 20. Sketch-map of the Norfolk Coast from Weybourne to Stiffkey, showing the great shingle beach from its articulation with the mainland (Kelling—Weybourne) to its extremity beyond Morston. The area enclosed is occupied by a system of salt marshes, of which two areas (R^1 and R^2) have been reclaimed.

smallest fishing craft, which now reach it by a long and devious channel during the periods of the spring tides only.

The portion of this region here to be described includes the shingle bank and the sand dunes which have formed at its western extremity, together with the system of salt marshes which lie on its lee or southern side—in other words the complex of shingle, sand and salt marsh that lies between Blakeney harbour and the North Sea (fig. 21).

The Weybourne-Blakeney shingle-bank has an average width at low water of 150 yards (*c.* 137 m.), and is directly continuous with the shingle which fringes the mainland between Sheringham and Weybourne. For a mile beyond Weybourne in the westerly direction the same relations obtain, the coastline running W. by N. At Kelling, however, the coast slightly alters its direction and runs due west, while the shingle bank maintains its former course. The interval between the shingle bank and the mainland is occupied by a strip of marshland which has an average width of about a mile (1·6 km.), though this width is considerably exceeded at the distal end of the estuary, beyond Blakeney, where it communicates with the sea. From Kelling to Cley ($3\frac{1}{2}$ miles = *c.* 5·64 km.) the marshland (fig. 20, R^1) has been reclaimed by means of a dyke stretching from the mainland to the shingle bank—the latter forming the seaward defence of the reclaimed ground. At Cley the River Glaven flows out, its channel running nearly due north for a mile till it reaches the shingle bank, when it turns abruptly west, reaching the open sea $3\frac{1}{2}$ miles further on at Blakeney Point. On its course it receives navigable creeks from the little ports of Blakeney and Morston, while just at its outlet the stream is reinforced by the waters of the Stiffkey River—the only other considerable stream draining into Blakeney harbour.

The "saltings."

Between Cley and Blakeney another large tract of marshland has been dyked off (fig. 20 R^2), but in this case the reclaimed area falls short of the shingle bank to allow of the outflow of the Glaven (Cley channel).

The main shingle-bank.

The length of the main shingle-bank from the point where it leaves the mainland (Kelling) to its distal extremity (Blakeney Point) is about 7 miles (*c.* 11·25 km.). As already stated it has an average width of about 150 yards (*c.* 137 m.) at low tide and 100—120 yards (*c.* 90—110 m.) at high water. The crest, which is on the seaward side, is some 20 feet (6 m.) above mean sea level and 6 to 8 feet (1·8 to 2·4 m.) above spring tide level. The seaward slope is relatively steep (20°), while on the landward side the shingle slopes gently to the saltings (Plate XXXVI a). The bank is composed of loose shingle, the stones—mainly flints—reaching a diameter of 1½ to 2 inches (3·25 to 5 cm.), and derived in all probability from the waste of the cliffs to the east, whence they are brought by the current which sets from east to west along this part of the coast.

Historical evidence shows that the distal extremity of the bank has for centuries steadily advanced in a westerly direction, so that Blakeney Point is to be regarded as its "growing point."

Lateral shingle-banks.

A striking feature of the shingle-bank is the existence of numerous lateral banks or "hooks," inserted on its landward face, and ranging in length from 200 to 800 yards (*c.* 183 to 730 m.). These hooks, which occur at irregular intervals, have evidently been produced from time to time as lateral displacements of the growing point. Traces of about six hooks are to be found traversing an area of salt marsh, locally called "the Marams[1]," one or more at "the Hood," half-a-mile further west, and a whole

[1] A misnomer, since marram grass (*Ammophila*) plays no part in this area.

series near the extremity of the main bank at Blakeney Point (Fig. 21 and Plate XXXV), where these lateral banks form a system of some complexity. The lateral shingle banks vary in height; the crests of some of them rise well above spring tide level, while others are quite covered by all tides other than the neaps.

Salt-marsh units.

When the successive lateral shingle banks are placed close together, the little bays enclosed between them silt up with mud, and the conditions become suitable for the production of salt marsh. Each bay forms an independent topographical unit of salt marsh with its own creek communicating with the main channel of the Glaven (Fig. 21 and Plate XXXV).

As a consequence of this progressive production of laterals, and the silting up of the bays, salt-marsh units of various ages are found in different stages of succession of the salt-marsh formation, younger stages existing near the point, older ones further east, as at "the Marams." The existence of this authentic sequence of salt-marsh associations gives this area a peculiar interest for the ecologist.

The sand dunes.

Contrasting with these salt marshes, on the lee side of the shingle bank are the sand dunes which have accumulated on parts of the main bank and laterals. These are restricted to two spots, viz. "the Hood" and Blakeney Point itself, where the shingle is largely covered with sand blown up from the extensive banks and flats beyond the harbour mouth.

The Blakeney Point complex consists of several ranges of sand hills parallel to the shore line, the higher summits reaching a height of some 30 feet (9 m.) above mean sea level. Nearer the sea are several systems of embryonic dunes arising round nuclei formed of seedlings of *Ammophila* (Plate XXXVI b). These systems illustrate all stages of dune development, starting with the aggregation of a

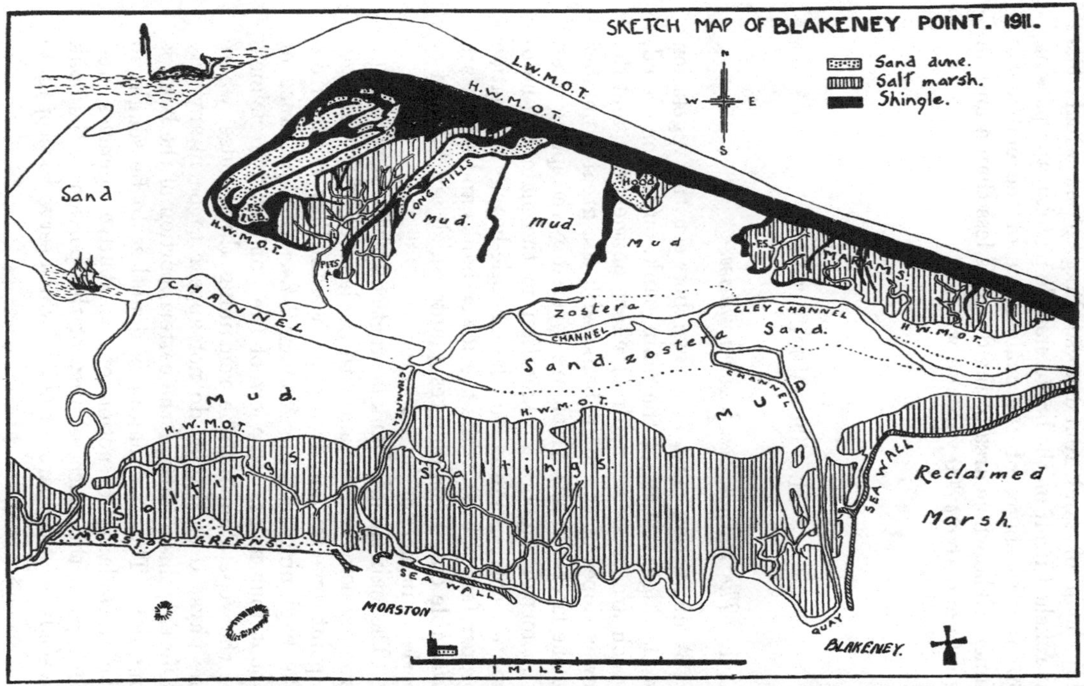

FIG. 21. Sketch-map of Blakeney Point, Blakeney Harbour, and the adjacent mainland, showing the three types of habitat. The principal topographic features only are given. Drawn by T. G. Hill.

tiny heap of sand round a seedling plant which has established itself on the broad shingle plateau, and passing through all stages of the coalescence of the embryonic dunes. The whole area of the dunes is less than a mile in length and about 300 yards in width.

II. The Vegetation

1. *The Shingle-beach communities.*

A distinction must be drawn between the vegetation of the higher parts of the main bank, on the one hand, which are outside the direct effect of the tides, and the lower slopes of the main bank with the greater number of the lateral banks which are covered by the spring tides. The former support a vegetation in the main non-halophytic, corresponding to the shingle-beach association proper (see p. 354), while the lower banks and slopes are available only to halophytes, plants which are tolerant of periodic immersion in sea water.

The main shingle-bank is practically bare of vegetation from its point of junction with the land at Weybourne to a point some few hundred yards west of Cley, not far from the beginning of "the Marams," *i.e.* for a distance of about four miles. In view of the comparative richness of the vegetation for the remaining three miles, where the physical conditions do not appear to be in any way different, this nudity of the eastern portion of the bank is singular. The explanation is doubtless to be sought in the irregularity of action of distributive agencies, and perhaps in the inaccessibility of this portion of the bank to seeds or other parts of plants which serve to propagate the species.

PLATE XXXV

Phot. F. W. Oliver

Looking east from near Blakeney Point. On the extreme left the main shingle bank overlaid with sand dunes. In the centre, three lateral shingle banks extending southwards towards the estuary and marked by lines of *Suæda fruticosa*. On the right, mudflats covered by *Salicornia europæa* and unattached *Pelvetia*. Beyond, the waters of the estuary.

From "the Marams" to the end of the bank at Blakeney Point there is a good deal of vegetation present, though it is always patchy, corresponding no doubt in the main to the facilities for spreading enjoyed by a restricted number of pioneers. The plants occur chiefly on the leeward slope; much more sparingly on the exposed slope between the crest of the bank and high-water mark.

The number of species present in this community is quite small. The following is a complete list.

Glaucium flavum	la	Calystegia Soldanella	vr
Silene maritima	a	Atriplex patula	l
Arenaria peploides	va	Suæda fruticosa	va
Sagina procumbens	l	Rumex crispus	
Tussilago Farfara	vr	var. trigranulatus	va
Sonchus arvensis	l	Poa annua	l
S. oleraceus		Agropyron junceum	l
Mertensia maritima	vr	Ammophila arenaria	l
Myosotis collina	l	Elymus arenarius	vr
Statice maritima		Sedum acre	la
Limonium binervosum			

Arenaria peploides and *Silene maritima* are the most abundant plants, forming extensive carpets. *Rumex crispus* var. *trigranulatus* is also extremely common, its seedlings arising in great numbers from among the mats of *Arenaria* and *Silene*. But the most conspicuous plant of these shingle-beaches is *Suæda fruticosa*, which often forms dense thickets 2 feet (·6 m.) high and flourishes equally well on the higher, and on the lower parts of the beach where it is covered by the spring tides (Plate XXXVI a). The species is rare and local in its British distribution, being confined to a few localities on the east and south coasts of England. This is apparently its most northerly station.

Of *Elymus arenarius* but one or two tufts exist on the shingle; of *Calystegia Soldanella* two small patches, about a third of a mile (*c.* ·5 km.) apart; of *Tussilago Farfara* one plant only has been found.

Mertensia maritima (the Oyster-plant) exists in one patch only, comprising (in 1910) some twenty plants within a few yards of one another. This species has a decidedly northern (Arctic) distribution in Europe. It occurs in a number of localities on the Scottish coast and also in Wales and Ireland, but it had not been previously found on the east coast of England south of Holy Island (Northumberland).

Seedlings of *Ammophila arenaria* arise directly on the shingle near Blakeney Point, where they play an important part in the establishment of embryo sand dunes as well as in their later development (Plate XXXVI b).

Apart from the close carpets of *Arenaria peploides* and the thickets of *Suæda fruticosa* the association is an open one, the plants growing very scattered with bare shingle between them. No closed association is formed on the habitat in this locality.

The actual soil in which the roots of these shingle plants exist occurs in the interstices of the shingle and is provided by the mouldering (disintegration) of the tidal drift, which is scattered widely over the bank by the wind, and by the decay of other plants. Of the physical conditions of plant life on shingle not much is known, but it is evident that the comparatively large spaces between the stones enable the roots and rhizomes to extend to great distances in all directions.

2. *The Sand dune formation.*

Apart from the continuous production of new dunes on the edge of the shingle plateau at Blakeney Point, where all stages of this process may be studied, the dune vegetation is not very noteworthy. The principal agent in the development of the dunes is *Ammophila arenaria*, though *Arenaria peploides* plays a similar *rôle* in the accumulation of small heaps of sand.

Phot. F. W. Oliver

a. Inner edge of main shingle beach abutting on "saltings," looking west. Bushes of *Suæda fruticosa* on inner slope of beach. Isolated plants of *Rumex crispus*, var. *trigranulatus*, and mats of *Silene maritima* and *Arenaria peploides* on upper part of beach. Beyond second creek in saltings a lateral shingle beach is marked by a line of *Suæda fruticosa*.

Phot. F. W. Oliver

b. Seedling plants of *Ammophila arenaria* established on shingle at Blakeney Point, and beginning to accumulate sand.

Blakeney Shingle-beach communities.

The associations are represented as follows:

(i) *Association of strand plants.*

Cakile maritima	r	Salsola Kali	r
Arenaria peploides	a	Agropyron junceum	a

(ii) *Association of Ammophila arenaria* (on mobile sand).

Eryngium maritimum	r	Ammophila arenaria	d
Carex arenaria	f	Festuca rubra var. arenaria	f

(iii) *Association of fixed dunes.*

Cerastium semidecandrum	Calystegia Soldanella
Erodium cicutarium	Phleum arenarium
Galium verum	Corynephorus canescens
Senecio Jacobæa	Polypodium vulgare
Cnicus lanceolatus	Tortula ruraloides
C. arvensis	Peltigera canina
Hieracium Pilosella	Cladonia sp.
Anagallis arvensis	

3. *The Salt-marsh formation.*

The description will be restricted to the salt marshes occupying the sheltered bays between the "hooks" or lateral shingle banks on the lee side of the main bank. They fall naturally into two groups: (*a*) a younger series, associated with Blakeney Point, and lying behind the main area of sand dunes: (*b*) an older series forming the area of "saltings" which bears the misleading name of "the Marams."

(*a*) *The Blakeney Point series* (*Association of Salicornia europæa*).

These are mainly contained between the high lateral shingle bank (partly covered with sand dunes) known as the Long Hills and the inner edge of Blakeney Point (Plate XXXV). These marshes are flat and muddy, and are traversed by meandering creeks; their average level is about 6 to 7 feet (*c.* 2 m.) above mean tide level.

The main area is covered with a very pure and densely

crowded association of *Salicornia europæa*, bright green in colour; with this is mingled, especially in the neighbourhood of the creeks, a certain amount of *Aster Tripolium*, which however rarely flowers because it is freely eaten by rabbits from the neighbouring dunes.

The most singular feature of these *Salicornia*-flats is the presence of a continuous carpet of *Pelvetia canaliculata*, consisting of sterile detached plants lying on the mud between the *Salicorniæ* and forming a layer 3 to 4 inches (*c.* 7 to 10 cm.) deep. The *Pelvetia* is not displaced by the tide, but remains permanently *in situ*.

So far as can be ascertained, this detached non-fruiting form of the plant has not been recorded elsewhere. On some parts of the flats it is replaced by an analogous form of a species of *Fucus* (probably *F. spiralis*). A similar form has been found associated with *Spartina* on the salt marshes of the Solent, particularly near Keyhaven at the junction of the Hurst Castle shingle-bank with the mainland.

The margins of the Blakeney Point marshes, where they slope up to the shingle-banks, show a well-marked peripheral zonation, the width of the zones depending of course on the steepness of the slope. Bordering the *Salicornia-Pelvetia* community is a zone of *Atriplex portulacoides* and *Suæda maritima*, which gives place to a zone of *Suæda fruticosa* on the flanks of the shingle bank itself. Associated with this, especially where mud has been deposited on the shingle, is *Limonium bellidifolium* (*Statice reticulata*) which is rare in Britain, but characteristic of the East Anglian coast.

The next zone belongs to the shingle-beach association, and is characterised by *Arenaria peploides, Silene maritima*, etc.

(*b*) "*The Marams*" (*General salt-marsh association*).

The halophilous communities carpeting most of the units of this system are in a more advanced stage of development than those of the Blakeney Point series. Owing to a continuation of the silting process the general level of the mud surface is here 2 feet or more above that of the *Salicornia*-flats just described. The shingle-banks which form the framework of "the Marams" have been to a considerable extent overlaid with mud, but they retain their characteristic covering of *Suæda fruticosa.*

Most of the saltings are occupied by a mixed salt-marsh association in which *Salicornia* is comparatively rare.

The following plants occur:

Cochlearia anglica	Glaux maritima
Frankenia lævis	Plantago maritima
Spergularia marginata	Atriplex portulacoides
Aster Tripolium	Salicornia europæa
Artemisia maritima	Suæda maritima
Limonium vulgare	Triglochin maritimum
Statice (Armeria) maritima	Glyceria maritima

Artemisia maritima often forms a conspicuous zone on the borders of the saltings where they merge into the shingle.

Atriplex portulacoides is scattered nearly everywhere, conspicuously on the banks of creeks. In at least one case *Atriplex portulacoides* is distributed round the edge of a marsh unit which has its centre densely clothed with *Salicornia stricta, Plantago maritima* and *Aster Tripolium,* thus furnishing a transitional stage between the *Salicornia*-association of the Blakeney Point flats and the mixed association characteristic of "the Marams."

(*c*) *Association of Glyceria maritima.*

Where silting has proceeded further still, so that the level of the mud is covered only by the highest spring tides, an association dominated by *Glyceria maritima,* forming a close turf, is formed. This does not occur within the limited area described but may be studied on the other side of the river beyond its mouth in the direction of Stiffkey. The seaward edge of this area is being eroded by the sea, so that a low mud cliff is formed, and detached columns of mud, capped by the *Glyceria*-turf, occur along its course.

(*d*) *Association of Juncus maritimus.*

At the back of the *Glyceria*-association where it abuts on the mainland is a well-marked zone of *Juncus maritimus* and *J. Gerardi*. The former species also occurs in some quantity on the Hood.

BIBLIOGRAPHY OF PAPERS ON BRITISH VEGETATION

(including only recent work on the lines of this book)

ADAMSON, R. S. "The Ecology of a Cambridgeshire Wood." *Proc. Linn. Soc.* 1911. (In the Press.)

HARDY, MARCEL. "Esquisse de la Géographie et de la Végétation des Highlands d'Écosse" (with vegetation map). Paris, 1904.

LEWIS, FRANCIS J. (*a*) "Geographical Distribution of Vegetation of the Basins of the Rivers Eden, Tees, Wear and Tyne." Part I (with vegetation map). *Geographical Journal*, March, 1904.

(*b*) Part II (with vegetation map). *Geographical Journal*, 1904.

MOSS, C. E. "Peat Moors of the Pennines: their Age, Origin, and Utilization." *Geographical Journal*, May, 1904.

MOSS, C. E. "Geographical Distribution of Vegetation in Somerset: Bath and Bridgwater District" (with vegetation map). *Royal Geographical Society*, London, 1906.

MOSS, C. E. "Vegetation of the Peak District" (with vegetation maps). Cambridge. (In the Press.)

MOSS, C. E., RANKIN, W. M., and TANSLEY, A. G. "The Woodlands of England." *New Phytologist*, Vol. IX, 1910, p. 113.

PETHYBRIDGE, G. H., and PRAEGER, R. LL. "The Vegetation of the District lying south of Dublin" (with vegetation map). *Proc. R. Irish Acad.* Vol. XXV, Section B. 6, 1905.

PRAEGER, R. LL. "A Tourist's Flora of the West of Ireland," Dublin, 1909.

SMITH, R. (*a*) "Botanical Survey of Scotland," Edinburgh District (with vegetation map). *Scottish Geographical Magazine*, 1900.

(*b*) "Botanical Survey of Scotland," Part II, North Perthshire (with vegetation map). *Scottish Geographical Magazine*, Aug. 1900.

SMITH, W. G. "Botanical Survey of Scotland," Parts III and IV, Forfar and Fife (with vegetation maps). *Scottish Geographical Magazine*, Vols. XX—XXI, 1904–5.

SMITH, W. G. and MOSS, C. E. "Geographical Distribution of Vegetation in Yorkshire" (with vegetation map), Part I, Leeds and Halifax District. *Geographical Journal*, April, 1903.

SMITH, W. G. and RANKIN, W. MUNN. "Geographical Distribution of Vegetation in Yorkshire," Part II, Harrogate and Skipton District (with vegetation map). *Geographical Journal*, Aug. 1903.

WATSON, W. "The Distribution of Bryophytes in the Woodlands of Somerset." *New Phytologist*, Vol. VIII, p. 90, 1909.

WOODHEAD, T. W. "Ecology of Woodland Plants in the Neighbourhood of Huddersfield." *Journ. Linn. Soc., Bot.* Vol. XXXVII, p. 333, 1906.

YAPP, R. H. "Wicken Fen" (Sketches of Vegetation at Home and Abroad. IV). *New Phytologist*, Vol. VII, p. 61, 1908.

Diagram illustrating the relationships
are in capitals, the most importa
genetic derivation. The upland
lowland toward the bottom. T
those with acid humus on the le

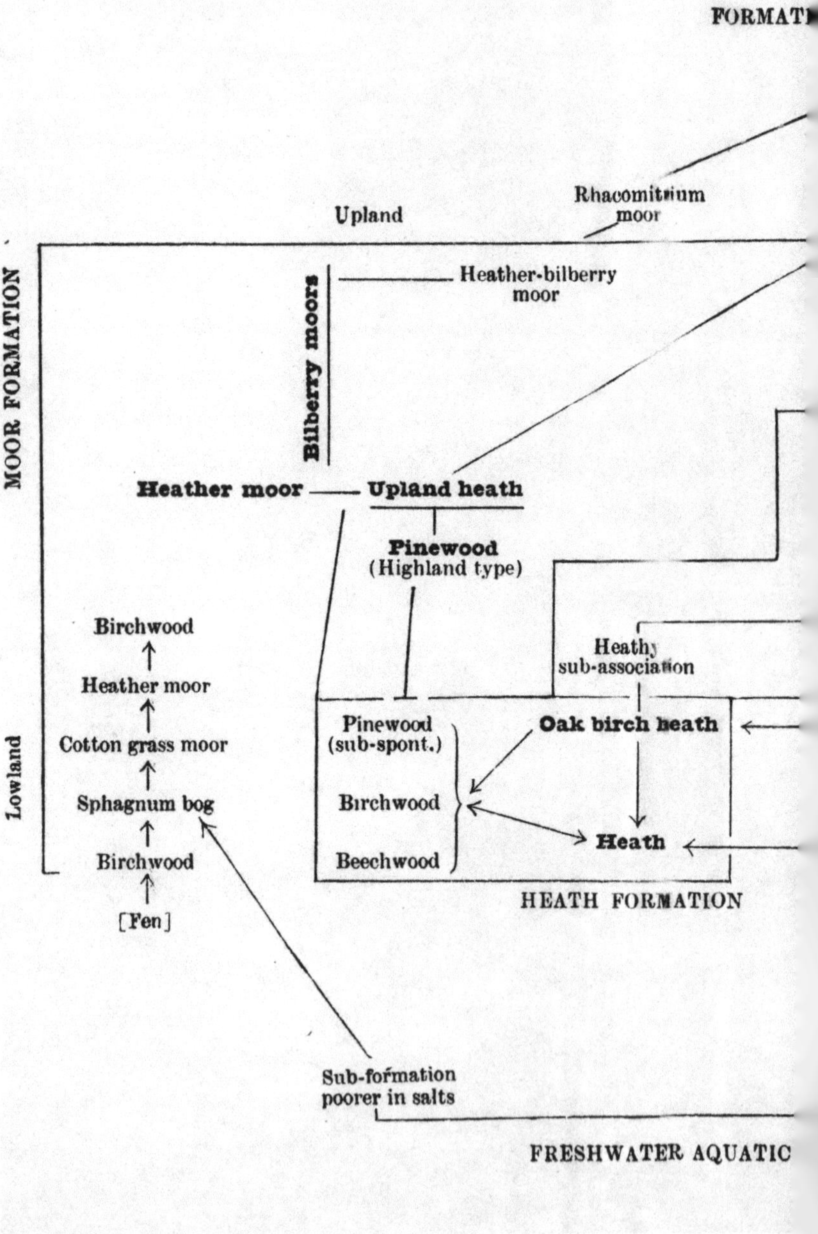

elationships of the fourteen British plant-formations recog
ost important associations in Clarendon type. The lines
he upland associations are on the whole towards the t
ottom. The calcareous soils are on the right, the ne
on the left.

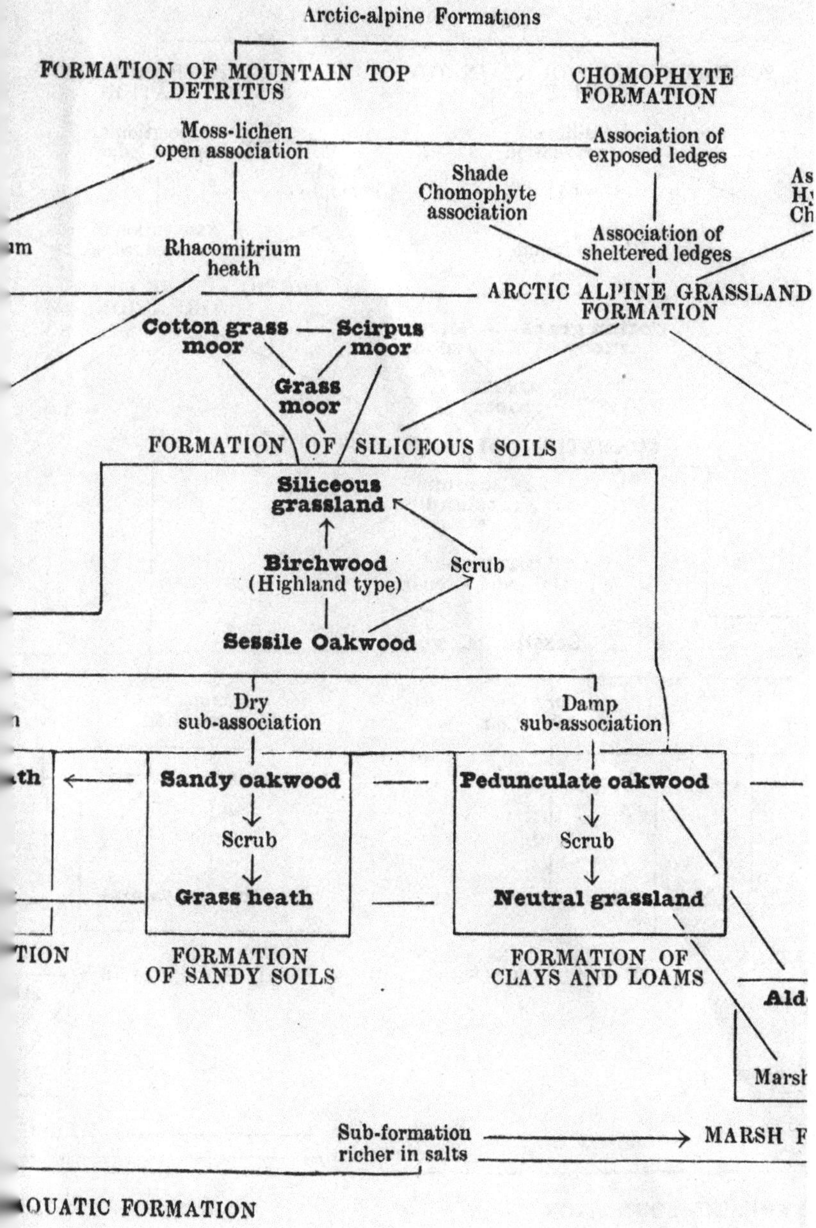

relationships of the fourteen British plant-formations recog
nost important associations in Clarendon type. The lines
The upland associations are on the whole towards the t
bottom. The calcareous soils are on the right, the ne
us on the left.

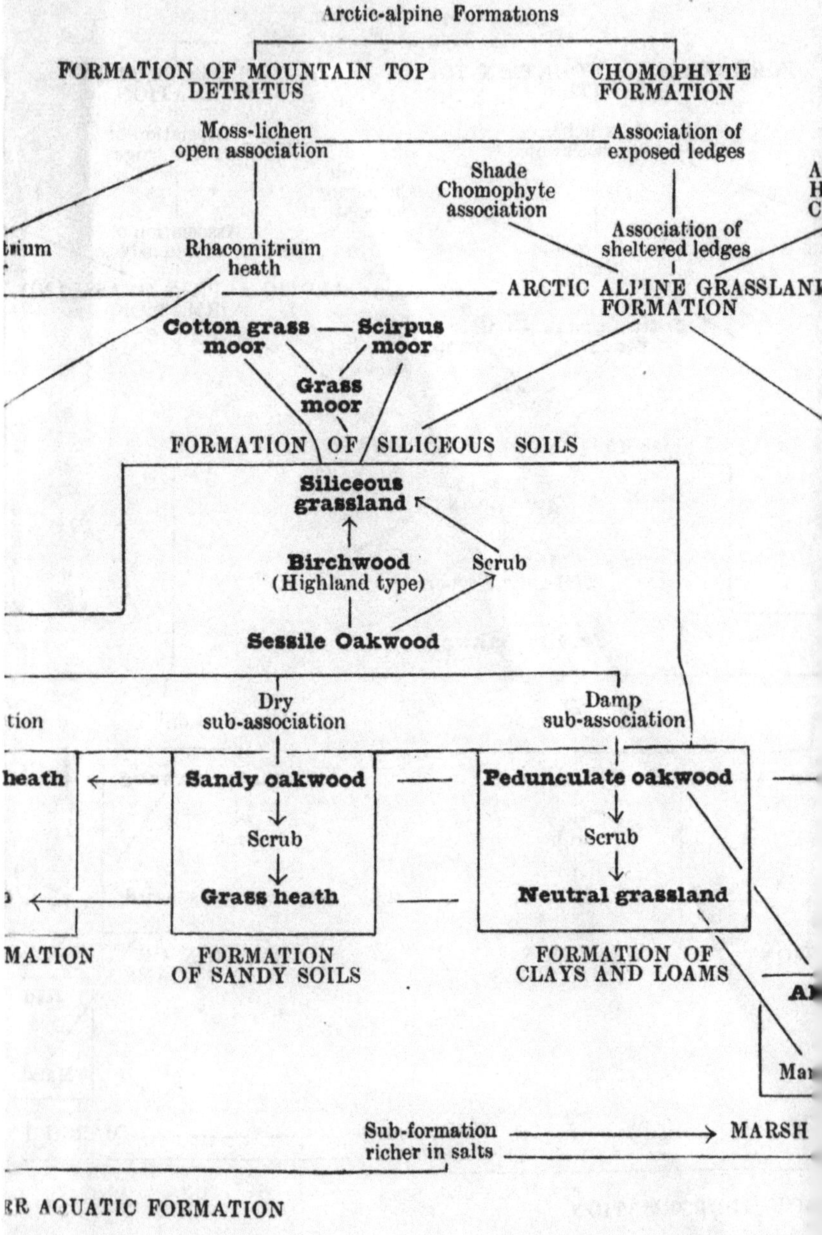

ıtions recognised. Formation-names
The lines show affinity, the arrows
·ards the top of the diagram, the
ht, the neutral in the centre and

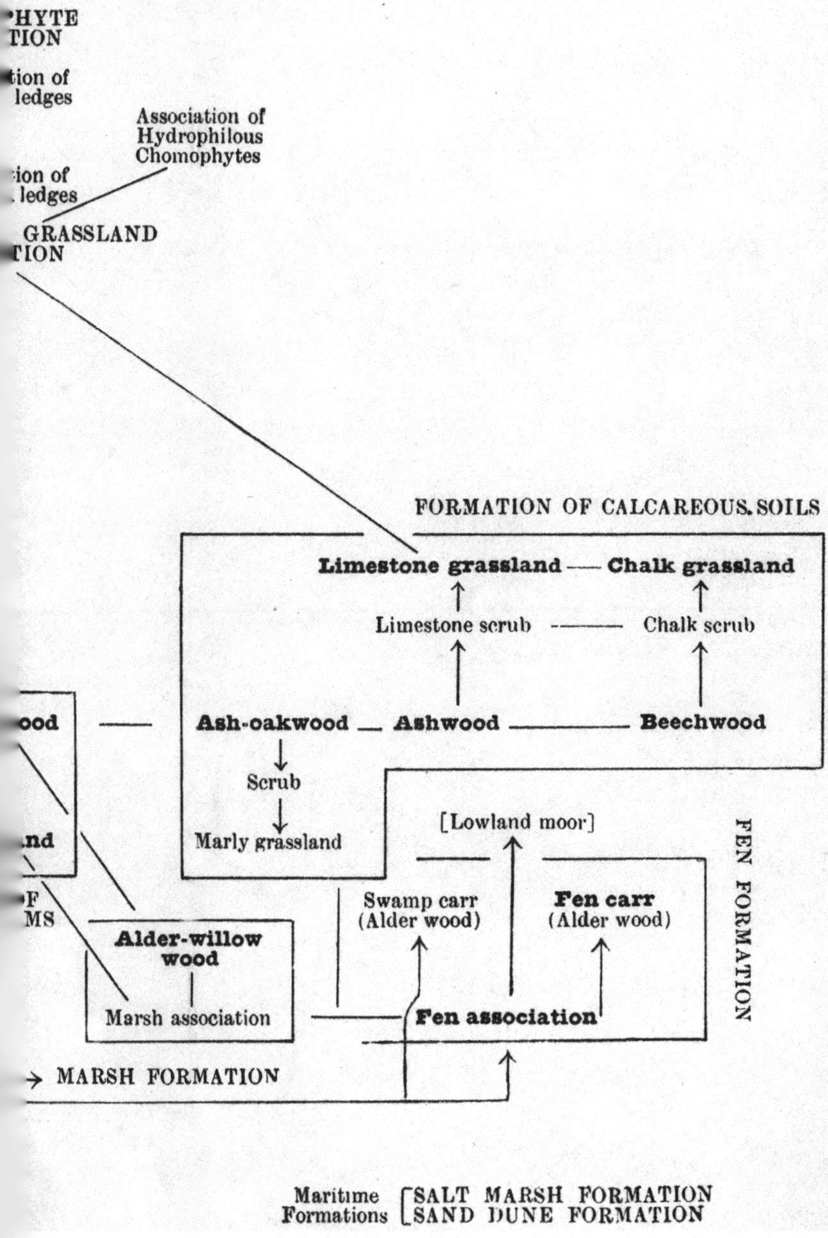

INDEX OF PLANT NAMES

GENERAL INDEX

The page-numbers printed in heavy type refer to the pages on which descriptions are given of the different formations, associations, etc.

Crevices 319
Cromer (Norfolk) 354
Crop rotation 54
Cross Fell 270
Crowden 136
Crystalline Rocks 32, 49
Cultivation 58, 63
zone of **293–296**
Cumberland 21, 51 n.
Cumbrian district 48
mountains 18, 21
Current 191, 194
tidal, 219, 224
Cushions (of sub-alpine species) 289
of mosses 298
Cutting back (by torrents) 281
Cyanophyceæ (*see* Blue-green Algæ)
Cyclonic depressions 23
Cyperaceæ 230 (*see* Sedges)

Dales of Derbyshire 147
streamless 150
Dartmoor 18, 20, 110
Dead water (formation of broads) 222
Decaying organic matter 188
Decoy Broad 220
Dee (Scottish) 292
heaths of Dee valley 113, 115
Deep ploughing 143
Deer 141, 299
forests 44
red 44
Definite succession 286
Degenerate woods 80, 110
oak and birchwoods 136
oakwood 80, 83
Degeneration of moorland **280**
of woodland 7, 108
Delamere Forest (Cheshire) 51, 61, 102, 111
Delta-formation 221, 222
Denmark 24, 64, 66, 163, 210
Denudation 281
channels 271
Depth of seas 15
of snow on Ben Nevis 309 n.
Derbyshire 148, 149, 151, 153, 156, 157, 158, 266, 270, 274, 279
ashwoods of 147
limestone 157
north 136
Derelict pasture 186
Desert regions 339
Desiccation (of moor) 248, 274
(of peat) 281
Desmid areas 200
flora 199, 201
plankton 193, 200
Desmids 199, 202, 203
Destruction (of forests) **66–67**
(of sand dunes) 282
Detritus 36
mountain-top **302–303**
plant-formation of 308, **310–316**
rock 305
soil 321
Devon 48, 61
coast 51
chalk of 161, 162
Spartina stricta in 337
Devon, ashwoods on chalk of East 167, 182
Ulex Gallii in 134
Quercetum sessilifloræ of 137
south 50
Devonian, Middle 50
limestones 61, 146
rocks 48, 49
sandstones and grits (heaths on) 110
Dew 4
Diabases 50
Diatoms 196, 197, 199, 200, 201, 202, 203
centric 197, 203
pennate 197
Dickie, G. 290 n.
Disforestation 174
Dissolved mineral salts 188
Distribution of chief forms of vegetation **62–74**
of geological formations 37
of plant-formations 27
of rainfall **21–23**
of soils **60–61**
of temperatures **24–26**
Ditches 188, 252
Dolomitic conglomerate 147
Domesday Book 108
Dominance (of heather) 277
Dominant plant-forms 104
species 11, 14, 173 (*see also lists throughout*)

Formation of mountain-top detritus 303, **310–316**
of sandy soils **92–97**
of siliceous soils **122–143**
salt marsh 6, 9, **330–339**, **363–366**
sand dune 6, 9, **339–352**, **362–363**
Formby 349, 352
Forth, Firth of 42, 43
Foul waters, sub-formation of **188–190**
Foulshaw Moss 248, 249, 250
France 54, 64, 66
Northern 98
South-western 24
Western 16, 98
Fraxinetum excelsioris **147–153**, **166–167**
Free floating-leaf association 226
Freshwater aquatic formation **188–196**, 214, **223–229**
marsh 335
Phytoplankton **196–203**
vegetation 187
Friesland, meres of 216
Fritton Decoy 220
Frost 26
action of 305

Galtymore 46
Galty range 46
Galway, Co. 45, 46, 47
desmids in lakes of 203
Quercetum sessiliflorae of 137
Gamekeeper 78
Game-preservation 69
preserves 38
Gamlingay Wood 184, 185
Gardiner, W. 292 n.
Gault 53, 55, 75
clay 61
Genera 1
Genetic relations of plant communities 83, 121, 137, 178, 245, 250, 329
varieties of carr association 229
Geographical botany 1
Geography, plant 1
scientific 3
Geological changes 8
formations (*see* Formations)
Geological structure
of British Isles 32
of England and Wales **47–48**
of Ireland 44
of Scotland 38
survey 304
memoirs 214, 305
German beechwoods 164 n.
heaths 98
Germany, forest area of 64
semi-natural vegetation of 64
North-west 90, 98, 210
Gilpin estuary 253
valley 248
Glacial boulder clay 47
débris 253
deposits 37, 40, 41, 45, 216
of Scotland 39
drift 37
loamy 43
period 57, 91, 97
terraces 296
Glaciated mountain plateaux 305
Glaciation 269
Glasgow 42
Glasswort association 332
Glaven, River 354, 356, 357, 358
Glen 117
Callater 292
Doll 292
More 40, 41
Lyon 293
Glendalough (co. Wicklow), Quercetum sessiliflorae of 138
Gloucestershire 28, 52, 162
Glyceria maritima, association of (*see* Glycerietum)
Glyceria-turf 366
Glycerietum maritimae 333, **338–339**, **365–366**
Gorleston 216, 220
Gosport 336
Graebner, P. 90
Gramineae 230
Grange-over-Sands 248
Granite-débris 311 n.
Granites 40, 41, 44, 45, 47
heath on 110
Grass heath 90, 97, 101, 104, 111, 112
association **94–96**, 107, 109, 175

Scirpetum cæspitosi **271–272**, 273, 280, 284
Schistose rocks 41, 298
  areas 325
Schists 40, 49, 122
  mica 47, 318, 319
  quartzitic 42
  sericite 305
Schleswig-Holstein 66
Schneeflecken 36
Schneetälchen 316, 321
Schröter, Prof. C. 2, 211 n
Scotland 16, 17, 23, 57, 60, 63, 64, 65, 66, 67, 71, 73, 98, 164
  Atlantic side of 89
  birchwoods of 121, 1 1
  Central 41
  East 23, 26–31, 40
  North 29, 31, 267
  scarcity of heath on western side of 98
  soils of **38–44**
  West 29, 31, 286
Scottish coast, *Mertensia maritima* on 362
  heaths 98, **113–116**
  Highlands 17, 21, 272, 273, 288, 289
  Lowlands 39
  lochs (lakes) 193, **202**
  moors 109, 208, 209
  mountains 16
  pinewoods 118
  Southern Uplands 17, 21, 28, 30, 42, 44, 113, 282
Screes 152, 319, 320
  block- 319
  limestone 153
  succession on 320
Scrub 7, 10, 78, 80, 83, 94, 112, 171, 186, 236, 341
  associations **83–84**, **94**, **130–131**, **153–154**, **171–173**
  calcareous 153, 163, 169
  chalk **171–173**
  climatic 130
  limestone **153–154**
  on dunes 341
  progressive 131, 173, 236
  retrogressive 131, 153, 236
Scully, Dr R. W. 139 n.
Sea-coast, vegetation of the **330–366**
Sea-coast, couch-grass association **342**, **348**
  effect of proximity to 26, 330
  erosion 347
  rush association **334**
  salt 330
  vegetation of the 185, 330–331 n.
  walls 59
  wrack association 337
Seadrift 353
Seas, depth of 15
Seasonal distribution of rainfall 20
Seathwaite (highest rainfall in Europe) 21
Secondary rocks 17, 50, 75
Sedge-communities (of Arctic-alpine hydrophilous chomophytes) 327
Sedges 174, 229
Seed, self-sown 78
Seedlings in closed associations 130
  Ammophila 358, 360
  ash 169
  beech 168–169
  birch 141
  pine 117, 120
  tree 66
  yew 170
Seeds in peat 253
  dormant 80
  myrmecochorous 106
  of ash 168, 169
  of beech 168, 169
  of pine 120
  of yew 170
  on shingle-beach 360
Selborne (Hants.), beechwoods of 163
Selsey peninsula 336
Seminatural grassland 72
  permanent grassland 207
  plant-communities 63, 69
  vegetation 9, 62, 64
  woods 71, 107
Sericite schist 305
Serpentine tract 306
Shade of beechwood 102, 164, 166, 167, 168, 172
  of birchwood 100
  of *Calluna* 104, 112, 277
  of dry oakwood 94
  of hazel coppice 79
  of oakwood 77, 79, 80

CAMBRIDGE: PRINTED BY JOHN CLAY, M.A. AT THE UNIVERSITY PRESS.

Printed in the United States
By Bookmasters